THE ENVIRONMENT IN GLOBAL SUSTAINABILITY GOVERNANCE

Perceptions, Actors, Innovations

Edited by
Lena Partzsch

BRISTOL
UNIVERSITY
PRESS

First published in Great Britain in 2024 by

Bristol University Press
University of Bristol
1–9 Old Park Hill
Bristol
BS2 8BB
UK
t: +44 (0)117 374 6645
e: bup-info@bristol.ac.uk

Details of international sales and distribution partners are available at bristoluniversitypress.co.uk

British Library Cataloguing in Publication Data
A catalogue record for this book is available from the British Library

ISBN 978-1-5292-2800-7 paperback
ISBN 978-1-5292-2801-4 ePub
ISBN 978-1-5292-2802-1 ePdf

Cover design: Andy Ward
Front cover image: stocksy/Giada Canu
Bristol University Press uses environmentally responsible print partners.
Printed and bound in Great Britain by CPI Group (UK) Ltd, Croydon, CR0 4YY

FSC
www.fsc.org
MIX
Paper | Supporting
responsible forestry
FSC® C013604

Contents

List of Figures, Tables and Boxes

Figures

Tables

Boxes

List of Abbreviations

AOSIS	Alliance of Small Island States
ABMTs	area-based management tools
ABNJ	Areas beyond national jurisdiction
AILPA	Australia Illegal Logging Prohibition Act
ATMs	automated teller machines
BBNJ	biodiversity beyond national jurisdiction
BECCS	Bioenergy with carbon capture and storage
BSBR	Bosque Seco Biosphere Reserve
CSOs	businesses and civil society organizations
C40	C40 Cities Climate Leadership Group
CB&TT	capacity building and marine technology transfer
CCS	carbon capture and storage
CIFOR	Center for International Forestry Research
CDR	carbon dioxide removal
NEPAD	New Partnership for Africa's Development
CDM	Clean Development Mechanism
CPI	Climate Policy Initiative
CEE	collaborative event ethnography
CIPEL	Commission for the Protection of the Waters of Lake Geneva
CLCS	Commission on the Limits of the Continental Shelf
CSD	Commission on Sustainable Development
CHP	common heritage of humankind principle
CFS	Committee on World Food Security
CBDR	common but differentiated responsibilities (CBDR)
CSA	community-supported agriculture
COP	Conference of the Parties
CBD	Convention on Biological Diversity
CITES	Convention on Trade in Endangered Species
EBSAs	ecologically and biologically significant marine areas
ECOWAS	Economic Community of West African States (ECOWAS)
EV	electric vehicle

ETS	Emissions Trading System
EPRA	Energy and Petroleum Regulatory Authority
EIAs	environmental impact assessments
ENGOs	environmental NGOs
EPE	ecofeminist political economy
EUTR	EU Timber Regulation
EEZs	exclusive economic zones
EPR	Extended producer responsibility
XR	Extinction Rebellion
FAO	UN Food and Agriculture Organization
FGNDs	feminist green new deals
FAD	food availability decline
FLR	Forest Landscape Restoration
FLEGT Action Plan	Forest Law Enforcement, Governance and Trade
FSC	Forest Stewardship Council
FPIC	free and prior informed consent
FOS	freedom of the seas
FFF	Fridays for Future
GHG	greenhouse gas
GDP	gross domestic product
HDI	Human Development Index
HRSDSW	human right to safe drinking water and sanitation
IHME	Institute for Health Metrics and Evaluation
IWRM	integrated water resource management
IGCs	Intergovernmental Conferences
IPCC	Intergovernmental Panel on Climate Change
IPF	Intergovernmental Panel on Forests
IPBES	Intergovernmental Platform on Biodiversity and Ecosystem Services
	Intergovernmental Science-Policy Platform on Biodiversity and Ecosystem Services
ICS	International Council of Science
ICESCR	International Covenant on Economic, Social and Cultural Rights
IEA	International Energy Agency
IFOAM	International Federation of Organic Agriculture Movements
ILUC	indirect land-use change
IMO	International Maritime Organization
IOC-UNESCO	International Oceanographic Commission of UNESCO
ISA	International Seabed Authority
IRENA	International Renewable Energy Agency

ITTA	International Tropical Timber Agreement
ITTO)	International Tropical Timber Organization
IUCN	International Union for the Conservation of Nature and Natural Resources
JMP	United Nations Joint Monitoring Programme for Water Supply and Sanitation1
LDC	Least Developed Countries
LNG	liquefied natural gas
LPG	liquefied petroleum gas
LGMA	Local Governments and Municipal Authorities Constituency
ICLEI	Local Governments for Sustainability
MBS	Mancomunidad of Municipalities in the Southwest Loja Province 'Bosque Seco'
MGRs	marine genetic resources
MPAs	marine protected areas
MAAE	Ministry of Environment and Water (Ecuador)
MRV	monitoring, reporting and validating (MRV)
MSPs	multistakeholder partnerships
NFPs	national forest programs
NDCs	Nationally Determined Contributions
NBS	nature-based solutions
NYDF	New York Declaration on Forests
NLBI	Non-Legally-Binding Instrument on All Types of Forests
OECD	Organisation for Economic Co-operation and Development
PSSAs	particularly sensitive sea areas
PCT	personal carbon trading
	Programme for the Endorsement of Forest Certification
PfA	Proposals for Action
RAN	Rainforest Action Network
REDD+	Reducing Emissions from Deforestation and Forest Degradation and the Role of Conservation, Sustainable Management of Forests and Enhancement of Forest Carbon Stocks in Developing Countries
RFMOs	regional fisheries management organizations
RISE	Regulatory Indicators for Sustainable Energy
ROVs	remotely operating vehicles
RTF	right to food
RTW	right to water

SIDS	Small Island Developing States
SER	Society for Ecological Restoration
SB	SocioBosque Program
SCP	sustainable consumption and production
SD	sustainable development
SDGs	Sustainable Development Goals
SEforAll	Sustainable Energy for All
SFM	sustainable forest management
SDSN	Sustainable Development Solutions Network
TPES	total primary energy supply
UCLG	United Cities and Local Governments
UN	United Nations
2030 Agenda	UN 2030 Agenda for Sustainable Development
UNACLA	UN Advisory Committee of Local Authorities
UNCED	UN Conference on Environment and Development
UNICEF	UN Children's Fund
UNCCD	UN Convention to Combat Desertification
UNCLOS	UN Convention on the Law of the Sea
	UN Decade of Ocean Science for Sustainable Development (2021–30)
UNDESA	United Nations Department of Economic and Social Affairs
UNDOALOS	UN Division for Ocean Affairs and the Law of the Sea
ECOSOC	UN Economic and Social Council
UNEP	UN Environment Programme
UNFF	UN Forum on Forests
UNFCCC	UN Framework Convention on Climate Change
UNESCO	UN Educational, Scientific and Cultural Organisation
UNISDR	UN International Strategy for Disaster Reduction Secretariat
MDGs	UN Millennium Development Goals
UNDRR	UN Office for Disaster Risk Reduction
UNSD	UN Statistics Division
UBS	universal basic services
VLRs	voluntary local reviews
VNRs	voluntary national reviews
WEF	water–energy–food nexus
WEE	women's economic empowerment
WEDO	Women's Environment & Development Organization

WBG	World Bank Group
WFTO	World Fair Trade Organization
WHO	World Health Organization
WWF	World Wide Fund for Nature
WSSD	World Summit on Sustainable Development

Notes on the Contributors

Eva Alfredsson, PhD, is Researcher at the Division of Sustainable Development, Environmental Science and Engineering (SEED) at KTH Royal Institute of Technology. She is also a Senior Analyst at the Swedish Agency for Growth Policy Analysis, under the Ministry of Enterprise, Energy and Communications. Her work focuses on analysis of sustainable economic development and green structural change, and she is interested in how to transition from the current unsustainable economy to one that is inclusive and sustainable. She is a member of Future Earth's Working Group on the Political Economy of Sustainable Consumption and Production and a Senior Adviser to the think tank Global Challenges.

Elanur Alsaç is master's student in political science at Freie Universität Berlin and holds a bachelor's degree in social sciences from Humboldt-Universität zu Berlin.

Myrodis Athanassiou received his bachelor's degree in political science at the Technische Universität Darmstadt and is on the master of arts programme in international relations at Humboldt-Universität zu Berlin, Universität Potsdam and Freie Universität Berlin.

Jana Beier is studying for a master's degree in international relations at Freie Universität Berlin, Humboldt-Universität Berlin and University of Potsdam. Her academic interests lie in the field of socially just and sustainable economic policy.

Ettore Benetti is a student on the MA programme in political science at Freie Universität Berlin, and previously graduated in politics, philosophy and economics from LUISS Guido Carli University in Rome.

Mareike Blum, Dr., is researcher at the Mercator Research Institute on Global Commons and Climate Change (MCC) in Berlin, where she examines the potential of mini-publics in co-producing policy advice for the German Energy Transition. Her dissertation at the University of Freiburg was on the (de)legitimation of global carbon markets, with a focus on legitimacy and discursive (de)legitimation processes in forest climate governance.

Maria Brockhaus is Professor and Chair of International Forest Policy at the University of Helsinki, and a member of HELSUS, the university's Sustainability Centre. Her main research themes are forest and land use policy and governance, with a particular focus on power and politics in climate change mitigation and adaptation. She has published extensively on the political economy of tropical deforestation, policy and institutional change, and related policy networks and discourses.

Johanna Carrasco Saravia is a bachelor's student in political science and government at Pontificia Universidad Católica del Perú, Lima.

Ekaterina Chertkovskaya, PhD, is a researcher based at Lund University, working on degrowth and critical organization studies. Her research addresses contemporary crises and explores paths for social-ecological transformation. She has been writing on corporate violence, problems with work and employability, and the plastic crisis, on the one hand, and focusing on degrowth as a vision for transformation, its political economy and alternative models of work and organizing, on the other.

Wonyoung Cho graduated from Chung-Ang University in the Republic of Korea with a BA in political science and is an MA student in political science at Freie Universität Berlin.

Paúl Cisneros, PhD, is Professor of Public Policy at the Escuela de Gobierno y Administración Pública of the Instituto Nacional de Altos Estudios Nacionales in Ecuador. He leads a research group on the implementation of Agenda 2030 and the SDGs. His research deals with policy and governance in the domains of mining, environment and sustainability.

Maurie Cohen, PhD, is Professor of Sustainability Studies and Chair of the Department of Humanities and Social Sciences at the New Jersey Institute of Technology. He is also a member of the management team of the Future Earth Knowledge–Action Network on Systems of Sustainable Consumption and Production and the Editor of the journal *Sustainability: Science, Practice, and Policy*. His most recent book is *Sustainability* (Polity Press, 2021).

Sofia Cordero Ponce, PhD, is Lecturer at the School of International Relations of the Instituto Nacional de Altos Estudios Nacionales in Ecuador and at the Facultad Latinoamericana de Ciencias Sociales (FLACSO) Sede Ecuador. Her research deals with democracy in Bolivia, political movements and the participation of Indigenous populations, and sustainable development.

Nopenyo E. Dabla is Senior Investment Officer in the Renewable Energy Department of the African Development Bank. As a renewable energy for development specialist, he has worked in other positions such as the lead of the International Renewable Energy Agency's engagement in sub-Saharan Africa, as well as in national and international NGOs promoting renewable energy in rural electrification and rural development approaches.

Emmanuel Dahan is currently completing an MA in political science with a focus on the social-ecological transformation at Freie Universität Berlin where he also completed his BA in political science.

Julie Duval is a master's student in public management at the University of Geneva and a student research assistant at the Department of Environmental Social Sciences at Eawag (Swiss Federal Institute of Water Science and Technology). She is interested in environmental governance, public policy and public administration.

Manuel Fischer, PD Dr., is Research Group Leader in Policy Analysis and Environmental Governance (PEGO group) at the Department of Environmental Social Sciences at Eawag (Swiss Federal Institute of Water Science and Technology) and an Associate Professor at the Institute of Political Science of the University of Bern. His research deals with policy and governance networks in the domains of water, environment and sustainability.

Liam Gavin is an Erasmus student at Freie Universität Berlin, from the University of Sussex where he is studying for a BA in international relations.

Andreas C. Goldthau, Dr., is Director of the Willy Brandt School of Public Policy at the University of Erfurt, where he holds the Franz Haniel Chair for Public Policy at the Faculty of Economics, Law and Social Sciences. He also leads the Research Group Energy Transition and the Global South at the Institute for Advanced Sustainability Studies, Potsdam.

Javier Gonzales-Iwanciw is climate change adaptation policy researcher and consultant linked to the Institute of Science and Social Research at Universidad Nur in Bolivia. His research and publications focus on the governance of adaptation, integrative approaches, collaboration and multilevel learning, implementation of SDG 13 and the Paris Agreement.

Sofie Jokerst completed her bachelor's degree in political science at the Technical University of Munich and is currently studying for a master's degree in political science at the Otto Suhr Institute of Freie Universität Berlin.

Elizaveta Kapinos is a master's student at Otto Suhr Institute of Freie Universität Berlin. She gained her bachelor's degree in international relations at the National Research University Higher School of Economics in Moscow.

Mawa Karambiri, PhD, is Policy Scientist at the World Agroforestry Centre (ICRAF), a member of the Helsinki Sustainability Centre (HELSUS) and a Visiting Scientist at the University of Helsinki. Her work is about land restoration, gender, local democracy, community forestry and forest policy translation to the local level.

Daniela Kleinschmit, Dr., is Professor for Forest and Environmental Policy at the University of Freiburg. Her main research interest focuses on land use policy and governance, including political communication. She is Vice-President of the International Union of Forest Research (IUFRO) and actively supports diverse science–policy interactions. As Vice-Dean, Dean and Vice-President of University Freiburg, she has been actively engaged in research management and governance.

Montserrat Koloffon Rosas is a PhD student within the Formas-funded Transformative Partnerships 2030 research project. She is based at the Department of Environmental Policy Analysis of the Institute for Environmental studies (IVM) at the Vrije Universiteit Amsterdam, where she is also a Junior Lecturer for the Political Science Global Environmental Governance track. Her research explores the effectiveness of international (environmental) regimes through the lens of complexity science.

Anna Kosovac, PhD, is Lecturer in the Faculty of Arts at the University of Melbourne in Australia. Her research focuses on urban water governance, urban politics and risk. In addition to a PhD, she holds degrees as a master of international relations and a bachelor of civil engineering from the University of Melbourne. She has a decade of experience working in the public sector on environmental project programmes and over seven years in academia. She has previously worked together with various international organizations such as the Global Covenant of Mayors, UN-Habitat, ICLEI and the Chicago Council on Global Affairs on topics related to city governance, environmental urban management and city diplomacy.

Laura Kräh is a master's student in political science at Freie Universität Berlin. She completed her bachelor's degree in public governance across borders at the University of Muenster, University of Twente and Universidad de Guadalajara.

Markus Kröger is Professor of Global Development Studies at the University of Helsinki. His research interests include natural resource politics, extractivism, political economy, social movements, and climatic-ecological crises and solutions. He has conducted extensive field research on agrarian, mining and forestry politics in South America, India and Finland. He is the author of four books and a series of articles and book chapters on these and other topics. In forest policy research, he has studied especially the global expansion and politics of tree plantations.

Ruth Krötz is studying international relations in the joint master's program of Freie Universität Berlin, Humboldt University of Berlin and the University of Potsdam. Her bachelor studies on political science were at the University of Leipzig and Sciences Po Lyon.

Maike Laengenfelder is a master's student of international relations with a focus on environmental politics at Freie Universität Berlin, Humboldt Universität Berlin and the University of Potsdam. She received her bachelor's degree in political science at the Otto Suhr Institute of Freie Universität Berlin.

Caroline Landolt is a double degree master's student in political science at Freie Universität and HEC Paris. She did her BA in business management in France.

Rosa-Lena Lange is studying international relations and political science in Berlin. Before that, she finished her bachelor's degree in liberal arts and sciences at the University in Freiburg.

Estefanía Lawrance Crespo is a bachelor of political science and public administration graduate from the University Autónoma of Madrid and a master's student in social policy, labour and welfare state at the University Autónoma of Barcelona. She is currently an exchange student at the Otto Suhr Institute of Freie Universität Berlin

Kathrin Lehmann received her bachelor's degree in political science from the Otto Suhr Institute of Freie Universität Berlin, where she is now studying for her master's degree.

Sylvia Lorek, Dr., is head of the Sustainable Europe Research Institute, Germany. She researched and published broadly in the fields of consumer economics and sustainable consumption. In 2021 she coauthored the book *Consumption Corridors: Living a Good Life within Sustainable Limits* (Routledge).

Maren Lorenzen-Fischer is pursuing her master's degree in the dual degree program on international affairs, with a focus on human rights and humanitarian aid, at Sciences Po Paris and Freie Universität Berlin, having completed her bachelor degree in European studies at the University of Passau.

Sherilyn MacGregor, PhD, is Professor of Environmental Politics at the University of Manchester. Her research focuses on the relationships between environmental sustainability and social justice, applying insights from intersectional ecofeminist and other critical political theories.

Aino Ursula Mäki is a PhD candidate in the Department of Politics at the University of Manchester. Their thesis is focused on the concept of housewifization. They studied politics with international relations at the University of York and gained a master's degree in international political economy at the University of Manchester.

Jens Marquardt, Dr., is Research Associate at the Institute of Political Science at the Technical University of Darmstadt. He has previously worked on non-state climate action at the Department of Political Science at Stockholm University after conducting research on the relation between climate science and politics at Harvard University's program on science, technology and society. His research interests include environmental governance, power relations and the politicization of climate change. Jens is the author of *How Power Shapes Energy Transitions in Southeast Asia* (Routledge, 2017) and coeditor of *Governing Climate Change in Southeast Asia* (Taylor & Francis, 2021), and has published articles about non-state climate action, populism and the imagination.

Marco Aurélio Mayer Duarte Neto is an MA candidate in the Human Rights and Multi-level Governance Program at Università degli Studi di Padova (UNIPD), Italy, and a holds a bachelor of law degree from Universidade Federal da Paraíba (UFPB), Brazil.

Lyla Mehta, PhD, is Professor at the Institute of Development Studies, UK, and a Visiting Professor at Noragric, Norwegian University of Life Sciences. She trained as a sociologist (University of Vienna) and has a PhD in development studies (University of Sussex). She uses water and sanitation to focus on the politics of scarcity, gender, human rights and access to resources, resource grabbing, power and policy processes in rural, peri-urban and urban contexts. Her work also deals with climate change and uncertainty and forced displacement. She has carried out extensive research and fieldwork in India and southern Africa and is currently leading a Belmont–Norface–EU–ISC project on 'Transformations as praxis' in South Asia. Her most recent books are *Water, Food Security, Nutrition and Social Justice* (Routledge, 2020) and *The Politics of Climate Change and Uncertainty in India* (Routledge 2021).

Felix Nütz is a master's student in political science at Freie Universität Berlin where he also received his bachelor's degree in political science.

Silvia Panini is a dual degree master's student at Sciences Po Paris and Freie Universität Berlin in European affairs and foreign policy. She completed her BA in international relations and foreign affairs at the University of Bologna (Italy).

Lena Partzsch, Dr., is Professor of Comparative Politics with a focus on Environmental and Climate Politics. Since 2018, she has been an Extraordinary Professor at the University of Freiburg. The University of Münster awarded her the Habilitation (postdoctoral qualification) in 2014 and she received her PhD from Freie Universität Berlin in 2007. Her research deals with sustainability transitions in the Global North and South. Her most recent book is *Alternatives to Multilateralism: New Forms of Social and Environmental Governance* (MIT Press, 2020).

Philipp Pattberg, Dr., is Professor of Transnational Environmental Governance and Head of the Environmental Policy Analysis Department at the Institute for Environmental Studies (IVM). He also serves as Director of the Amsterdam Sustainability Institute (ASI), a platform for interdisciplinary research collaboration between all faculties of the Vrije Universiteit Amsterdam. He has published widely on global environmental politics with a focus on climate change governance, biodiversity, forest and ocean governance, transnational relations, multistakeholder partnerships, network theory and institutional analysis.

Daniel Pejic is Research Fellow at the Melbourne Centre for Cities at the University of Melbourne. His research explores the role of cities in global affairs, focusing particularly on city diplomacy and cities in multilateral systems. His work has been published in journals such as the *International Journal of Urban and Regional Research*, *Journal of International Affairs*, *The Hague Journal of Diplomacy* and *Scientific American*. Daniel has also held a number of professional research roles and leadership positions, working to communicate and translate evidence into policy for international organizations, non-government organizations and both state and federal governments in Australia.

Sabaheta Ramcilovic-Suominen, PhD, is Associate Research Professor and Academy of Finland Research Fellow at Natural Resources Institute Finland, Luke. She studies global sustainability commitments and their socioecological and political effects, including power relations and socioecological justice in Ghana and Laos, among other countries. More recently she has drawn on decolonial and degrowth theories and thinking to help advance anticolonial and just responses to the global socioecological and inequality crises.

Sabine Reinecke, Dr., is senior researcher and lecturer at the University of Freiburg and a freelance consultant for knowledge coordination. Her research interests are global environmental governance, sustainable development, and the sociology of scientific knowledge, especially in the context of forests and climate and recent initiatives on forest landscape restoration and REDD+.

Claudia Ringler, PhD, is Deputy Division Director of the Environment and Production Technology Division at the International Food Policy Research Institute (IFPRI), where her research focuses on global water and food security, gender–water and energy and gender–climate change linkages, and the synergies of climate change adaptation and mitigation. She coauthored of a recent book on *Water for Food Security, Nutrition and Social Justice* (Routledge, 2019). She has a PhD in agricultural economics from University of Bonn and an MA in international development economics from Yale University.

Arthur Saillard studied political science in Moscow and Berlin and is currently pursuing a double degree in management and public policy at HEC Paris and Freie Universität Berlin.

Miranda Schreurs, PhD, is Chair of Climate and Environmental Policy at the Bavarian School of Public Policy, Technical University of Munich. She investigates environmental movements, green politics and climate policy making both comparatively and internationally.

Sandra Schwindenhammer, PD Dr., is Assistant Professor at Justus Liebig University Giessen and Principal Investigator for the project SUSKULT – Development of a Sustainable Cultivation System for Food in Resilient Metropolitan Regions (2019–24), funded by the German Federal Ministry of Education and Research. Her work focuses on international norms, norm entrepreneurship, and food and agricultural governance.

Alice B.M. Vadrot, PhD, is Associate Professor for International Relations and the Environment in the Department of Political Science at the University of Vienna. She is the Principal Investigator on the Europ ean Research Council –funded project MARIPOLDATA, which combines ethnography, bibliometrics and oral history to study the role of science and knowledge in marine biodiversity negotiations. Since 2019, she has been a member of the Young Academy of the Austria Academy of Sciences and Senior Research Fellow of the Earth System Governance Platform.

Shiney Varghese is Senior Policy Analyst at the Institute for Agriculture and Trade Policy, US, and a member of the High Level Panel of Experts to the UN Committee on World Food Security. She is currently working on global initiatives on water governance and food security, focusing on initiatives and investments related to water and climate crises, and their implications for the food security of smallholder producers, especially women, and possible solutions that emphasize equity, environmental justice and sustainability. She has presented and published extensively on these issues.

Note on the Figures
We wish to acknowledge that the text describing some of the images included in this book refers to colours in the images, despite the images being printed in black and white in the print version of this book. The decision to retain these colour descriptions was made in order to provide greater visual clarity to readers. The images are available in full colour in the electronic version of this book; furthermore, a PDF containing all the colour images in their original form can be viewed at: https://bristoluniversitypress.co.uk/the-environment-in-global-sustainability-governance

1

Introduction: The Integration of Development and Environmental Agendas

Lena Partzsch

The United Nations (UN) General Assembly adopted Agenda 2030 in 2015 with a set of wide-ranging goals that articulate the desired outcome of sustainable development. These so-called Sustainable Development Goals (SDGs) were the result of two processes: the Millennium Development Goals (MDGs) of 2000, and the results of the 2012 Rio+20 Summit, which augmented Agenda 21 of the 1992 Rio Earth Summit. Hence the SDGs are an effort to integrate the development and environmental agendas. Humans are dramatically accelerating global environmental change, and some scholars consider the SDGs to be example of development approaches being increasingly 'in tune with the biosphere, of reconnecting development to the biosphere preconditions' (Folke et al, 2016: 5). However, others argue that the SDGs mask ongoing contestations over sustainable development (Sachs, 2017; Bengtsson et al, 2018; Elder and Olsen, 2019). Humanity is already outside the 'safe operating space' for at least four of nine 'planetary boundaries': climate change, biodiversity, land-system change and biogeochemical flows (nitrogen and phosphorus imbalance) (Rockström et al, 2009; Steffen et al, 2015). Moreover, as a result of the coronavirus pandemic and the Russian invasion of Ukraine, crucial measures of environmental protection are being postponed, have been watered down or risk being completely abandoned.

This volume discusses the Agenda's environmental content and takes a critical account of sustainability governance over the last decades. Each chapter provides an accessible and comprehensive introduction to and assessment of sustainability governance in a field that is crucial for the environment. Authors address three fundamental questions:

1. How have perceptions of the environment changed in sustainability governance and research since the 1992 Earth Summit?
2. Which actors and institutions have mattered most for governance efforts over the last three decades?
3. Which alternative and innovative forms of governance exist and deserve more research attention for a transition to environmentally salient sustainability?

With the High-Level Political Forum (HLPF), UN member states have created a body that is mandated to implement the SDGs. However, the Forum does not have an enforcement function comparable to executive or judicial agencies at the level of the nation-state (Bernstein, 2017). The SDGs represent 'global governance through goal-setting' (Kanie et al, 2017), each government being responsible for implementation in its own territory. Classically, governments and other state actors have governance authority over a defined nation-state territory. Since the 1990s, however, those who have been progressive about governance action are non-state actors, including businesses and civil society organizations (CSOs) (Partzsch, 2020). At the same time, failure to implement green goals in one country frequently has consequences for the people and the natural environment beyond this individual nation-state (Gupta and Nilsson, 2017). Therefore the term *governance* is used in this book to refer to hierarchical and non-hierarchical steering activities by state and non-state actors, including transnational activities.

The first part of this chapter outlines tensions between environmental governance and socio-economic development. The central conceptual contribution of this volume concerns these tensions. Agenda 2030 has been characterized as universal, transformative and integrative (Kanie et al, 2017). Simultaneously, as its effective implementation depends on diverse actors' priorities, there is a risk of uneven attention given to the environmental dimension. Confronted with the overshoot in planetary boundaries (Steffen et al, 2015), more and more environmental scientists are demanding that a balancing approach be given up in favour of ecosystem protection (Griggs et al, 2013; Folke et al, 2016). Countries with a high income in terms of gross domestic product (GDP) per capita, in particular, are expected to prioritize environmental over economic goals at this stage of implementation (Forestier and Kim, 2020). While some demand greater prioritization of environmental goals, others welcome Agenda 2030 for pursuing environmental goals in connection with social and economic goals, seeing planetary boundaries and human development as mutually dependent (Raworth, 2017; Swilling, 2020).

Against the backdrop of this debate, the second part of this chapter is dedicated to the environmental content of Agenda 2030. Studying the weight given to environmental concerns on the global agenda begins with identifying the SDGs that are crucial for a transition to environmentally salient sustainability. There is a broad consensus that the environmental core of Agenda 2030 consists of SDG 6 (Clean water and sanitation), SDG 13 (Climate action), SDG 14 (Life below water) and SDG 15 (Life on land) (eg Folke et al, 2016). These green goals interact with other SDGs in positive and negative ways, either helping to boost ecosystems or compromising other concerns (Bowen et al, 2017; Nilsen, 2020). For example, environmental synergies between SDG 13 and SDG 7 (Affordable and clean energy) are emphasized in the context of renewable energy promotion (eg Wackernagel et al, 2017), depending on the energy sources, the expansion of energy infrastructure can also result in a trade-off with climate change mitigation, hence compromising environmental concerns (Bowen et al, 2017: 91).

Like SDG 7, several SDGs can be expected to have environmental synergies and trade-offs, in particular, as shown later in this chapter, with SDG 2 (Zero hunger), SDG 5 (Gender equality), SDG 8 (Decent work and economic growth) and SDG 12 (Responsible consumption and production) (Le Blanc, 2015; Bowen et al, 2017; Nilsen, 2020). The goals that are relevant for an environmentally sound implementation of Agenda 2030, in particular, are SDG 11 (Sustainable cities and communities) and SDG 17 (Partnerships for the goals). Empowering cities is increasingly seen as a straightforward approach to realizing sustainability on the ground (Bansard et al, 2017; Kosovac et al, 2020). Likewise, multi-stakeholder partnerships have become mainstream implementation mechanisms for attaining the SDGs. However, the latter's effectiveness is increasingly being called into question (Kalfagianni et al, 2020).

The third and final part of this chapter introduces the diverse chapters of this volume. Each chapter focuses on one of the abovementioned SDGs, explaining environmental synergies and trade-offs with other goals. It has been argued that raising living standards is compatible with green growth, but the chapters demonstrate inevitable tensions between the different dimensions of sustainability. As a transition to environmental sustainability is overdue, this chapter and volume aim to bring forward informed debates on alternative and innovative forms of governance that exist and deserve more research attention.

1.1 Prioritizing or balancing environmental protection?

The preamble of Agenda 2030 highlights that the SDGs are 'integrated and indivisible and balance the three dimensions of sustainable development: the economic, social and environmental' (UN, 2022). Agenda 2030 hence takes up the Brundtland Commission's three pillars concept. Emphasizing the need to integrate the environmental, social and economic dimensions of sustainable development, the World Commission on Environment and Development's (WCED) report *Our Common Future* (WCED, 1987) laid the groundwork for the landmark Rio Earth Summit and the adoption of Agenda 21, the Rio Declaration, and to the establishment of the Commission on Sustainable Development (CSD) in 1992 (Hajer et al, 2015).

While the three pillars concept has prevailed in the Rio process for the last three decades, development practice has continued to be dominated by the economic growth paradigm, giving little or no attention to environmental concerns. The Human Development Index (HDI) was created in 1990 to emphasize that people and their capabilities should be the ultimate criteria for assessing the development of a country, not economic growth alone (UNDP, 1990). Still, the most recent Human Development Report ranks high-polluting countries as best in development (Norway, Ireland, Switzerland) (UNDP, 2020). Only one of eight MDGs was dedicated to the environment (UN, 2015). Hence, using an integrated approach was far from self-evident in the negotiations leading to the adoption of the SDGs. Moreover, on the positive side, there was a surprising demand for universal goals that would apply to all countries (Donald and Way, 2016; Sachs, 2017).

Prior to the adoption of Agenda 2030, a number of environmental scientists called for abandoning the three-pillar concept in favour of an approach 'that meets the needs of the present while safeguarding Earth's life-support system, on which the welfare of current and future generations depends'[1] (Griggs et al, 2013: 306; see also Elder and Olsen, 2019). In this vein, developed countries insisted that the SDGs take greater account of the environmental dimension of sustainability compared to the MDGs (Kamau et al, 2018; Elder and Olsen, 2019). By contrast, stakeholders, especially from developing countries, emphasized in the negotiation process that planetary boundaries and human development are mutually dependent. They welcomed Agenda 2030's overcoming of 'the environmental bias that plagued the latter years of the Commission on Sustainable Development' (Bernstein, 2017: 223).

Fukuda-Parr and McNeill (2019: 9–10) explain that the North/South divide regarding the weight given to environmental concerns in the Agenda 2030 negotiations was related to political settings rather than only to discrepancies between developed and developing countries. The MDG follow-up process was dominated by donor countries. Think tanks, particularly from the UK and the US, were prominent in producing analyses and organizing discussion events on a 'Post-2015 Development Agenda' (Fukuda-Parr and McNeill, 2019: 10). By contrast, the Rio+ 20 conference was hosted by Brazil, and in this context Colombia initiated the idea of the SDGs (Fukuda-Parr and McNeill, 2019: 10).

The constituency and many of the policy makers in the Rio process were from the environmental community, including environmental ministries, academics, activists, think tanks and business. These actors ensured that the process leading to the SDGs was not perceived as donor driven (Kamau et al, 2018; Fukuda-Parr and McNeill, 2019). In response to concessions made by the environmental community, nevertheless, Sachs (2017: 2580) criticizes Agenda 2030, from an environmental perspective, for falling behind Agenda 21 of the 1992 Rio Earth Summit by failing to acknowledge the physical limits of growth. There is no mention anywhere in Agenda 2030 of planetary boundaries.

As Gupta and Vegelin (2016: 440) point out, Agenda 2030 has a rhetorical commitment to 'sustainable development' (mentioning it 85 times), but does not mention 'inclusive development'. These authors find that, while Agenda 2030 succeeds in integrating economic development and social well-being, it fares less well in respect to ecological viability. Sharpening this point, Brühl (2018) argues that the SDGs serve neither the development nor the environmental agenda. As will be discussed later, the SDGs stick to a vision of economic development serving society in a context of unlimited growth. The focus is on technology transfer and scientific solutions to address environmental problems (Braunmühl, 2017; Sachs, 2017). The integrative nature of Agenda 2030 is supposed to address interactions, in theory, but the complexity of the systems involved, limited knowledge and competing interests challenge its implementation in reality. Synergies and trade-offs between the SDGs are unavoidable and become most obvious in how subtargets were defined, as outlined later in this chapter for the environmental dimension of each goal.

The SDGs are not legally binding. No sanctions and few formal mechanisms are in place to ensure that targets and outcomes are achieved (Bowen et al, 2017: 93). On the one hand, flexibility has enabled broad participation and support for the SDGs (Gupta and Nilsson, 2017). On the other hand, there is a risk of less attention being given to the environmental dimension in particular. Governments have already been shown to prioritize economic and social over environmental goals (Tosun and Leininger, 2017; Forestier and Kim, 2020). While some speak of a bottom–up approach (eg Gupta and Nilsson, 2017; Forestier and Kim, 2020), others criticize the 'cockpit-ism' of Agenda 2030, 'the illusion that top-down steering by governments and intergovernmental organizations alone can address global problems' (Hajer et al, 2015: 1652). Although this is essentially relevant, only a few scholars are explicit about which SDGs are environmentally significant in respective debates.

1.2 Environmental content of Agenda 2030

The implementation of all SDGs is of relevance to the environment (Elder and Olsen, 2019), but only a few goals and subtargets explicitly consider

Figure 1.1: The global goals for sustainable development

Source: Folke et al (2016). Published under the terms of CC BY-NC 4.0.

biosphere protection. So, which SDGs constitute the environmental dimension of Agenda 2030? Folke et al (2016) provide the most popular SDG categorization according to Brundtland's three pillars concept. They consider SDG 6 (Clean water and sanitation), SDG 13 (Climate action), SDG 14 (Life below water) and SDG 15 (Life on land) to demonstrate the biosphere. Their tripartite figure became famous as the wedding cake model (see Figure 1.1) (Elder and Olsen, 2019: 72), where the base layer consists of the four environmental SDGs, covered by a middle layer of society and a top layer of the economy (Folke et al, 2016). On the one hand, Folke et al emphasize the artificiality and arbitrary nature of the distinction between natural and social systems and, on the other hand, they argue that the global ecological system integrates 'all living beings and their relationships, humans and human actions included, as well as their dynamic interplay with the atmosphere, water cycle, biogeochemical cycles and the dynamics of the Earth system as a whole' (Folke et al, 2016: 1).

Folke et al's article had been preceded by Waage et al's (2015) SDG figure, which consists of three concentric circles (Figure 1.2). Here, only SDGs 13–15 represent the natural environment; these three goals form the

Figure 1.2: Sustainable Development Goals

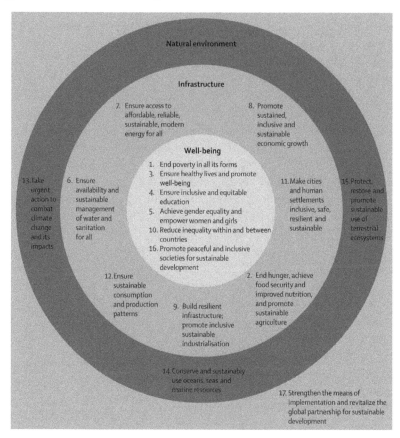

Source: Waage et al (2015). Open access article published under the terms of CC BY.

outer circle surrounding the economic and social goals (except for SDG 17: discussed later). Several authors consider these three SDGs as 'the green targets' (Bengtsson et al, 2018: 1539), and every categorization has assigned these three goals to the environmental dimension of Agenda 2030. By contrast, however, there is no consensus on SDG 6 being an environmental goal, as suggested by Folke et al. At the same time, there is no explicit controversy. Scholars tend to discuss the SDGs in silos. There are separate debates for each policy field, ranging from contradictory language to respect of planetary boundaries.

SDG 6 itself has a primary focus on expanding infrastructure to 'achieve universal and equitable access to safe and affordable drinking' (target 6.1) and 'adequate and equitable sanitation and hygiene for all' (target 6.2). The goal very much follows the wording of MDG 7.C, which aimed to 'halve the proportion of people without sustainable access to safe drinking water and basic sanitation' by 2015. While infrastructure was expanded and MDG 7.C was even accomplished years ahead of schedule, the environmental performance of this goal was particularly poor (UN, 2015). Not classifying SDG 6 as an environmental goal can be seen as an acknowledgement of this neglect. However, in addition to expanding water access, SDG 6 now aims to improve 'water-use efficiency ... and ensure sustainable withdrawals and supply of freshwater to address water scarcity' (target 6.4.), to implement 'integrated water resources management at all levels' (target 6.5) and to 'protect and restore water-related ecosystems' (target 6.6). Hence, it makes sense to consider SDG 6 as a core biosphere and hence green goal of Agenda 2030, despite potentially ambiguous subtargets (Elder and Olsen, 2019).

In addition to SDGs 6 and 13–15, a range of publications categorize SDG 12 (Responsible consumption and production) as an environmental goal (Elder and Olsen, 2019: 71; Hickel, 2019: 874). Target 12.2 sets targets on production and consumption patterns including 'sustainable management and efficient use of natural resources'. Although the goal deals with efficiency improvements rather than sufficiency in the sense of self-limitation and renunciation (Sachs, 2017: 2581; Hagedorn and Wilts, 2019: 124;), SDG 12 clearly focuses on biosphere protection. In the context of this goal, several scholars have pointed out that sustainable consumption and production (SCP) has suffered from being addressed only as an add-on (Bengtsson et al, 2018). However, with Agenda 2030, actors in many areas now have to work with SCP-related targets under their goals. SDG 12 is the most connected to other goals through subtargets (14 other goals in total: see Le Blanc, 2015: 180).

Following Le Blanc (2015: 181, 182), SDG 12 is most strongly linked to SDG 4 (Quality education) and further to SDG 5 (Gender equality). Target 12.8 wants 'people everywhere (to) have the relevant information and awareness for ... lifestyles in harmony with nature'. However, there is no explicit mention of nature or the environment in SDG 4. By contrast, target 5a states: 'Undertake reforms to give women equal rights to economic resources, as well as access to ownership and control over land and other forms of property, financial services, inheritance and natural resources, in accordance with national laws' (Gunawan et al, 2020; UN SDG, 2022). Hence, only SDG 12 and SDG 5 constitute environmentally relevant goals on the basis of their subtargets.

Moreover, Sachs (2017: 2575) and Wackernagel et al (2017: 518) consider SDG 11 (Sustainable cities and communities) as an environmental goal. In total, Sachs names SGD 11, 12 (like Elder and Olsen, 2019, and Hickel 2019), 13, 14 and 15 as the 'five goals to ecological vulnerability', while excluding SDG 6 (in line with Waage et al, 2015, but different from Folke et al, 2016, and others). SDG 11 sets targets for enhancing the quality of life by providing access to open and green spaces for all, sustainable transport systems, sustainable urbanization, sustainable human settlement planning and improved air quality and waste management within sustainable and resilient cities (UN, 2022). Urbanization has significant impacts on the environment, and cities are at the frontline of global change (Kosovac et al, 2020). Cities and local governments have an underestimated potential in improving the health of citizens (SDG 3), and bring to the fore inequalities (SDG 10) in global and national contexts (Koch, 2020). In consequence, Kosovac et al (2020) find that cities are not only an 'an environmental affair' but play a central role in development. SDG 11 is hence not a green goal per se but is relevant to an environmentally sound implementation of Agenda 2030. Although this is also true of SDG 16 (Peace, justice and strong institutions), this latter goal does not include any reference to nature or the environment. Accordingly, no authors were found who consider SDG 16 an environmental goal.

Wackernagel et al (2017) provide a list of seven 'resource relevant goals'. Besides SDGs 6 and 11–15, these authors name SDG 7 (Affordable and clean energy). Similar to SDG 6, SDG 7 is primarily focused on infrastructure expansion for 'universal access to affordable, reliable and modern energy services' (target 7.1), while increasing 'the share of renewables in the global energy mix' (target 7.2). However, renewable energy production does not necessarily mean protection of natural resources. Biomass-based energy, if grown on deforested land, may have a higher carbon footprint than fossil fuels (SDG 13). In addition, bioenergy productivity is widely assumed to counteract food security (SDG 2) through land competition (Nilsson et al, 2016: 321). Efforts to accelerate SDG 7 through modern agriculture ultimately impact the environment (UNEP, 2020). Even worse, using coal to improve energy access would accelerate climate change and acidify the oceans, undermining environmental sustainability (SDGs 13 and 14), in addition to impairing social well-being by exacerbating damage to health from air pollution (disrupting SDG 3) and so on (see also Nilsson et al, 2016; Nilsen, 2020). In contrast to SDG 6, there is no consideration of scarce resources and environmental restoration in SDG 7. Hence, this goal should not be categorized as a green goal per se but as environmentally relevant due to synergies and trade-offs.

Finally, Gupta and Vegelin (2016: 441–2) provide the longest list, with a total of 11 SDGs, for which they identify environmental subtargets: SDG 1 (No poverty), SDG 2 (Zero hunger), SDGs 6–8 (Decent work and economic growth) and 9 (Industry, innovation and infrastructure) and again SDGs 11–15. In addition to the scholars mentioned before, Gupta and Vegelin record that target 1.4 aims for 'ownership and control over land and ... natural resources'. Target 2.4 mentions the need to 'increase (agricultural) productivity and production, that help maintain ecosystems, that strengthen capacity for adaptation to climate change, extreme weather, drought, flooding and other disasters and that progressively improve land and soil quality'. However, Agenda 2030 does not require biosphere protection as a necessary condition to accomplish SDGs 1 and 2. Therefore, while the two goals should not be classified as green goals, at least SDG 2 should be considered as highly environmentally relevant through its subtarget 2.4 and potential trade-off with the green goals (Breitmeier et al, 2021).

SDG 8 (Decent work and economic growth) is the goal that has caused most debates regarding its environmentally destructive impact (Hickel, 2019; Nilsen, 2020). The goal calls for both 'sustained' and 'sustainable' economic growth and employment. Developing countries insisted on headline goals of economic growth (SDG 8) in combination with industrialization (SDG 9) (Elder and Olsen, 2019: 77). Target 8.1 is the only one with a specific numerical objective. On the one hand, it defines per capita economic growth of 'at least 7 per cent gross domestic product growth per annum in the least developed countries'. On the other hand, target 8.4 aims to 'improve progressively, through 2030, global resource efficiency in consumption and production and endeavour to decouple economic growth from environmental degradation' (UN, 2022). Considering this latter subtarget, Gupta and Vegelin (2016: 441–2) argue that SDG 8 serves environmental protection. By contrast, Hickel (2019: 875) calculates that the defined minimum growth rate translates into aggregate global GDP growth of 3 per cent per year. If the global economy grows at such a rate, so he argues, the world would need to achieve emissions reductions of 4 per cent and a decoupling (or decarbonization) of 7.29 per cent per year. Otherwise, it would not be possible to keep global warming to well below 2 °C above preindustrial levels, as defined by the Paris Agreement (Hickel, 2019: 882). Hickel outlines that such decoupling is not feasible on a global scale. In consequence, he demands that target 8.1 on GDP growth be removed (Hickel, 2019: 881). Therefore SDG 8 should definitely not be considered a green goal, but it is highly relevant in terms of environmental trade-offs.

In SDG 9, 'sustainable industrialization' (target 9.2) and 'increased resource-use efficiency' (target 9.4) recognize the concept of limited ecospace (Gupta and Vegelin, 2016: 440). However, developing countries successfully advocated a soft formulation here (Elder and Olsen, 2019: 77), and target 9.2 aims to 'significantly raise industry's share of employment and gross domestic product in line with national circumstances, and double its share in least developed countries'. Hence, GDP growth, defined by SDG 8, should be primarily industrial (Hickel, 2019: 874). Biosphere protection is at best a secondary goal, as SDG 9 does not recognize planetary boundaries. SDG 9, especially in combination with SDG 8, becomes highly environmentally relevant in regard to trade-offs.

Finally, there is SDG 17 (Partnerships for the goals), which is the only goal outside of the circle in Waage et al's (2015) figure. For Folke et al (2016), it is considered an integrative goal that forms the middle axis of their 'wedding cake'. SDG 17 considers mainly economic means for the implementation of Agenda 2030 and policy coherence for 'global macroeconomic stability' (target 17.14). Hickel (2019: 882) highlights that target 17.19 states: 'By 2030, build on existing initiatives to develop measurements of progress on sustainable development that *complement* gross domestic product, and support statistical capacity-building in developing countries' (emphasis added). The term 'complement' here reveals that GDP remains the dominant indicator of progress.

Target 17.7 aims to 'promote the development, transfer, dissemination and diffusion of environmentally sound technologies to developing countries on favourable terms, including on concessional and preferential terms, as mutually agreed' (UNSDG, 2022). However, countries may also choose environmentally destructive technologies. As in the case of SDG 11, it is therefore crucial for the environment how SDG 17 is implemented. Both goals are particularly relevant for an environmentally sound implementation of Agenda 2030.

In sum, four green goals constitute the core of the environmental content of Agenda 2030: SDGs 6, 13–15 (first part of this volume). In addition, through their subtargets, there are the goals that consider environmental trade-offs and synergies: SDGs 5, 7, 8 (and 9) and 12 (second part of this volume). Finally, two goals that are especially relevant for an environmentally sound implementation of Agenda 2030 – SDGs 11 and 17 – also need to be considered (third part of this volume) (see Table 1.1).

1.3 Organization of this volume

The previous section made it clear that the SDGs are far from redefining an economic paradigm based on a 'safe operating space' (Rockström et al, 2009) and from advocating a cultural change towards cooperative economics and politics for the commons (Braunmühl, 2017; Sachs, 2017). Governments

Table 1.1: The environmental content of Agenda 2030

Environmental SDGs (the green goals)	SDGs with environmental trade-offs and synergies	SDGs relevant for an environmentally sound implementation
SDG 6 Clean water and sanitation	SDG 2 Zero hunger	SDG 11 Sustainable cities and communities
SDG 13 Climate action	SDG 5 Gender equality	SDG 17 Partnerships for the goals
SDG 14 Life below water	SDG 7 Affordable and clean energy	
SDG 15 Life on land	SDG 8 Decent work and economic growth	
	SDG 12 Responsible consumption and production	

have mandated the United Nations to follow up, monitor and review all commitments related to sustainable development, as well as to mobilize means of implementation. The 2012 UN Conference on Sustainable Development created the High-Level Political Forum on Sustainable Development to orchestrate all efforts, replacing the CSD (Ocampo and Gómez-Arteaga, 2016).

While many scholars are currently busy discussing indicators to articulate goal achievements and rank countries' efforts towards sustainability (for a critique see Fukuda-Parr and McNeill, 2019), this volume focuses on the Agenda's environmental content and takes a critical account of sustainability governance over the last few decades. The aim is to provide a political science introduction to the most relevant topics of global environmental governance. To facilitate access to the topics, chapters of this volume are each followed by an interview with the authors. These interviews were conducted by master's students. In addition, the authors gave public lectures on their chapters and engaged in discussion. The recordings of these lectures are available online.[2]

The book is divided into three sections. The first part deals with the green goals, that is, SDGs 6 and 13–15. Chapter 2 starts with climate action, as this is the most institutionalized field of global environmental governance. Jens Marquardt and Miranda Schreurs outline interlinkages between SDG 13 and the Paris Agreement on climate change. They demonstrate that both recognize that climate change and development need to be addressed together not only to avoid harmful trade-offs and high costs, particularly for poorer countries, but also to exploit the benefits that come from strengthening these linkages. On a more critical note, in Chapter 3 Daniela Kleinschmit et al link discourses on 'life on land' (SDG 15) to questions of change in governance arrangement since the Rio Summit in 1992. They reveal and criticize dominant 'selling nature to save it' storylines, especially, with regard to Global North/South asymmetries.

In Chapter 4, Alice B.M. Vadrot continues to outline developments regarding 'life below water' (SDG 14). She shows that, although SDG 14 precedes SDG 15 in Agenda 2030, ocean governance is less institutionalized compared to biodiversity and forest governance. Vadrot shows how SDG 14 builds on previous efforts to negotiate a new legally binding instrument for the protection and sustainable use of marine biodiversity beyond national jurisdiction (BBNJ). Her chapter uses the BBNJ case to demonstrate how different principles, norms and legal systems that are applied to different maritime zones and marine resources continue to challenge the protection of the ocean. Following this, in Chapter 5 Manuel Fischer et al discuss the development of water sustainability principles and related institutions and actors since the 1992 International Conference on Water and the Environment in Dublin. The output from this conference, the Dublin

Declaration, was presented at the Rio Earth Summit a few months later that year, where the UN Framework Convention on Climate Change (UNFCCC) was adopted. The authors use three case studies in Bolivia, Ecuador and Switzerland to demonstrate that global water management (SDG 6) has both synergies and trade-offs with climate action (SDG 13). Their chapter outlines types of innovative governance arrangements that local municipalities are using to sustainably address water and climate issues in the Global North and South.

The second part of the book contains chapters on the SDGs with subtargets that signify environmental synergies and trade-offs. Here, Lyla Mehta et al look at water (SDG 6) too, while outlining linkages to access to land and food (SDG 2) and reducing inequality (SDG 10). The authors argue for a reframing of the debate concerning production processes, waste and food consumption while proposing alternative strategies to improve land and water productivity, putting the interests of marginalized and disenfranchised groups upfront. Mehta et al highlight that land and water rights often go hand in hand, and are marked by gender, caste, racial and other exclusions. In Chapter 7, Sandra Schwindenhammer and Lena Partzsch demonstrate the robustness of food security in conjunction with paradigms of productivism and technological innovation in global agri-food governance. With the fourth subtarget, SDG 2 (Zero hunger) requires ecosystem protection, while there is no commonly used indicator yet for monitoring. In consequence, SDG 2 is likely to invoke multiple synergies and trade-offs with the green goals of Agenda 2030. In a similar vein, in Chapter 8, Nopenyo E. Dabla and Andreas C. Goldthau note that accelerating SDG 7 (Affordable and clean energy) by increasing the share of renewables in the global energy mix would mitigate climate change and, hence, increase environmental sustainability. At the same time, the authors also show caution against some forms of energy production such as biomass-based renewables may counteract climate mitigation efforts (SDG 13) if grown on deforested land. In addition, biomass-based energy expansion increases competition for land with agriculture and nature.

In Chapter 9, Ekaterina Chertkovskaya problematizes SDG 8 (Decent work and economic growth) which contradicts the green targets due to the impossibility of decoupling GDP growth from material and energy throughput. Her chapter also pays attention to some of the ways in which growth is expected to be achieved, such as expansion of industrial activity (SDG 9). In Chapter 10, Sherilyn MacGregor and Aino Ursula Mäki present a critical assessment of how objectives of ecofeminism and gender equality (SDG 5) are articulated in Agenda 2030. Their chapter outlines risks and possibilities associated with linking developmental and environmental goals with the pursuit of gender. Following an ecofeminist interrogation, the authors suggest that rectifying gender injustice requires both an intersectional approach and political goals for commoning care work to redress the structural dimensions of gendered and racialized inequality. In Chapter 11, Sylvia Lorek et al critically analyse the notion of sustainable consumption and production (SDG 12). While much of the debate about sustainable consumption and production has revolved around efficiency and technological innovation, less attention has been given to the dimension of social innovation, such as how social power relations and actor roles are changing (or could change) in the process of making consumption and production more sustainable.

Finally, the third part of this volume is about the relevant goals for an environmentally sound implementation of Agenda 2030. It looks at the role that cities (SDG 11) and partnerships (SDG 17) might play as incubators of scalable and transferable social innovations. In Chapter 12, Anna Kosovac and Daniel Pejic emphasize governance pressure to create an urban focus in Agenda 2030. Agenda 21 (adopted in 1992) had already stated that 'by the turn of the century, the majority of the world's population will be living in cities' (para 7.3) and 'urban settlements, particularly in developing countries, are showing many of the symptoms of the global environment and development crisis ... if properly managed can develop the capacity to ... improve the living conditions of their residents and manage natural resources in a sustainable way'. Kosovac and Pejic highlight, however, that the most common theme aligned with city mentions in UN documents since Rio is not the environment. They discuss the intricacies of SDG 11 (Sustainable cities and communities), and the goal's intersections with other SDGs. In Chapter 13, Montserrat Koloffon Rosas and Philipp Pattberg explain that SDG 17 (Partnerships for the goals) calls for partnerships as the main vehicle of delivering sustainable development globally. Empirically focusing on partnerships that work between SDG 13 (Climate action) and SDG 15 (Life on land), their chapter scrutinizes synergies and conflicts between partnerships working in different fields, analyses the level of integration of development interests in environmental partnerships and suggests avenues for governance reform.

The volume concludes with a synthesis chapter that highlights the prevailing, but controversial perception of the environment as a global commodity. Looking at actors and institutions, it outlines the highly fragmented and polycentric landscape of global sustainability governance. Planetary boundaries do not contradict development goals per se. However, innovative and alternative forms of governance that integrate environmental, social and economic goals are limited to voluntary actions. There are alarming signs that governments are generally trading off the environment in their implementation of these goals. Therefore, what comes after the SDGs and whether humans want to continue along chosen paths need to be considered seriously.

Notes

[1] Griggs et al reframe the definition of the 1987 Brundtland Report here, which invented the three-pillar concept. The original definition is: "Sustainable development is development that meets the needs of the present without compromising the ability of future generations to meet their own needs" (WCED, 1987: 41).

[2] The authors gave public lectures on their chapters and engaged in discussion with students. The recordings of these lectures are available online at www.youtube.com/playlist?list= PLHT9ScVgSX3mnVpPFkPwekHloNVb-WOpu [accessed 31 May 2023].

References

Bansard, J.S., Pattberg, P. and Widerberg, O. (2017) 'Cities to the rescue? Assessing the performance of transnational municipal networks in global climate governance', *International Environmental Agreements: Politics, Law and Economics*, 17(2): 229–46.

Bengtsson, M., Alfredsson, E., Cohen, M., Lorek, S. and Schroeder, P. (2018) 'Transforming systems of consumption and production for achieving the Sustainable Development Goals: moving beyond efficiency', *Sustainability Science*, 13(6): 1533–47.

Bernstein, S. (2017) 'The United Nations and the governance of Sustainable Development Goals', in N. Kanie and F. Biermann (eds) *Governing through Goals*, Cambridge, MA: MIT Press, pp 213–40.

Bowen, K.J., Cradock-Henry, N.A., Koch, F., Patterson, J., Häyhä, T. and Vogt, J. (2017) 'Implementing the 'Sustainable Development Goals': towards addressing three key governance challenges – collective action, trade-offs, and accountability', *Current Opinion in Environmental Sustainability*, 90–96.

Braunmühl, C. von (2017) 'Feministische Diskurse zu Entwicklungspolitik und Entwicklungstheorie', in H.-J. Burchardt, S. Peters and N. Weinmann (eds) *Entwicklungstheorie von heute – Entwicklungspolitik von morgen*, Baden-Baden: Nomos, pp 133–50.

Breitmeier, H., Schwindenhammer, S., Checa, A., Manderbach, J. and Tanzer, M. (2021) 'Aligned sustainability understandings? Global inter-institutional arrangements and the implementation of SDG 2', *Politics and Governance*, 9(1): 141–51.

Brühl, T. (2018) 'Die Zusammenführung von Entwicklungs- und Umweltagenda: Hat sie Vorteile?', in T. Debiel (ed) *Entwicklungspolitik in Zeiten der SDGs: Essays zum 80. Geburtstag von Franz Nuscheler*, Duisburg: Institut für Entwicklung und Frieden, pp 18–31.

Donald, K. and Way, S.-A. (2016). 'Accountability for the sustainable development goals: A lost opportunity?' *Ethics & International Affairs*, 30(2): 201–13.

Elder, M. and Olsen, S.H. (2019) 'The design of environmental priorities in the SDGs', *Global Policy*, 10(S1): 70–82.

Folke, C., Biggs, R., Norström, A.V., Reyers, B. and Rockström, J. (2016) 'Social-ecological resilience and biosphere-based sustainability science', *Ecology and Society*, 21(3): art 41.

Forestier, O. and Kim, R.E. (2020) 'Cherry-picking the Sustainable Development Goals: goal prioritization by national governments and implications for global governance', *Sustainable Development*, 28(5): 1269–78.

Fukuda-Parr, S. and McNeill, D. (2019) 'Knowledge and politics in setting and measuring the SDGs: introduction to special issue', *Global Policy*, 10(S1): 5–15.

Griggs, D., Stafford-Smith, M., Gaffney, O., Rockström, J., Öhman, M.C., Shyamsundar, P. et al (2013) 'Sustainable development goals for people and planet', *Nature*, 495(7441): 305–7.

Gunawan, J., Permatasari, P. and Tilt, C. (2020) 'Sustainable development goal disclosures: do they support responsible consumption and production?', *Journal of Cleaner Production*, 246: 118989.

Gupta, J. and Nilsson, M. (2017) 'Toward a multi-level action framework for Sustainable Development Goals', in N. Kanie and F. Biermann (eds) *Governing through Goals*, Cambridge, MA, MIT Press, pp 275–94.

Gupta, J. and Vegelin, C. (2016) 'Sustainable development goals and inclusive development', *International Environmental Agreements: Politics, Law and Economics*, 16(3): 433–48.

Hagedorn, W. and Wilts, H. (2019) 'Who should waste less? Food waste prevention and rebound effects in the context of the Sustainable Development Goals', *GAIA – Ecological Perspectives for Science and Society*, 28(2): 119–25.

Hajer, M., Nilsson, M., Raworth, K., Bakker, P., Berkhout, F., De Boer, Y. et al (2015) 'Beyond cockpit-ism: four insights to enhance the transformative potential of the Sustainable Development Goals', *Sustainability*, 7(2): 1651–60.

Hickel, J. (2019) 'The contradiction of the Sustainable Development Goals: growth versus ecology on a finite planet', *Sustainable Development*, 27(5): 873–84.

Kalfagianni, A., Partzsch, L. and Widerberg, O. (2020) 'Transnational institutions and networks', in F. Biermann and E.K. Rakhyun (eds) *Architectures of Earth System Governance: Institutional Complexity and Structural Transformation*, Cambridge: Cambridge University Press, pp 75–96.

Kamau, M., Chasek, P.S. and O'Connor, D.C. (2018) *Transforming Multilateral Diplomacy: The Inside Story of the Sustainable Development Goals*, Abingdon: Routledge.

Kanie, N., Bernstein, S., Biermann, F. and Haas, P.M. (2017) 'Introduction: Global governance through goal setting', in N. Kanie and F. Biermann (eds) *Governing through Goals*, Cambridge, MA, MIT Press, pp 1–27.

Koch, F. (2020) 'Cities as transnational climate change actors: Applying a Global South perspective', *Third World Quarterly*, 36(3): 1–19.

Kosovac, A., Acuto, M. and Jones, T.L. (2020) 'Acknowledging urbanization: a survey of the role of cities in UN frameworks', *Global Policy*, 11(3): 293–304.

Le Blanc, D. (2015) 'Towards integration at last? The Sustainable Development Goals as a network of targets', *Sustainable Development*, 23(3): 176–87.

Nilsen, H.R. (2020) 'Staying within planetary boundaries as a premise for sustainability: on the responsibility to address counteracting Sustainable Development Goals', *Etikk i praksis – Nordic Journal of Applied Ethics*, 14(1): 29–44.

Nilsson, M., Griggs, D. and Visbeck, M. (2016) 'Policy: map the interactions between Sustainable Development Goals', *Nature News*, 534(7607): 320.

Ocampo, J.A. and Gómez-Arteaga, N. (2016) 'Accountability in international governance and the 2030 Development Agenda', *Global Policy*, 7(3): 305–14.

Partzsch, L. (2020) *Alternatives to Multilateralism: New Forms of Social and Environmental Governance*, Cambridge MA: MIT Press.

Raworth, K. (2017) *Doughnut Economics: Seven Ways to Think Like a 21st-century Economist*, London: Random House.

Rockström, J., Steffen, W., Noone, K., Persson, Å., Chapin, F.S., Lambin, E.F. et al (2009) 'A safe operating space for humanity', *Nature*, 461(7263): 472–75.

Sachs, W. (2017) 'The Sustainable Development Goals and *Laudato si*': varieties of post-development?, *Third World Quarterly*, 38(12): 2573–87.

Steffen, W., Richardson, K., Rockström, J., Cornell, S.E., Fetzer, I., Bennett, E.M. et al (2015) 'Planetary boundaries: guiding human development on a changing planet', *Science*, 347(6223).

Swilling, M. (2020) *The Age of Sustainability: Just Transitions in a Complex World*, Abingdon: Routledge.

Tosun, J. and Leininger, J. (2017) 'Governing the interlinkages between the Sustainable Development Goals: approaches to attain policy integration', *Global Challenges*, 1(9): 1700036.

UN (United Nations) (2015) *The Millennium Development Goals Report 2015*, New York: United Nations.

UN (2022) Sustainable Development Goals, Available from: www.un.org/sustainabledevelopment [accessed 31 May 2023].

UNDP (United Nations Development Programme) (1990) *Human Development Report*, New York: Oxford University Press.

UNDP (2020) *Human Development Report 2020: The Next Frontier*, Available from: http://hdr.undp.org/sites/default/files/hdr2020.pdf [accessed 31 May 2023].

UNEP (UN Environment Programme) (2020) 'Monitoring progess', Available from: https://www.unenvironment.org/explore-topics/sustainable-development-goals/what-we-do/monitoring-progress [accessed 31 May 2023].

Waage, J., Yap, C., Bell, S., Levy, C., Mace, G., Pegram, T. et al (2015) 'Governing the UN Sustainable Development Goals: interactions, infrastructures, and institutions', *Lancet Global Health*, 3(5): e251–e252.

Wackernagel, M., Hanscom, L. and Lin, D. (2017) 'Making the Sustainable Development Goals consistent with sustainability', *Frontiers in Energy Research*, 5: 518.

WCED (World Commission on Environment and Development) (1987) *Our Common Future: World Commission on Environment and Development*, Oxford: Oxford University Press.

PART I

The Green Goals

2

Governing the Climate Crisis: Three Challenges for SDG 13

Jens Marquardt and Miranda Schreurs

The 13th Sustainable Development Goal of the United Nations Sustainable Development Agenda addresses climate change, considered by many to be one of the most existential threats to humanity (UN, 2021), and described 'as the defining issue of our time' (UN, 2019). SDG 13 shapes not only other environmental SDGs like forest protection (see Chapter 3 on SDG 15 by Kleinschmit et al) and marine ecosystems (see Chapter 4 on SDG 14 by Vadrot); but also broader socio-economic targets such as clean energy access (see Chapter 8 on SDG 7 by Dabla and Goldthau) and sustainable production and consumption (see Chapter 11 SDG 12 by Lorek et al). SDG 13 is thereby confronted with what Partzsch (see Introduction in this volume) highlights as a critical tension in Agenda 2030: the need to balance environmental protection with other socio-economic priorities.

Attempts to govern climate change have developed over decades, long before Agenda 2030 came into being as a global sustainable development framework. Since measurements of carbon dioxide in the atmosphere began at the Mauna Loa Observatory in 1958, there has been a steady rise in their levels. In the early 2020s, they had risen to approximately 415 parts per million (ppm) (Lindsey, 2021) compared to a pre-industrial level of about 280 ppm (IPCC, 2018). Such high levels were last seen during the Pliocene era, over four million years ago (NOAA, 2021). Rising greenhouse gases in the atmosphere are acting like a blanket around the planet, trapping inbound solar radiation and warming the earth at unprecedented rates in the last 2,000 years (NASA, 2021). As a result, the average surface temperature on the planet has risen about 1.2 °C (2.16 °F) since pre-industrial levels.

Left unchecked, global warming has the potential to make many parts of the planet uninhabitable. It will contribute to widespread species dieback and extinction, intensify hunger, speed the spread of deadly diseases and add fire to the flames of ethnic and religious conflicts. The most vulnerable will be left struggling to survive. The massive loss of glaciers and the melting of Antarctic ice will not be reversible for tens or even hundreds of thousands of years. Yet, there is still time to act to prevent the worst impacts of climate change. As the UN Secretary-General António Guterres (2019) stated, 'the climate emergency is a race we are losing, but it is a race we can win'. This, however, requires ambitious and urgent collective action.

The international community began dealing with the global climate crisis in the early 1990s. In 1994 the United Nations Framework Convention on Climate Change (UNFCCC) took effect, paving the way for a global climate change regime. If the world community acts and makes deep cuts in greenhouse gas (GHG) emissions, the climate crisis could be the trigger to lead humanity towards developing a greener and more just world (Schlosberg and Collins, 2014; Porter et al, 2020). Combating climate change could create synergies with and reinforce all the other SDGs (Fuso Nerini et al, 2019; Venkatramanan et al, 2021). For example, shifting away from fossil fuels towards affordable and clean energy sources (SDG 7) will result in a more environmentally friendly energy supply and can support energy independence, create local jobs and trigger community empowerment and more democratic energy systems (Ram et al, 2022; Wahlund and Palm, 2022). Adopting climate-friendly approaches to urban design, changing lifestyles to become more sustainable and promoting shared mobility concepts would certainly also make cities far more liveable (Mendizabal et al, 2018).

This chapter examines why combating climate change internationally has been so cumbersome despite the many ecological, social and economic benefits that can be anticipated with early action. A myriad of forces have delayed, prevented or in some cases reversed ambitious climate action. While there are certainly technological barriers that still need to be overcome, and the immediate financial costs of climate action are considerable, arguably the real opponents of climate action have been powerful vested industries. Particularly determined efforts to slow and block policy reforms have come from fossil fuel industries and the scientists and politicians they have supported (Oreskes and Conway, 2010). This chapter focuses on three key challenges that shape not only SDG 13 specifically, but also the broader climate governance architecture more broadly. The urgent need for more ambitious climate action is confronted by various *governance challenges* including the voluntary focus of the international climate governance framework; the *responsibility challenges* that are tied to the quest to pursue ambitious climate action while simultaneously addressing development needs and social inequities; and the *political challenges* stemming from the issue linkages between climate change and other sociopolitical concerns. This chapter addresses these challenges as well as the movements and actors calling for climate action now.

2.1 Governance challenges: tackling climate change internationally

Any attempts to solve a complex and 'wicked problem' (Lazarus, 2009) such as climate change are confronted with governance challenges related to coordination, the unequal distribution of power and knowledge imbalances. These challenges have shaped the climate talks under the UNFCCC for decades. SDG 13 acknowledges the UNFCCC as the primary forum for negotiating the global response to climate change, with the Paris Agreement as the key guiding document. The climate governance framework established in Paris in 2015 rests on governments' voluntary commitments to act. The process of arriving even at this weak consensus on the need for transformative action was slow and frustrating, an indication of just how powerful incumbent industries and fossil fuel exporting countries remain. To date, national policy plans and actions still do not add up to the level of action needed to prevent dangerous increases in global average temperatures, despite some signs that a critical juncture may have been reached as climate awareness strengthens and renewable energy rapidly expands. What follows is a brief summary of the targets under SDG 13 and their links to the international climate governance landscape.

2.1.1 Governing by goals: SDG 13

Largely referring to what was agreed upon in Paris, SDG 13 outlines an ambition to limit global warming to 1.5 °C above pre-industrial levels by the end of this century. SDG 13 consists of the following key targets:

- 13.1 recognizes that many people and regions worldwide are already facing the devastating effects of climate change. Resilience and adaptive capacity to climate-related disasters should be strengthened.
- 13.2 calls for integrating climate change measures into broader political agendas, national policies, strategies and planning.
- 13.3 considers education, human capacity and knowledge as prerequisites for tackling the climate crisis. Climate change education should be mainstreamed into national education policies and curricula.
- 13.a reiterates the commitment made by developed countries to jointly mobilize $100 billion annually by 2020 under the UNFCCC to address climate change mitigation in the Global South (a target that has still not been achieved as of 2022).

- 13.b calls for mechanisms to raise the capacity for effective climate change-related planning and management in the most vulnerable countries, focusing particularly on women, youth and marginalized communities.

The targets for SDG 13 are thin in scope and broadly formulated. This means that climate action has been largely defined by decisions made during the international climate negotiations.

2.1.2 The long road to Paris

Climate scientists issued some of their earliest warnings about global warming in the 1970s. The first World Climate Conference in 1979 led to the creation of the Intergovernmental Panel on Climate Change (IPCC) in 1988. The IPCC provides governments with regular reports on the state of scientific, technical and socio-economic knowledge on climate change (Bolin, 2007). Politically, climate change has been on the international agenda since at least the 1992 United Nations Conference on Environment and Development (UNCED) when the world community recognized climate change as a matter of global concern and established the UNFCCC. Based on this convention, a Conference of the Parties (COP) takes place generally once a year. In 1997 the Kyoto Protocol was signed as the first international agreement addressing climate change, although it came into effect only in 2005 after enough national parliaments had ratified it. The Kyoto Protocol obliged the wealthier countries of the world to reduce their combined GHG emissions by 5.2 per cent below 1990 levels by the period 2008–12 (Bohringer, 2003). The protocol's effectiveness was, however, greatly limited by the failure of the United States to ratify the agreement and by Canada's decision to pull out of it just before it was due to expire (Schott and Schreurs, 2020). Efforts to negotiate a successor agreement dragged on for years. Hopes were high that a global agreement would be reached in Copenhagen in 2009, but delegates failed to bridge their differences.

In parallel to the negotiations under the UNFCCC, nations began discussing sustainability and development issues. At the World Summit on Sustainable Development in Johannesburg in 2002, over 100 heads of state and government committed to achieving development while protecting the environment, thereby recognizing climate change effects in sectors like water and agriculture (UN, 2002). In the subsequent formation of the sustainable development goals, developed countries were required to accept that global progress on climate change also required action on other goals, such as poverty alleviation, education and gender equality (Udapudi and Sakkarnaikar, 2015).

Shortly after the SDGs were announced in 2015, parties to the UNFCCC adopted the Paris Agreement on climate change. The Paris Agreement called on the global community to act to reduce GHGs, as well as to prepare for and adapt to the consequences of climate change, including sea level rise and more frequent and extreme weather events. The Paris Agreement set a goal to prevent a rise in global average temperatures to 2 °C above pre-industrial levels as this was the level beyond which scientists concurred that tipping points could be reached beyond which the natural climate system could be irreversibly altered, putting humanity at great risk (Knutti et al, 2016). For small island states and low-lying countries this target was insufficient; they pressurized instead for a 1.5 °C upper temperature limit. Even this level brings with it serious risks, for example from rising sea levels. Unable to reach a consensus, the Paris Agreement calls on its signatories to hold the increase to well below 2 °C and strive to stay within 1.5 °C (UNFCCC, 2015).

2.1.3 Voluntary initiatives

Given the lack of global acceptance of an agreement with legally binding targets, at the Paris climate negotiations (COP 21) hopes were placed on a 'hybrid multilateralism' (Kuyper et al, 2018). It assumed that, under the prevailing political realities, progress on the climate crisis would be politically achievable only through collaborative action, voluntary commitments, win–win solutions, and the development of synergies. Political stalemate at the climate negotiations in Copenhagen (COP 15) in 2009 required negotiations to make a shift from aiming for a *regulatory* regime to accepting a *catalytic and facilitative* model. Thus, the Paris Agreement incorporated voluntary commitments of action (nationally determined contributions, or NDCs) that are deposited with the UNFCCC Secretariat. The agreement further incorporates a regular review mechanism intended to pressurize states to examine the effectiveness of their measures and to assess the latest scientific findings and climate developments so that they can adjust their climate commitments and ambitions accordingly. Finally, the agreement includes non- and substate actors far more directly than previous regimes (Hale, 2016). The UNFCCC Global Climate Action Portal identifies the climate action pledges and commitments of more than 29,000 different non- and substate actors. Countries' voluntary national reviews of their SDG efforts pay considerable attention to these kinds of climate change actions (Elder and Bartalini, 2019).

While critics question the effectiveness of a system that relies so heavily on private actors as standard setters and where accountability mechanisms remain weak (Streck, 2020), there are also positive dimensions to this 'era of nonstate climate leadership' (MacLean, 2020), with its polycentric and voluntary characteristics (Ostrom, 2009).

An example is the United Kingdom's effort to pull together a club of countries to agree to phase out coal, the dirtiest of the fossil fuels. By

November 2021, more than 40 countries had joined this informal club, although critics point out that the biggest coal users are not on board and phase-out dates remain too late (Harvey et al, 2021). Students' and citizens' groups have spearheaded divestment campaigns, urging pension funds, governments and financial institutions to divest from fossil fuels. In response, the Norwegian Pension Fund, the world's largest sovereign wealth fund, decided to divest from fossil fuels (Ambrose, 2019). Forbes reports that the divestment movement is a $14.5 trillion movement with over a thousand major investors (Carlin, 2021).

A growing number of countries, companies and organizations, including universities, have committed to net zero carbon targets. The UN-backed Race to Zero campaign highlights several such initiatives (UNFCCC, 2021). A growing number of countries have also adopted climate neutrality goals (Wallach, 2021), and companies have taken on the net zero challenge, including even various energy companies such as BP, Repsol and Sasol (Geck, 2021). A 2019 survey found that 13 of 132 energy companies had formulated their own net zero targets (Dietz et al, 2019). The UN, however, warns that much more needs to be done to stay within a 1.5 °C warming (IPCC, 2021).

2.1.4 Nationally Determined Contributions and the emissions gap

Several countries have notched their NDCs upwards, but the commitments that have been made are still far short of what is needed to stop the world from entering a real temperature danger zone (Climate Analytics and Next Climate Institute, 2021a). In November 2018 the United Nations issued an emissions gap report indicating that the G20 countries, the largest economies in the world whose combined emissions accounted for almost 80 per cent of global emissions, were not doing enough to rein in emissions growth, putting their 2030 pledges at risk (UNEP, 2018). The IPCC sent out a stark warning in the same year that the time frame available to stay within a 1.5 °C target was rapidly closing (IPCC, 2019). The Climate Action Tracker, a scientific analysis by a consortium of climate research organizations, estimated that the pledges and commitments made at the Paris COP in 2015 were leading the world in the direction of a 2.5 °C to 2.9 °C temperature increase, even if all pledges were to be fully implemented (Climate Analytics and Next Climate Institute, 2019). In 2020 at COP26 in Glasgow, Scotland, an increasingly concerned public put pressure on policy makers to strengthen their pledges. In November 2021 the Climate Action Tracker assessed that, even if all of the pledges countries made for 2030 are to be fulfilled, there is still a 50 per cent chance that global temperatures will be 2.4 °C higher than pre-industrial levels and a 95 per cent chance that the 1.5 °C target will be missed (Climate Analytics and Next Climate Institute, 2021b).

2.2 Responsibility challenges: unequally distributed emissions

Tackling the climate crisis hinges on questions of fairness, equity and responsibility. The world's biggest GHG emitters are spread across the Global North and South and are responsible for more than 50 per cent of global emissions (Friedrich et al, 2020). The climate commitments made by the four largest emitters – China, the US, the EU and India – are considered briefly later. In addition, the situations in Brazil, Indonesia and Tuvalu are introduced as snapshots of the highly heterogenous group of countries that make up the Global South. For these countries, climate action competes with other development needs and priorities, raising in turn, responsibility issues for wealthier countries and those with high historical and current emissions.

2.2.1 Large emitters from the Global North: the US and the EU

Carbon emissions mainly arise from industrial production and fossil fuel consumption. Thus, countries that were the first to industrialize have historically contributed most to global warming.

Historically, the *United States* has emitted more greenhouse gases into the atmosphere than any other country. Today, it is the world's second largest emitter after China but the largest emitter from a cumulative historical perspective. Depending on the administration in power, the US has either sought to lead on global climate action or to block multilateral climate agreements. While the William J. Clinton administration (Democrat) signed the Kyoto Protocol, the George W. Bush administration (Republican) rejected the Kyoto Protocol and the Donald J. Trump administration (Republican) pulled the country out of the Paris Agreement. President Barack Obama was unable to convince Congress to pass meaningful climate legislation and thus was largely limited to seeking change through executive action. These executive actions introduced by Obama were largely annulled by Trump. A sharply divided country has stood in the way of finding a consensus on climate action (Schreurs, 2019; Fiorino, 2022).

In the meantime, President Joe Biden brought the US back into the Paris Agreement in 2021. His administration also worked with Congress to pass major climate legislation. The first big success came in the form of the Infrastructure Law, which passed with bipartisan support. This law will channel funds for infrastructure projects, including public transport, rail, electric vehicle (EV) chargers, clean energy transmission and grids, and cleaning up brownfield sites and abandoned mines. There is a strong focus on ensuring environmental justice in the allocation of funding. The second, more complicated and precarious win came with the passage of a special form of budgetary legislation known as a reconciliation bill. The Inflation Reduction Act of 2022 does not sound like a climate bill but is actually the largest climate bill the US Congress has ever passed. It includes funding for renewable energy, batteries, forestry (for climate resilience) and electric vehicles, and sets major GHG emission reduction targets (by about a billion metric tons in 2030).

Given the federal system in the US and as a result of decades of inconsistency in federal action on climate change, subnational actors have also stepped up to the plate. States like California, Oregon, Washington and New York have succeeded in introducing important policies and measures to reduce emissions within their states (Stokes, 2020). US emissions dropped by 7.3 per cent between 1990 and 2020 (US EPA, 2022).

The *European Union* is often perceived as a global leader in international climate negotiations. The block of 27 countries (28 until the UK's exit in 2020) championed the Kyoto Protocol and later the Paris Agreement after the US retreat. The EU met its goals to reduce its GHG emissions by 20 per cent of 1990 levels by 2020, and exceeded its target to achieve 20 per cent renewables in its energy mix. Targets for 2030 announced in 2014 have subsequently been tightened in response to warnings from the IPCC. The EU raised its carbon dioxide reduction ambition from 40 to 55 per cent of 1990 levels. The European Commission is promoting local climate action, for example with an initiative to realize 100 climate-neutral cities by 2030. In 2019 the EU announced the European Green Deal, which aims at climate neutrality by 2050, the development of a circular economy, major improvements in building efficiency, sharp reductions in the use of chemical pesticides in agriculture, large-scale reforestation and leadership in research and development of climate-friendly technologies (Bloomfield and Steward, 2020). The Fit for 55 package outlines steps to be taken by 2030, including raising the ambition and reach of the Emissions Trading System (ETS) to include not only major industries but also the airline and marine sectors. It also calls for updating member states' national targets in areas not covered by the ETS (European Council, 2022).

In reaction to Russia's illegal and devastating invasion of Ukraine, the EU has made extraordinary efforts to reduce dependency on Russian fossil fuels, speed the development of renewables and enhance energy efficiency. The war has become a catalyst for speeding up action on renewables and energy efficiency. At the same time, at least in the short term, Europe is returning to more use of coal to meet gaps in its energy supplies as a consequence of the loss of Russian fossil fuel sources. There is also fear that soaring energy prices could lead to social unrest, which has encouraged European governments to cooperate more on energy and to introduce a variety of measures to cushion consumers and small and medium industries from exploding fuel costs.

2.2.2 Heavyweights in the Global South: China and India

When emissions from land-use change and forestry are included, the Global South is estimated to contribute about 63 per cent of today's total GHG emissions (Fuhr, 2021). These emissions are heavily concentrated. The ten biggest emitters from the Global South are responsible for around 78 per cent of the group's emissions; the remaining 120 countries account for only 22 per cent. China and India alone are responsible for about 60 per cent of all emissions from the Global South. Emissions pathways, climate-related visions and strategies to tackle climate change differ significantly across the Global South.

China overtook the US as the world's largest emitter in the mid-2000s. Given its population of close to 1.4 billion people and an economy that has experienced rapid economic growth since 1980, China's position in the international climate negotiations has shifted. In the early years, it positioned itself as a developing country, arguing that the responsibility for climate change lay primarily with North America, Europe and Japan. China was not required to reduce its emissions under the Kyoto Protocol, and instead became the recipient of technological assistance under the Clean Development Mechanism, a policy instrument designed to allow developed countries to obtain credits towards their own emission reductions by taking actions to reduce or prevent emissions in developing countries (Zhang and Yan, 2015). With its GHG emissions now reaching about 30 per cent of the global total, China has had to accept greater responsibility. At COP 15 in 2009, China was perceived as a blocker of a global climate agreement. Offering some hope to the global community, a decade later in September 2020, China's President Xi Jinping announced that the country would aim to peak its emissions by 2030 and to achieve carbon neutrality by 2060.

China has taken major strides to reduce the carbon intensity of its economy, shuttering highly polluting factories, closing many small and hazardous coal mines, and investing heavily in a modernization of production systems. Between 1978 and 2018, China's economy grew by 176 per cent and its population by 16 per cent, but its CO_2 emissions increased a much smaller sixfold because of sharp declines in energy and carbon intensity (Zheng et al, 2020). China has invested heavily in renewables, accounting for 45 per cent of global investments in renewables in 2020. It has the world's largest renewable energy generation capacity (over 900 GW at the end of 2020). In 2021, China's renewable energy investments outpaced investments in fossil fuels under the Belt and Road Initiative for the first time. While China still dominates global investments in overseas coal power plants (REN21, 2021), the government recently banned the financing of such projects. In addition to promoting hydrogen fuels, circular economy concepts, electric vehicles, and digital technologies, the 14th Five-Year Plan envisions substantial reliance on what it calls the clean and efficient use of fossil fuels (NDRC, 2021).

India has overtaken China as the world's most populated country and its population is expected to be over 1.6 billion by 2050 (PTI, 2019). This will put additional burdens on an economy in transition that is still struggling to supply its entire population with their basic needs. Demand for energy and resources will expand significantly in the decades ahead. Yet, average per capita CO_2 emissions remain low, at an average of 1.69 tons per year in 2019, compared with a global average of 4.39 tons.

Reflecting its stage of development and its belief that developed countries should carry the weight of responsibility for addressing climate change, India initially resisted setting a substantial climate neutrality target. This changed in November 2021, when President Narendra Modi announced that India would aim to become carbon neutral by 2070 (McGrath, 2021). Further measures aim to reduce the emissions intensity of GDP by 33–35 per cent from the 2005 level and to obtain 40 per cent of cumulative installed electric power capacity from non-fossil fuel energy resources (renewables and nuclear power), both by 2030. Taking the position of many developing countries, India demands financial and technical assistance from developed countries and the Green Climate Fund, which was set up by the Paris Agreement to help developing countries adapt to and mitigate against climate change. India is eager to become a player in the production and export of clean energy technologies. However, the country also continues to build coal-fired power plants (Varadhan and Sheldrick, 2021), claiming that these are necessary to meet its rapidly growing energy demands.

2.2.3 Divergent perspectives from the Global South

The Global South accounts for the majority of countries in the world. It is an economically, politically and culturally diverse group. Brazil, Indonesia and South Africa are examples of countries that have seen substantial economic progress and positive human development over the past few decades, albeit on the back of widespread environmental degradation and increasing social inequality. Given their current and future GHG trends, the global fight against climate change very much hinges on developments in these countries. On the other side of the spectrum, small island countries with small carbon footprints such as Tuvalu, Fiji and the Maldives are struggling to adapt to climate change. Their stories raise troubling climate justice and equity concerns.

Brazil has one of the largest tracks of rainforest in the world. Yet, large areas have been deforested in response to demands for agricultural land and timber exports, as well as through corruption and illegal logging. Deforestation rates surged under the far right presidency of Jair Bolsonaro. His successor,

Lula da Silva, has pledged to protect the Amazon and its peoples. In April 2022 Brazil published an updated NDC to underline its commitment to reduce GHG emissions by 37 per cent by 2025, and by 50 per cent by 2030 (compared to 2005 levels), and to attain climate neutrality by 2050. The plan notes the country's already high share of renewables, which accounted for 48.4 per cent of total energy demand, 84.8 per cent of electricity and 25 per cent of transport fuel. Brazil is a world leader in the development and consumption of biofuels in heating and transport (Martinelli et al, 2022). However, there are controversies involving the extent to which biofuel strategies contribute to social inequities, biodiversity loss and loss of arable land for agriculture. As part of its NDC, Brazil has committed to ending illegal logging by 2028.

Indonesia is the world's fourth most populous nation, with more than 270 million inhabitants. This lower middle-income country is the largest economy in Southeast Asia and the eighth largest worldwide. With its per capita consumption-based CO_2 emissions of 2.21 tons, excluding land-use emissions, Indonesia's contributions to global warming might be viewed as moderate (World Bank, 2022). However, Indonesia is actually one of the world's largest emitters of GHGs, mainly as a result of the high level of emissions stemming from land use, land-use change and the energy sector, which together are responsible for 80 per cent of the country's emissions. Indonesia originally agreed to reduce emissions by 26 per cent (unconditionally) compared to a 2030 business as usual (BAU) scenario and by up to 41 per cent below the 2030 BAU level, depending on international assistance for finance, technology transfer and capacity building (Wijaya et al, 2017). Decarbonization of its economy was to follow a phased approach involving improvements to land-use policies, energy conservation and renewable energy development (Dunne, 2019). Activists rejected the government's plans as not ambitious enough, and criticized the government for planning to categorize coal gasification, brown hydrogen (developed from fossil fuels) and nuclear energy as 'renewable energy' (Jong, 2021). They pointed out that emissions 'might even double by 2030' compared to 2014 levels if more ambitious actions were not taken (Climate Analytics and Next Climate Institute, 2022). The Indonesian government has in the meantime committed to doing more. In July 2021 it submitted an updated NDC with a 29 per cent unconditional emission reduction target for 2030 and a net zero emissions target for 2060 or sooner (compared to earlier discussions of a 2070 date).

South Africa is Africa's largest economy. Situated in a drought belt, the country regularly experiences severe water shortages. In 2019 Cape Town almost ran out of water (Heggie, 2021). After decades of apartheid, major income inequalities still plague the country, with over half the population living in poverty and a quarter experiencing food poverty. Poor communities are particularly hard hit by climate change. South Africa's national climate adaptation strategy notes that climate change threatens its ability to meet the SDGs (DEFF (Department of Environment, Forestry and Fisheries), South Africa, 2020). It also points out that women experience climate change challenges differently from men. At COP26, South Africa called for developed countries to honour their pledges to provide developing countries with financial and technical support for climate adaptation (Creecy, 2021).

Finally, *Tuvalu* and the other 43 UN member states in the Alliance of Small Island States (AOSIS) contribute only marginally to global GHG emissions but are most seriously threatened by climate change. For Tuvalu, a small island nation in the Pacific with fewer than 12,000 inhabitants, sea level rise and extreme storms will almost certainly mean that its citizens will need to find a new home. AOSIS played a critical role in the Paris climate negotiations in demanding the inclusion of a 1.5 °C target, characterizing this as an existential issue for them (Ourbak and Magnan, 2018). While some areas may be able to get by with climate adaptation strategies, others will suffer an irreversible loss of territory.

2.3 Political challenges: climate change deniers versus climate activists

Climate change touches all sectors of society. There are countless interests and a plethora of different views about how best to address it. Two antagonistic poles are presented here. One is populated by climate change deniers and sceptics who are backed by powerful and wealthy industries and philanthropists. The other is represented by climate movements and their members, many of whom are young and worried about what the future might hold. Climate change is thus a highly politicized field.

2.3.1 Climate change deniers and sceptics

Climate action has been slowed by climate change deniers and sceptics, who question either the extent to which humans are contributing to global warming or whether global warming is happening at all. They found powerful supporters in the likes of former US Senator James Inhofe, former US President Donald Trump and former Czech President Vaclav Klaus. Many prominent climate deniers have either headed up conservative think tanks (the Heartland Institute, the Cato Institute, the Competitive Enterprise Institute) or had their research financed by fossil fuel companies. The movie *Before the Flood* (2016) describes these linkages and raises awareness of the financing behind many climate change deniers (see also Thornton 2023). While climate change denialism is not equally strong in all parts of the world, various far right movements have taken up these arguments. Germany's far right party, Alternative für Deutschland, has for example, questioned the wisdom of the German energy transition and has campaigned against renewable energy projects. Others, such as France's Rassemblement National and Spain's Vox (Onishi, 2019; de Nadal, 2021; Serhan, 2021), reject the need for international solutions. In the wake of the Russian war on Ukraine, there is growing evidence that numerous European far right movements have received funding from the Kremlin (Datta, 2022).

2.3.2 New climate movements

In sharp contrast to the climate deniers, youth climate action is on the rise and has led to new climate movements such as Fridays for Future (FFF) and Extinction Rebellion (XR). These movements have used a broad spectrum of strategies from press conferences to street protests, to more radical civil disobedience, to demand more ambitious climate action. They combine vocal public protest with dedicated policy work, agenda setting and lobbying (Sovacool and Dunlap, 2022). Slowly, the balance on climate action appears to be tipping in their favour, although not with the speed or intensity they rightfully demand.

Greta Thunberg initiated a worldwide movement in August 2018 when she announced the first Skolstrejk för klimatet (school strike for climate). Her efforts gained unprecedented social media attention, helping to launch the global Fridays for Future movement. With support from climate scientists, the movement regards itself as bipartisan and politically neutral. It demands radical and immediate climate action that acknowledges and meaningfully responds to the mounting evidence of climate change (Marquardt, 2020). Mass protests around the world were severely disrupted during the pandemic, but the movement continued to lobby for climate action through various online formats. During COP26 in Glasgow, Thunberg criticized the official UNFCCC negotiations as 'blah, blah, blah' and joined an alternative summit instead.

The Sunshine Movement, which was launched in 2017 in the US, has similarly organized a wide range of protests and carried out policy work. During the 2018 US midterm elections, the group attacked candidates with ties to the fossil fuel industry and supported candidates who were in favour of renewables, such as Alexandria Ocasio-Cortez who promoted a Green New Deal bill. Movement protesters gathered noisily near politicians' residences to wake them up. On the other side of the Atlantic, in the United Kingdom Extinction Rebellion emerged from a network of environmental initiatives 'to spark and sustain a spirit of creative rebellion' (Extinction Rebellion, 2018) and call for immediate action against climate change. More radical than FFF, XR employs more disruptive tactics. Driven by a strong sense of urgency in light of depleting carbon budgets and the rapidly dwindling time left to avoid dangerous anthropogenic climate change, XR engages in disruptive modes of protest such as street sit-ins and blockades of carbon-intensive infrastructures. These more radical strategies and tactics of civil disobedience have been adopted by other movements such as Ende Gelände or Letzte Generation (Sovacool and Dunlap, 2022).

2.4 Three challenges for SDG 13

Numerous actors at multiple levels have sought to advance climate action or delay progress (Jänicke et al, 2015). This has made the governance of climate change complex and conflictual. Three cautionary tales arise when it comes to combating climate change. They translate into key challenges for promoting and implementing SDG 13.

2.4.1 Governance challenge: urgency versus voluntary governance

Climate governance is characterized by a high degree of tension between different actors both within and across countries and raises many concerns about what is being handed down to future generations. Can the growing need for immediate and urgent action to prevent the most devastating effects of climate change be met by voluntary pledges? Will states rachet up their commitments enough over the coming years to make a real difference? Researchers see a serious credibility gap between national announcements and climate trends (Climate Analytics and Next Climate Institute, 2021a). Jernnäs (2021: 60) describes the post-Paris climate governance architecture as merely facilitative and incapable of tackling a global collective action problem like climate change. In her words, it aims to 'meet urgency with voluntarism.' Yet, the flexible regime also holds a chance for increased action as a result of intensifying pressure from below. At least in democratic societies, citizens can hold governments accountable, demand that they implement their NDCs, contest weak commitments and advocate for more ambitious action (Marquardt and Bäckstrand, 2022). As a promising example, the global climate youth movement has not only put climate change back on the agenda of high-level politics, but also shaped elections and domestic political debates. Various forms of climate activism, protest and civil disobedience can be expected to foster societal debates, articulate climate justice concerns and give a voice to marginalized positions (Martiskainen et al, 2020). These activities alone will not be sufficient to solve the climate crisis, and may even be dangerous in some more authoritarian systems, but they can and have had important impacts. Nevertheless, they still need to be accompanied by climate-friendly policies and shifts towards sustainability across all economic sectors and at all levels of government.

2.4.2 Responsibility challenge: climate concerns versus development needs

Mitigating climate change has been framed as the responsibility of the rich countries. Historically, Europe, North America and Japan have contributed the most to global warming, especially in terms of consumption-based emissions (Liddle, 2018). Yet, emission trends are changing the responsibility discussion. Under the 1997 Kyoto Protocol only industrialized countries were obliged to reduce their GHG emissions; in contrast, the 2015 Paris Agreement covers both industrialized and developing countries (Obergassel et al, 2016). Today, developing countries account for about 63 per cent of GHG emissions (Fuhr, 2021), which means that SDG 13 cannot be seriously tackled without considering the development challenges attached to it. Climate change and sustainable development must be considered together to avoid harmful trade-offs especially in the Global South, but also to take advantage of the benefits that can come from fostering interconnections between them. A number of developing countries have responded by setting less ambitious unconditional and more ambitious conditional targets. The latter depend on financial and technical support from the Global North, which has pledged financial support but has to date failed to fully meet its promises.[1]

The climate crisis highlights the need for deeper transformation as it points to the links among existing economic structures, global inequalities and the maldistribution of resources. Various tools have been developed to explore the interlinkages between climate action and the other 16 SDGs. According to Gonzales-Zuñiga et al (2018: 4), the 'synergies outweigh the trade-offs found for most of the SDGs'. Fuso Nerini and colleagues (2019: 675) identify synergies particularly with regard to SDG 2 (Zero hunger), SDG 7 (Affordable and clean energy) and SDG 9 (Industry, innovation and infrastructure) (IGES, 2021).

2.4.3 Political challenges: depoliticization versus politicization

The rise of climate movements and street protests as well as climate-related populism and denial point to a third critical challenge, namely the sociopolitical conflicts, tensions and cleavages attached to climate action. Climate governance has long been a struggle between attempts at politicizing and depoliticizing the climate issue. Some scholars argue that framing climate change in ecomodernist terms has led to a post-political condition where climate change is understood as an ecological but less as a political problem that can be managed and solved through technological innovations (Swyngedouw, 2011). Such a framing has come increasingly under pressure. Right-wing populists have discovered climate politics as a major societal battleground as it reflects a broader ideological dispute between an environmentally friendly elite and the population at large (Marquardt and Lederer, 2022).

Since the early years of this century, the climate issue has turned into a cultural cleavage, shaped by competing world views and ideologies. The more obvious it becomes that tackling climate change is not only about environmental protection but also about fundamental changes in society, the wider the gap has grown between supporters and opponents of climate science and the greater the partisan divide over climate change (Hoffman, 2011). While right-wing populists employ modes of climate science denialism, climate policy nationalism and climate policy conservatism (Vihma et al, 2020: 22), left-wing activists and movements such as XR or FFF, as well as progressive left-wing parties, highlight climate justice concerns and global inequalities to mobilize for more ambitious climate action. These interventions can give voice to typically marginalized interests as well as future generations. Machin (2020) describes this form of engagement as 'ecological agonism', where democratic disagreement over climate change provides an opportunity to develop alternatives, disrupt business as usual policy making, and foster civic participation. Acting for the climate thus means working for democracy and shaping the society humans want to live in. Along those lines, scholars like Willis et al (2022) promote a switch from elitist democratic practices to deliberation-based reforms such as deliberative mini publics to effectively but democratically address climate change.

2.5 Conclusion

Realizing SDG 13 will be challenging given the urgency of the climate crisis, competing development priorities and political struggles. Climate change was included in the SDGs because it had to be. But, in reality, the development of climate change goals, targets and funding decisions has been largely left to the international climate negotiations and resulting treaties and agreements. Many climate-relevant measures can, however, be found in other SDGs such as SDG 7 (Affordable and clean energy) and SDG 12 (Responsible consumption and production). The impacts of climate change on SDG 2 (Zero hunger) and SDG 14 (Life below water) are explicitly addressed in the *Sustainable Development Goals Report 2022* (UN, 2022).

Post-Paris climate governance has led to contestation and pressure from below not only to achieve more ambitious climate mitigation targets and to adapt to climate change, but also to work towards more just, fair and democratic climate politics both globally and domestically (Marquardt et al, 2022). Yet, there is little doubt that the Paris Agreement is still too limited to keep the climate crisis in check (Allan, 2019). The interests of those countries most affected by climate change often fail to gain sufficient attention, but there are also signs that climate awareness is deepening. Competition for climate technology leadership among the biggest emitters is increasingly visible. Thus, while the world will most likely miss the 1.5 °C target set out in Paris and incorporated in SDG 13, legislative changes are happening, new technologies and processes are being adopted, and sustainable lifestyles are becoming more popular. With signs of climate change all around us, protests and initiatives to demand action and transform societies will certainly intensify. There is still room for some optimism but there is no time to lose. Climate action is needed now.

Note

[1] In Copenhagen (2009) industrialized countries promised to allocate $100 billion a year by 2020 to help the Global South adapt to climate change. In 2021 that target had still not been met.

References

Allan, J.I. (2019) 'Dangerous incrementalism of the Paris Agreement', *Global Environmental Politics*, 19(1): 4–11.

Ambrose, J. (2019) 'World's biggest sovereign wealth fund to ditch fossil fuels', *The Guardian*, 12 June, Available from: www.theguardian.com/business/2019/jun/12/worlds-biggest-sovereign-wealth-fund-to-ditch-fossil-fuels [Accessed 20 August 2022].

Bloomfield, J. and Steward, F. (2020) 'The politics of the Green New Deal', *Political Quarterly*, 91(4): 770–79.

Bohringer, C. (2003) 'The Kyoto Protocol: a review and perspectives', *Oxford Review of Economic Policy*, 19(3): 451–66.

Bolin, B. (2007) *A History of the Science and Politics of Climate Change*, Cambridge: Cambridge University Press.

Carlin, D. (2021) '*The case for fossil fuel divestment*', Forbes, Available from: https://www.forbes.com/sites/davidcarlin/2021/02/20/the-case-for-fossil-fuel-divestment [Accessed 20 August 2022].

Climate Analytics and Next Climate Institute (2019) 'Climate governance in the Philippines', Available from: https://climateactiontracker.org/publications/climate-governance-in-the-philippines [Accessed 20 August 2022].

Climate Analytics and Next Climate Institute (2021a) 'Glasgow's 2030 credibility gap: net zero's lip service to climate action', Available from: https://climateactiontracker.org/publications/glasgows-2030-credibility-gap-net-zeros-lip-service-to-climate-action [Accessed 20 August 2022].

Climate Analytics and Next Climate Institute (2021b) 'The CAT thermometer', Available from: https://climateactiontracker.org/global/cat-thermometer [Accessed 20 August 2022].

Climate Analytics and Next Climate Institute (2022) 'Indonesia', Available from: https://climateactiontracker.org/countries/indonesia [Accessed 20 August 2022].

Creecy, B. (2021) 'Statement by H.E. Ms Barbara Creecy, Minister of Forestry, Fisheries and the Environment of South Africa', United Nations Framework Convention on Climate Change', Available from: https://unf ccc.int/sites/default/files/resource/SOUTH_AFRICA_cop26cmp16cma3 _HLS_EN.pdf [Accessed 20 August 2022].

Datta, N. (2022) 'We know Russia funds Europe's far Right. But what does it get in return?', openDemocracy, Available from: https://www.opende mocracy.net/en/5050/russia-ukraine-war-putin-europe-far-right-fund ing-conservatives/ [Accessed 20 August 2022].

de Nadal, L. (2021) 'Spain's VOX party and the threat of 'international environmental populism'', openDemocracy, Available from: https://www. opendemocracy.net/en/can-europe-make-it/spains-vox-party-and-the-thr eat-of-international-environmental-populism [Accessed 20 August 2022].

Dietz, S., Jahn, V., Noels, J., Stuart-Smith, R. and Hepburn, C. (2019) *A Survey of the Net Zero Positions of the World's Largest Energy Companies*, Available from: https://www.oxfordmartin.ox.ac.uk/downloads/reports/ A-survey-of-the-net-zero-positions-of-the-worlds-largest-energy-compan ies.pdf [Accessed 31 May 2023].

Dunne, D. (2019) 'The Carbon Brief Profile: Indonesia', Carbon Brief, Available from: www.carbonbrief.org/the-carbon-brief-profile-indonesia [Accessed 20 August 2022].

Elder, M. and Bartalini, A. (2019). *Assessment of the G20 Countries' Concrete SDG Implementation Efforts: Policies and Budgets Reported in Their 2016–2018 Voluntary National Reviews*, Hayama: Institute for Global Environmental Strategies, Available from: https://iges.or.jp/en/pub/assessment-g20- countries'-concrete-sdg [Accessed 20 August 2022].

European Council, Council of the European Union (2022) 'Fit for 55', Available from: www.consilium.europa.eu/en/policies/green-deal/fit-for- 55-the-eu-plan-for-a-green-transition [Accessed 20 August 2022].

Extinction Rebellion (2018) 'Extinction Rebellion', Guerrilla Foundation, Available from: https://guerrillafoundation.org/grantee/extinction-rebell ion [Accessed 20 August 2022].

Fiorino, D.J. (2022) 'Climate change and right-wing populism in the United States', *Environmental Politics*, 31(5): 801–19.

Friedrich, J., Ge, M. and Pickens, A. (2020) 'This interactive chart shows changes in the world's top 10 emitters', World Resources Institute, Available from: https://www.wri.org/insights/interactive-chart-shows-changes-wor lds-top-10-emitters [Accessed 20 August 2022].

Fuhr, H. (2021) 'The rise of the Global South and the rise in carbon emissions', *Third World Quarterly*, 42(11): 2724–46.

Fuso Nerini, F., Sovacool, B., Hughes, N., Cozzi, L., Cosgrave, E., Howells, M. et al (2019) 'Connecting climate action with other Sustainable Development Goals', *Nature Sustainability*, 2(8): 674–80.

Geck, M. (2021) 'Seven major companies that committed to net-zero emissions in 2021', Principles for Responsible Investment, Available from: www.unpri.org/pri-blog/seven-major-companies-that-committed-to-net-zero-emissions-in-2021/9197.article [Accessed 20 August 2022].

Gonzales-Zuñiga, S., Roeser, F., Rawlins, J., Luijten, J. and Granadillos, J. (2018) *SCAN (SDG & Climate Action Nexus) Tool*, Methodology Paper, Available from: https://ambitiontoaction.net/wp-content/uploads/2018/10/Methods_note_final.pdf [Accessed 20 August 2022].

Government of South Africa (2020) *South Africa's National Climate Change Adaptation Strategy*, Available at: www.dffe.gov.za/sites/default/files/docs/nationalclimatechange_adaptationstrategy_ue10november2019.pdf [Accessed 20 August 2022].

Guterres, A. (2019) 'Remarks at 2019 Climate Action Summit', United Nations, Available from: www.un.org/sg/en/content/sg/speeches/2019-09-23/remarks-2019-climate-action-summit [Accessed 20 August 2022].

Hale, T. (2016) "All Hands on Deck': The Paris Agreement and Nonstate Climate Action', *Global Environmental Politics*, 16(3): 12–22.

Harvey, F., Ambrose, J. and Greenfield, P. (2021) 'More than 40 countries agree to phase out coal-fired power', *The Guardian*, 4 November, Available from: https://www.theguardian.com/environment/2021/nov/03/more-than-40-countries-agree-to-phase-out-coal-fired-power [Accessed 20 August 2022].

Heggie, J. (2021) 'Day zero: where next?, *National Geographic*, Available from: www.nationalgeographic.com/science/article/partner-content-south-africa-danger-of-running-out-of-water [Accessed 20 August 2022].

Hoffman, A.J. (2011) 'talking past each other? Cultural framing of skeptical and convinced logics in the climate change debate'. *Organization & Environment*, 24(1): 3–33.

IGES (Institute for Global Environmental Strategies) (2021) *SDG Interlinkages Analysis & Visualisation Tool*, Available from: https://www.iges.or.jp/en/pub/sdg-interlinkages-web-tool-v4/en [Accessed 20 August 2022].

IPCC (Intergovernmental Panel on Climate Change) (2018) 'FAQ Chapter 1', Available from: www.ipcc.ch/sr15/faq/faq-chapter-1 [Accessed 20 August 2022].

IPCC (2019) 'Global warming of 1.5 °C', Available from: www.ipcc.ch/sr15 [Accessed 20 August 2022].

IPCC (2021) 'Climate change widespread, rapid, and intensifying', Available from: www.ipcc.ch/2021/08/09/ar6-wg1-20210809-pr/ [Accessed 20 April 2022].

Jänicke, M., Schreurs, M. and Töpfer, K. (2015) *The Potential of Multi-Level Global Climate Governance*, IASS Policy Brief 2/2015, Potsdam: Institute for Advanced Sustainability Studies.

Jernnäs, M. (2021) *Governing Climate Change under the Paris Regime Meeting Urgency with Voluntarism*, Linköping: Linköping University Press.

Jong, H.N. (2021) 'Indonesia's net-zero emissions goal not ambitious enough, activists say', Mongabay, Available from: https://news.mongabay.com/2021/04/indonesia-net-zero-emissions-target-coal-energy-2070 [Accessed 20 August 2022].

Knutti, R., Rogelj, J., Sedláček, J. and Fischer, E.M. (2016) 'A scientific critique of the two-degree climate change target', *Nature Geoscience*, 9(1): 13–18.

Kuyper, J.W., Linnér, B.O. and Schroeder, H. (2018) 'Non-state actors in hybrid global climate governance: justice, legitimacy, and effectiveness in a post-Paris era', *WIREs Climate Change*, 9(1): 1–18.

Lazarus, R.J. (2009) 'Super wicked problems and climate change: restraining the present to liberate the future', *Cornell Law Review*, 94(5): 1153–233.

Liddle, B. (2018) 'Consumption-based accounting and the trade-carbon emissions nexus', *Energy Economics*, 69: 71–8.

Lindsey, R. (2021) 'Climate change: atmospheric carbon dioxide', Climate. gov, Available from: www.climate.gov/news-features/understanding-climate/climate-change-atmospheric-carbon-dioxide [Accessed 20 August 2022].

Machin, A. (2020) 'Democracy, disagreement, disruption: agonism and the environmental state', *Environmental Politics*, 29(1): 155–72.

MacLean, J. (2020) 'Rethinking the role of nonstate actors in international climate governance', *Loyola University Chicago International Law Review*, 16(1): 21–43.

Marquardt, J. (2020) 'Fridays for Future's disruptive potential: an inconvenient youth between moderate and radical ideas', *Frontiers in Communication*, 5(48), 1–18.

Marquardt, J. and Bäckstrand, K. (2022) 'Democracy beyond the state: non-state actors and the legitimacy of climate governance', in B. Bornemann, H. Knappe and P. Nanz (eds) *The Routledge Handbook of Democracy and Sustainability*, London: Routledge, pp 237–53.

Marquardt, J. and Lederer, M. (2022) 'Politicizing climate change in times of populism: an introduction', *Environmental Politics*, 31(5): 735–54.

Marquardt, J., Fast, C. and Grimm, J. (2022) 'Non- and sub-state climate action after Paris: from a facilitative regime to a contested governance landscape', *WIREs Climate Change*, 13(5): e791.

Martinelli, F.S., Biber-Freudenberger, L., Stein, G. and Börner, J. (2022) 'Will Brazil's push for low-carbon biofuels contribute to achieving the SDGs? A systematic expert-based assessment', *Cleaner Environmental Systems*, 5: 10007.

Martiskainen, M., Axon, S., Sovacool, B.K., Sareen, S., Furszyfer Del Rio, D. and Axon, K. (2020) 'Contextualizing climate justice activism: knowledge, emotions, motivations, and actions among climate strikers in six cities', *Global Environmental Change*, 65: 102180.

McGrath, M. (2021) 'COP26: India PM Narendra Modi pledges net zero by 2070', *BBC News*, Available from: https://www.bbc.com/news/world-asia-india-59125143 [Accessed 20 August 2022].

Mendizabal, M., Heidrich, O., Feliu, E., García-Blanco, G. and Mendizabal, A. (2018) 'Stimulating urban transition and transformation to achieve sustainable and resilient cities', *Renewable and Sustainable Energy Reviews*, 94: 410–18.

NASA (National Aeronautics and Space Administration) (2021) 'How do we know climate change is real?', NASA Global Climate Change, Available from: https://climate.nasa.gov/evidence [Accessed 20 August 2022].

NDRC (National Development and Reform Commission) (2021) '*Outline of the People's Republic of China 14th Five-Year Plan for National Economic and Social Development and Long-Range Objectives for 2035*'. Beijing: NDrC, Available from: https://perma.cc/73AK-BUW2. [Accessed 3 July 2023].

NOAA (National Oceanic and Atmospheric Administration) (2021) 'Carbon dioxide peaks near 420 parts per million at Mauna Loa observatory', *NOAA Research News*, Available from: https://research.noaa.gov/article/ArtMID/587/ArticleID/2764/Coronavirus-response-barely-slows-rising-carbon-dioxide [Accessed 20 August 2022].

Obergassel, W., Arens, C., Hermwille, L., Kreibich, N., Mersmann, F., Ott, H.E. and Wang-Helmreich, H. (2016) *Phoenix from the Ashes – An Analysis of the Paris Agreement to the United Nations Framework Convention on Climate Change*, Wuppertal: Wuppertal Institut für Klima, Umwelt, Energie.

Onishi, N. (2019) 'France's Far Right wants to be an environmental party, too', *New York Times*, 17 October, Available from: www.nytimes.com/2019/10/17/world/europe/france-far-right-environment.html [Accessed 20 August 2022].

Oreskes, N. and Conway, E.M. (2010) *Merchants of Doubt: How a Handful of Scientists Obscured the Truth on Issues from Tobacco Smoke to Climate Change*, New York: Bloomsbury Press.

Ostrom, E. (2009) 'Beyond markets and States: polycentric governance of complex economic systems', *American Economic Review*, 100(3): 641–72.

Ourbak, T. and Magnan, A.K. (2018) 'The Paris Agreement and climate change negotiations: Small Islands, big players', *Regional Environmental Change*, 18(8): 2201–7.

Porter, L., Rickards, L., Verlie, B., Bosomworth, K., Moloney, S., Lay, B. et al (2020) 'Climate justice in a climate changed world', *Planning Theory & Practice*, 21(2): 293–321.

PTI (2019) 'India likely to add 273 million people between 2019 and 2050: UN Report', *Economic Times*, 17 June, Available from: https://economictimes.indiatimes.com/news/politics-and-nation/india-likely-to-add-273-million-people-between-2019-and-2050-un-report/articles how/69830509.cms?from=mdr [Accessed 20 August 2022].

Ram, M., Osorio-Aravena, J.C., Aghahosseini, A., Bogdanov, D. and Breyer, C. (2022) 'Job creation during a climate compliant global energy transition across the power, heat, transport, and desalination sectors by 2050', *Energy*, 238: 121690.

REN21 (2021) *Renewables 2021 Global Status Report*, Available from: https://www.ren21.net/reports/global-status-report.

Schlosberg, D. and Collins, L.B. (2014) 'From environmental to climate justice: climate change and the discourse of environmental justice', *WIREs Climate Change*, 5(3): 359–74.

Schott, S. and Schreurs, M. (2020). 'Climate politics and fossil fuel sector developments in Canada and Germany: potentials for greater transatlantic cooperation', *Canadian Journal of European and Russian Studies*, 14(2): 29–55.

Schreurs, M. (2019) 'Climate change politics in the United States, China and the Europea vn union: climate science and the framing of climate action', in J. Men, S. Schunz and D. Freeman (eds), *The Evolving Relationship between China, the European Union and the USA*, New York: Routledge, pp 192–212.

Serhan, Y. (2021) 'The Far-Right view on climate politics', *The Atlantic*, Available from: https://www.theatlantic.com/international/archive/2021/08/far-right-view-climate-ipcc/619709 [Accessed 20 August 2022].

Sovacool, B.K. and Dunlap, A. (2022) 'Anarchy, war, or revolt? Radical perspectives for climate protection, insurgency and civil disobedience in a low-carbon era', *Energy Research & Social Science*, 86: 102416.

Stokes, L.C. (2020) *Short Circuiting Policy: Interest Groups and the Battle over Clean Energy and Climate Policy in the American States*, New York: Oxford University Press.

Streck, C. (2020) 'Filling in for governments? The role of the private actors in the international climate regime', *Journal for European Environmental & Planning Law*, 17(1): 5–28.

Swyngedouw, E. (2011) 'Depoliticized environments: the end of nature, climate change and the post-political condition', *Royal Institute of Philosophy Supplement*, 69: 253–74.

Thornton, F. (2023) 'Top 11 Climate Deniers', Before the Flood, Available from: https://beforetheflood.com/top-climate-deniers/ [Accessed 3 July 2023].

Udapudi, S. and Sakkarnaikar, F.S. (2015) 'From Stockholm to Rio to Rio + 20: green economy and the road ahead', *American International Journal of Research in Humanities, Arts and Social Sciences*, 15(325): 65–73.

UN (United Nations) (2002) *Report of the World Summit on Sustainable Development, Johannesburg, South Africa, 26 August–4 September 2002*, A/CONF.199/20, New York: United Nations, Available from: https://undocs.org/en/A/CONF.199/20 [Accessed 20 August 2022].

UN (2019) Global issues: climate change, Available from: www.un.org/en/global-issues/climate-change [Accessed 20 August 2022].

UN (2021) 'Climate change 'biggest threat modern humans have ever faced', world-renowned naturalist tells Security Council, calls for greater global cooperation', Available from: www.un.org/press/en/2021/sc14445.doc.htm [Accessed 20 August 2022].

UN (2022) *The Sustainable Development Goals Report 2022*, New York: United Nations, Available from: https://unstats.un.org/sdgs/report/2022/The-Sustainable-Development-Goals-Report-2022.pdf [Accessed 20 August 2022].

UNEP (United Nations Environment Programme) (2018) *Emissions Gap Report 2018*, Nairobi, Available from: www.un.org/Depts/Cartographic/english/htmain.htm [Accessed 20 August 2022].

UNFCCC (United Nations Framework Convention on Climate Change) (2015) Paris Agreement, 21st Conference of the Parties, Paris: UNFCCC.

UNFCCC (2021) 'Race To Zero', Available from: https://unfccc.int/climate-action/race-to-zero-campaign [Accessed 20 August 2022].

US EPA (United States Environmental Protection Agency) (2022) *Inventory of U.S. Greenhouse Gases and Sinks, 1990–2020*, US Environmental Protection Agency, EPA 430_R-22_003. Available from: www.epa.gov/system/files/documents/2022-04/us-ghg-inventory-2022-main-text.pdf [Accessed 31 May 2023].

Varadhan, S. and Sheldrick, A. (2021) 'COP26 aims to banish coal: Asia is building hundreds of power plants to burn it', Reuters, Available from: www.reuters.com/business/energy/cop26-aims-banish-coal-asia-is-building-hundreds-power-plants-burn-it-2021-10-29 [Accessed 20 August 2022].

Venkatramanan, V., Shah, S. and Prasad, R. (eds) (2021) *Exploring Synergies and Trade-offs between Climate Change and the Sustainable Development Goals*, Singapore: Springer.

Vihma, A., Reischl, G., Andersen, A.N. and Berglund, S. (2020) *Climate Change and Populism: Comparing the Populist Parties' Climate Policies in Denmark, Finland and Sweden*, Helsinki: Finnish Institute of International Affairs.

Wallach, O. (2021) 'Carbon neutral goals by country', Visual Capitalist, Available from: https://www.visualcapitalist.com/sp/race-to-net-zero-carbon-neutral-goals-by-country/ [Accessed 3 July 2023].

Wahlund, M. and Palm, J. (2022) 'The role of energy democracy and energy citizenship for participatory energy transitions: a comprehensive review', *Energy Research & Social Science*, 87: 102482.

Wijaya, A., Chrysolite, H., Ge, M., Wibowo, C.K. and Pradana, A. (2017) *How Can Indonesia Achieve its Climate Change Mitigation Goal?*, Washington, DC: World Resources Institute.

Willis, R., Curato, N., & Smith, G. (2022) 'Deliberative democracy and the climate crisis', *WIREs Climate Change*, 13(2): e759.

World Bank (2022) 'Indonesia', Available from: https://data.worldbank.org/country/ID [Accessed 20 August 2022].

Zhang, C. and Yan, J. (2015) 'CDM's influence on technology transfers: a study of the implemented clean development mechanism projects in China', *Applied Energy*, 158: 355–65.

Zheng, X., Lu, Y., Yuan, J., Baninla, Y., Zhang, S., Stenseth, N. et al (2020) 'Drivers of change in China's energy-related CO_2 emissions', *Proceedings of the National Academy of Sciences*, 117(1): 29–36.

Interview with Miranda Schreurs and Jens Marquardt: Ending the North/South Divide in Climate Action

Johanna Carrasco Saravia and Caroline Landolt

Fossil fuel companies are powerful influencers in politics. Did they ever try to influence your research?

Schreurs: One day, I was asked if I was willing to do a paid interview. In exchange for money, I would have had to give up my right to determine what was written later in the report. The journal would have had the right to edit what I wrote. So I refused. It turned out that it was the atomic lobby. They were hoping they could get an American professor to take their side of the story. This also happens in more indirect ways through research projects and funding meant to support a particular kind of research, for instance about clean coal or fossil fuel's energy efficiency. In this way, the fossil fuel lobby has had a lot of influence on the research that was conducted over the years.

Considering the position of most developing countries as providers of raw material and natural resources, what role can these countries play in climate action?

Marquardt: Developing countries need more agency. The Yasuní-ITT Initiative in Ecuador was one example where the government said 'We have this beautiful rainforest here and underneath us, there is oil. We'll agree not to cut the trees and not to extract the oil if you give us 50% of the revenues that we could have expected from that.' There was a fund for a couple of years, but it didn't reach the target, so they cancelled it. That's a clear example of how the Global South and the Global North could work together by sharing responsibilities. This would have been a win–win solution, but it failed. I think the role of developing countries is to demand action, compensation and technologies. However, I also wouldn't say developing countries are homogeneous; they have often varying interests ranging from those of China to Small Island Developing States. It is important for this group to keep different responsibilities and climate justice issues on their agenda, working together with civil society organizations too.

With regards to civil society, what role can transnational environmental movements, such as Fridays for Future play in the design and implementation of climate policies?

Marquardt: I'm hopeful that these movements bring the politics back to climate action and demonstrate that tackling climate change is more than a question of technology, economic prosperity or science. Fridays for Future raises the voice of marginalized groups across the world and calls for climate justice. Their success depends on creating alliances with unions, Indigenous peoples, all the people who suffer from the current economic system and climate change. It's important to build coalitions and to get away from this Global North versus Global South narrative. We need to promote alternatives, such as *buen vivir* in South America, for example, which rests on solidarity and unity between humans and nature. It is important to think out of the box and not to reproduce the system that created the mess of the climate crisis.

3

Key Logics of International Forest Governance and SDG 15

*Daniela Kleinschmit, Mareike Blum, Maria Brockhaus,
Mawa Karambiri, Markus Kröger, Sabaheta Ramcilovic-Suominen
and Sabine Reinecke*

In 2015 the 2030 Agenda for Sustainable Development, building on the Millennium Development Goals, was adopted by the UN member states. The SDGs, and in particular SDG 17 (Partnerships for the goals), have emphasized global partnerships as a means of overcoming the shortcomings of the MDGs in relation to Goal 8 (To develop a global partnership for development) (UN System Task Team, 2012: 5). In these regards, partnerships at the global level between governments, as well with other stakeholders such as the private sector, are described as being essential (Lomazzi et al, 2014). This was in response to known power asymmetries in international arrangements, namely, a Northern-driven agenda and a bias in governance forms that favour international organizations, governmental actors and a strong private sector, while dismissing Southern and local participation as well as non-state actors and authorities (Menashy, 2019). SDG 17 is essential for, and thus should be integrated by, all other SDGs, including SDG 15 (Life on land).

While the other SDGs focus on economic development, social rights or cooperation, SDG 15 flags terrestrial ecosystems, and correspondingly forests, as being essential for sustainable development. Forest ecosystems are assigned a particular role not only in SDG 15 but also in other SDGs, building a 'complex relationship' between them (Baumgartner, 2019: 1). This is in part due to their geographic scope (with forest covering about one third of the world's land area) but also because of the goods and services they provide: timber and non-timber forest products; a fundamental basis for ecological processes (CO_2 mitigation, water supply and quality); habitat for plants and animals; basis for livelihood and human well-being; and a space for human culture and spiritual engagement. Hence, it is argued that the future of the world's forests is 'critical for sustainable development at all scales, from global to local' (Katila et al, 2019: 3). This overarching relevance is partly mirrored in SDG 15 with the goal to 'Protect, restore and promote sustainable use of terrestrial ecosystems, sustainably manage forests, combat desertification, and halt and reverse land degradation and halt biodiversity loss'.

Many of the targets and objectives of SDG 15, such as protection, restoration and sustainable use of ecosystems, have already been tackled in earlier international frameworks. These include the Convention on Biological Diversity (CBD) and its Aichi Biodiversity Targets, the UN Convention to Combat Desertification (UNCCD) and the Convention on Trade in Endangered Species (CITES), as well other international governance initiatives – particularly those concentrating on forest (and) landscape restoration (FLR), such as the Bonn Challenge or the New York Declaration on Forests (NYDF). These examples indicate the complexity in international forest governance, in which international governance is understood to be 'the formal and informal bundles of rules, roles, rights and relationships that define and regulate the social practices of state and nonstate actors in international affairs' (Slaughter et al, 1998: 371). These types of governance include UN-driven governmental processes such as the United Nations Forum on Forests (UNFF); private sector governance; market-oriented governance like Forest Stewardship Council (FSC) certification; and hybrid settings involving both state and non-state actors , such as the Bonn Challenge (Sotirov et al, 2020).

SDG 15 builds on this legacy of international forest governance with its diverse objectives, targets and forms of governance. Yet history has demonstrated its limited effectiveness. Trade-offs between different objectives, as well as power asymmetries, are often named as key obstacles for efficient and just international forest governance (Fleischman et al, 2021). Acknowledging the complexity of, as well as the power asymmetries in, international forest governance, this chapter has a twofold aim: first, to structure the complexity by identifying key logics that characterize international forest governance according to the issues, problems and proposed solutions (Kleinschmit et al, 2023); and, second, to better understand whether SDG 15 supports the overarching aim of the SDGs to overcome power asymmetries or whether the legacy of power asymmetries in international forest governance pertains in SDG 15.

To achieve these aims, this chapter maps key institutions and interests, considering inherent trade-offs and power asymmetries. It is assumed that institutions and the interests involved are central to international forest-related governance over the past three decades and thus might be (still) embedded in SDG 15. Institutional rules are understood not as neutral but rather as outcomes of political bargaining processes in which different beliefs and discourses compete, and only certain ones are adopted. Starting with the assumption that international forest governance is shaped by multiple, more and less powerful, and partly conflicting institutions, the chapter focuses on formal laws, regulations and procedures. Following Barnett and Duvall (2005: 3), it perceives institutional power to involve the control of some actors over the conditions for the actions of (socially distant) others (Barnett and Duvall, 2005). Interests can be broadly described as the preferences of political actors. This dimension is also an expression of power inasmuch as some actors may be more capable of attaining and exercising their interests than others, depending on the distribution of resources and power in a governance setting. While existing institutions are the (partly unintended) consequence of earlier and historically defined institutional decisions and sets of rules, these typically reflect the preferences of more privileged actors. Likewise, existing institutions shape the preferences and power relations of actors in a given governance setting. Hence, institutions and interests are intimately interlinked (see Schmidt, 2008). Key institutions and interests that have been recognized in the literature as being embedded in international forest governance are described in section 3.1, which sets out the basis for synthesizing the three logics presented in section 3.2. In the mapping of institutions and interests, trade-offs and power asymmetries are considered.

For decades, studies have uncovered power asymmetries in international environmental and forest governance. These reveal, on the one hand, a Northern-driven agenda that diminishes the needs and priorities of Southern countries (eg Karlsson, 2002). On the other hand, they understand the diversity of types of governance not only as an expression but also as a source of conflict about responsibilities, which in turn results in or furthers existing global power asymmetries (McDermott et al, 2019). While elite capture and power struggles are taking place in societies in North and South alike, the main interest in this chapter is the specific targets and objectives related to forests in the SDG 15, and the different (historic) roles and responsibilities of the Global North and South in the governance of forest and forest lands.

The substantive scope and main data sources of this overview covers multidisciplinary scientific knowledge on past and current developments published up to the end of 2022. In terms of geographical coverage, this chapter focuses on international forest governance. As described earlier, international forest governance is approached from a broader perspective, going beyond forestry-specific concerns and taking into account forest-related and interlinked concepts. But this chapter focuses only on multilateral governance processes. Scholars have identified significant differences in agendas, types of governance and actor constellations in international forest governance before, during and after the Rio Earth Summit in 1992 (eg Humphreys, 1996a; Sotirov et al, 2020). Therefore these three periods are differentiated.

3.1 International forest governance: a chronological overview

SDG 15 is embedded in a range of different and diversifying approaches to international forest governance and is characterized by powerful institutions and interests. This section provides a chronological overview of those institutions and interests that have been highlighted in the literature as being embedded in international forest governance. Building on a literature review, two interrelated general trends of international forest governance can be observed: (1) the diversification of interests addressing forests, extending beyond classical wood production–oriented forestry; and (2) the diversification of forms of international forest governance beyond classical state actors and international organizations. Though these developments have accelerated since the United Nations Conference on Environment and Development (UNCED) held in Rio de Janeiro in 1992 (referred to in this chapter as the Earth Summit or the Rio Conference), they did not develop at the time in the context of an institutional void but rather built on historical developments in international forest governance. Therefore, the overview starts with the period preceding the Rio conference, before focusing on Rio and its relevance for international forest governance and the SDGs, and finally addressing the developments that followed thereafter until the 2020s. These developments will be discussed in light of the targets and objectives of SDG 15 in section 3.2.

3.1.1 International forest governance before the Earth Summit

In the period before the Rio conference, governments (especially those in high-income countries) saw forests mainly as a site of wood production. This view is informed by centuries of colonial rule, colonial administrations and a network of pro-colonial private companies practising forest exploitation, often combined with expropriation of local populations in the tropics (eg Peluso and Vandergeest, 2001). At the time of independence, power was transferred to new postcolonial states that continued to pursue timber extraction, while public and private companies started to develop commercial forest plantations. These have often used the narrative of reforesting or restoring degraded forest lands (eg Feintrenie, 2014). In the late 1960s, in line with the wood production and global market logic, the UN Food and Agriculture Organization (FAO) created a Committee on Forest Development in the Tropics, emphasizing the development (meaning exploitation) of forests (Humphreys, 2004). In the late 1970s, negotiations on a commodity agreement on tropical timber commenced, ultimately resulting in the International Tropical Timber Agreement (ITTA) sponsored by the United Nations Conference on Trade and Development. Adopted in 1983 and ratified in 1985, the ITTA is an international, legally binding agreement focusing on tropical forests and trade in tropical timber. As a result of an intervention by the International Union for the Conservation of Nature and Natural Resources (IUCN), the agreement also makes reference (in one clause) to forest conservation and the long-term ecological balance (Humphreys, 2004). There are two broad categories of signatories to the ITTA, producing and consuming countries, with reference to their focus on the production or consumption of tropical timber. ITTA established the International Tropical Timber Organization (ITTO) as the body for policy development and decision-making processes in implementation.

In the 1980s, the practice of forest exploitation and the establishment of (large-scale tree) plantations, particularly in the tropics, became a prioritized issue of environmental campaigning and a hot topic in international public discourses supported by environmental NGOs (ENGOs) and rural people trying to prevent the clearing of natural forests for tree plantations (Gulbrandsen and Humphreys, 2006). This global attention was curbed by the emergence of global environmental issues such as ozone destruction and desertification (Humphreys, 1996a). Responding to this global attention, the United States, Canada and some European countries made several initiatives to negotiate a legally binding international forest instrument in the mid-1980s. Their main interest was to halt and reverse deforestation in the tropical regions and protect old-growth forests. This resulted in diverse attempts to propose the establishment of a global forest instrument in the shape of a global forest convention or a forest protocol for a climate change or biodiversity convention (Humphreys, 1996b). The global attention on and debate around tropical deforestation also affected the preparatory meetings of the UNCED conference starting in August 1990 in Nairobi, which focused on forests but also other areas such as trade, climate change and agriculture (Schally, 1994). While some countries were in favour of a global forest convention, others raised general concerns. From a powerful position in the international forest governance arena based on their forest resources, Malaysia, Brazil and other countries, in particular, criticized the biased focus on tropical forests (Pülzl, 2005).

International forest governance concentrated on wood and international markets in the period before the Earth Summit, and was spurred mainly by large and arguably powerful international state-driven institutions such as the ITTO, the UN and its affiliated agencies. The private sector pushing the international wood market, and also supported by the World Bank, has been a key player in international forest governance. In the 1980s, however, the attention of ENGOs at the global level was also directed towards forest conservation. This movement was flanked by an evolving discussion about certification in the 1990s, responding to both the environmental discussion in industrialized countries (focusing on the threats to tropical forests) and the demand of timber retailers and distributors for forest products from legal, well-managed forests (Meidinger, 2003).

The Convention on International Trade in Endangered Species of Wild Fauna and Flora, adopted in 1976, is another trade-oriented international institution. However, CITES does not focus only on wood products but also on endangered species of wild plants (including trees) and animals in general by restricting imports and exports. The three appendices to the Convention offer different levels of protection from prohibited trade. CITES is perceived as an early example of the power struggles in negotiating international rules and regulations, and the related problems of involving or considering (specific) non-state actors (Challender et al, 2015). The CITES regulation actually has implications for rural communities who may be dependent on CITES-listed wildlife for their livelihoods (eg Roe et al, 2002; Velázquez Gomar and Stringer, 2011). Some authors criticize CITES for being characterized by a preservation perspective of powerful Northern industrial states and conservationists in other (non-affected) regions of the world demanding a halt to the utilization of wildlife (Duffy, 2013). The significant attention, supported by media coverage, given to prestigious species (eg rhinos or elephants) is perceived as demonstrating the interests of non-state actors in international governance (Duffy, 2013).

3.1.2 Earth Summit in Rio 1992: *non-agreement and broadening the forest perspective*

The UNCED conference in Rio 1992 has fundamentally affected international forest governance over a long period, because it failed to agree on a global forest convention and broadened the range of issues to be addressed through forests, incorporating, for example, climate change and biodiversity.

At the Earth Summit, a non-binding statement was agreed upon as called for by the third preparatory committee (Prep Com): the 'Non-legally Binding Authoritative Statement of Principles for a Global Consensus on the Management, Conservation, and Sustainable Development of All Types of Forests' (the Forest Principles) in Agenda 21. The Forest Principles, together with the conventions described later, were endorsed by the United Nations General Assembly at the Rio Conference. The 15 principles deal with subjects as diverse as the sovereign right of states, women's participation and the need to provide financial resources (Zentilli, 1995). Chapter 11 of Agenda 21 (Combating deforestation), in particular, focuses on forest goals including sustainable management and conservation, as well as on utilization and assessments and systematic observation. Though not legally binding, Chapter 11 is the first international commitment of states to the sustainable development of forests (Humphreys, 2001).

Experts have identified various reasons for the failure to agree on a legally binding convention on forests, including the role of forests for international commerce (Lipschutz, 2000), strategic moves against increased standards for forest management (Dimitrov, 2003), the worries of tropical countries about sovereignty (Humphreys, 1996a, 1996b) and the mismatch between costs and benefits for countries from the Global North (Davenport, 2005). The output of the Rio Conference likely reflects a combination of all these factors.

Apart from the principles focusing specifically on forests, forests are addressed in the context of three conventions resulting from the Rio Conference: the Convention on Biological Diversity and the Framework Convention on Climate Change and the Convention to Combat Desertification (UNCCD). Both were formally negotiated by an Intergovernmental Negotiating Committee between 1991 and 1992. They ended up as legally binding framework conventions signed by a large number of states.

The CBD aims to combat the destruction of plant and animal species, and ecosystems at large (CBD, 1992). The CBD specifically addresses biodiversity in developing countries, and also considers the responsibility of contracting parties to recognize the knowledge of Indigenous and local communities; equitable sharing of the resulting benefits; and the sovereign authority to determine access to genetic resources and commercial benefits from biodiversity. Arts (2006: 183) recognizes the CBD as 'a delicate balance and complex compromise between Northern and Southern preferences, assets and interests'. Many NGOs, like World Wide Fund for Nature (WWF), Friends of the Earth and Greenpeace to name just a few, participated in and influenced the intergovernmental process of the CBD (Arts, 1998). Additionally, the principle of free and prior informed consent (FPIC) was upheld by the CBD to facilitate participation, foster transparency of information exchange and support benefit sharing. CBD has particular relevance for forests as it is one of the Convention's priorities (with its own programme of work) to ensure forests as a space for species protection.

The UNFCCC as a framework convention aims to achieve the stabilization of greenhouse gas concentrations in the atmosphere at a level that would prevent dangerous anthropogenic interference with the climate system within a timeframe sufficient to allow ecosystems to adapt naturally to climate change (UNFCCC, 1992). The treaty established three categories of signatory states with different responsibilities: developed countries (Annex 1 countries), developed countries with special financial responsibilities (covered in Annex 2, ie Annex 1 countries excluding countries in transition) and developing countries. Forests and the forest sector gained high relevance in the UNFCCC in relation to the mitigation of GHGs and adaptation to climate change.

The goal of the UNCCD is 'to combat desertification and mitigate the effects of drought ... through effective action at all levels, supported by international cooperation and partnership arrangements' (UNCCD, 1994). Forests did not feature heavily in the convention but were addressed because signatories committed themselves to include measures to conserve natural resources – for example, through ensuring integrated and sustainable management of forests – in national action programmes.

The limited problem-solving effectiveness of the conventions and principles agreed at Rio has since become evident (Sotirov et al, 2020), specifically concerning forests as deforestation and forest degradation persist at high levels. Thus, the Rio Conference was only the starting point for further initiatives, declarations, protocols, annexes or instruments both within and outside the UN system, with forests continually receiving increased attention.

3.1.3 Beyond the Rio Conference: shift towards 'new' types of governance

For those actors that have aimed for a global forest convention, the Rio Conference marked a starting point for continual frustration, because the soft laws delivered in further intergovernmental processes lacked legal bite (Levin et al, 2008). For others, the Rio Conference is seen as having catalysed a series of innovative market-oriented international forest governance approaches (eg Shaffer and Bodansky, 2012). The newer dynamics and trends in forest-related international governance within and beyond the treaties are presented in this section.

Forest under the UN

The UN Commission on Sustainable Development established first the Intergovernmental Panel on Forests (IPF), based on the Forest Principles, and after two years the Intergovernmental Forum on Forests (IFF) as its successor, with the task of analysing priority forest issues, which resulted in over 150 Proposals for Action (PfA). Based on this, the UN Economic and Social Council (ECOSOC) created the United Nations Forum on Forests in 2000, an intergovernmental body still in existence today, with a focus on sustainable forest management (SFM) and its diverse facets (for an overview from IPF to UNFF, see Humphreys, 2001). The UNFF was also tasked with laying the groundwork for a global forest convention. However, in 2007 this resulted in yet another soft law known as the Non-Legally-Binding Instrument on All Types of Forests (NLBI). Finance was flagged by many developing countries as the most crucial issue while 'most of the donor countries were not prepared to agree on the establishment of a strong financial mechanism' (Kunzmann, 2008: 987). The NLBI emphasizes, among other things, a reversal of the loss of forest cover and sustainable forest management. The latter is supported by the development of national forest programs (NFPs) with the aim of strengthening national sovereignty and allocating the responsibility for achieving global objectives to member states (Glück et al, 2003). NFPs shift power from the state in decision making to new forms of governance. Critical voices argue that the NLBI reaffirms neoliberal ideas of international forest governance highlighting the primary role of forests and their goods and services for the market (eg Humphreys, 2009).

More obviously in the interest of wood production and international markets are the updated ITTAs and other diverse governance processes concentrating on timber markets. ITTO's mandate was renewed by ITTAs in 1994 and 2006, the latter with the explicit aim to 'promote the expansion and diversification of international trade in tropical timber from sustainably managed and legally harvested forests' (International Tropical Timber Agreement, 2006: 3). Some authors understand the ITTO to be strongly supportive of SFM, for example through the Guidelines for the Sustainable Management of Tropical Natural Forests, including corresponding criteria and indicators (Linser et al, 2018), and through its support for SFM initiatives such as the African Timber Organization (Kadam et al, 2021). Others perceive ITTO to be dominated by trade interests emphasizing 'sustainable timber production' instead of SFM (Gulbrandsen and Humphreys, 2006).

Certification

Certification, described by Cashore et al (2004) as a prime example of 'non-state market-driven' governance, concentrates on timber production and international markets as well as on sustainable forestry. In contrast to state-centred traditional international governance, the authority of certification is diffuse and is located in the marketplace (Bernstein and Cashore, 2004). Early efforts towards forest certification started in the late 1980s, accompanying movements against threats to tropical rainforests (Kadam et al, 2021). Justified and boosted by not agreeing on a legally binding global forest convention in Rio, the FSC was formed in 1993 in Canada, and continues to be led by the WWF. At the centre of the FSC's approach are voluntary forest management standards, developed through a multistakeholder process at the country level. Researchers have noted the trade-off between interests involved in the certification scheme, notably between trade and conservation issues (eg Fernàndez-Blanco et al, 2019). Hence, it is not surprising that the FSC initially received only limited support from the forest industry (eg Cashore et al, 2004; Gulbrandsen, 2006). While this has changed subsequently, growing industry support has been criticized for lowering the forest management standards (Humphreys, 2012). The repeated criticism from ENGOs resulted in prominent NGOs withdrawing from the FSC and its domestic branches; for example, in 2018 Greenpeace International, one of the FSC's funders, did not renew its membership. Other certification schemes have evolved over time, for example the Programme for the Endorsement of Forest Certification, which was established in 1998 and broadened in 2002 (Auld et al, 2009). With reference to power relations, Pattberg (2006: 579) argues that in the forest certification arena 'southern actors have not benefited so much economically from private certification schemes, [but] they have been partially empowered through cognitive and integrative processes of governance'.

Legality verification

Timber legality trade restrictions and verification have been developed on the legacy of forest certification and in response to continued global concerns regarding forest degradation and deforestation (Acheampong and Maryudi, 2020). The emerging timber legality regime builds on a cooperation between state actors, NGOs and (multinational) corporations. The set of mechanisms developed over time consists of a range of instruments, including mandatory state-based regulations such as bilateral voluntary partnership agreements (VPAs) under the EU's Forest Law Enforcement, Governance and Trade (FLEGT) Action Plan adopted in 2003. VPAs are legally binding political and trade agreements between the EU as a 'consumer' and a partner 'producer' country with the aim of granting products a FLEGT licence and setting up an effective timber legality assurance system (Sotirov et al, 2020). Another mechanism developed as part of the FLEGT Action Plan is the EU Timber Regulation (EUTR) adopted in 2010. Similar demand-side consumer regulations outside the EU are the US Lacey Act (amended in 2008) and the Australia Illegal Logging Prohibition Act (AILPA) adopted in 2012 (eg Cashore and Stone, 2012). Perceptions of these consumer regulations vary a great deal, with some considering that they support sustainable forest use while others see them as instruments for green protectionism of consumer markets (Winkel et al, 2017).

Forests as climate change governance issue

Since the Rio Conference, forests have gained increasing attention in international climate governance – starting with the Kyoto Protocol, which addressed forests as sinks and reservoirs for GHGs and committed signatories to promoting sustainable forest management practices, afforestation and reforestation. The Clean Development Mechanism (CDM) allowed Annex 1 countries to offset GHG emissions through forestry activities in developing countries, consisting of afforestation and reforestation projects. The importance of standing forests for climate change mitigation and adaptation efforts was acknowledged in climate negotiations early on (Locatelli et al, 2008), even though at the intersection of land use, climate and sustainable development, forest-based mitigation seems to be dominant across levels of governance (Di Gregorio et al, 2019). Reducing Emissions from Deforestation and Forest Degradation and the Role of Conservation, Sustainable Management of Forests and Enhancement of Forest Carbon Stocks in Developing Countries (REDD+) is the most prominent climate governance instrument on forests. Envisioned originally as part of a global cap-and-trade carbon market, REDD+ has developed into a form of results-based payments scheme (Angelsen et al, 2017). It is supposed to create an incentive for forested developing countries to protect, better manage and sustainably use their forest resources, thus contributing to global efforts to limit climate change and in return receiving payments for verified/certified emission reductions and removals. Yet deforestation continues, for example in Brazil and the Democratic Republic of the Congo, and questions are raised about the effectiveness of REDD+. In addition, critiques point to risks of REDD+ being a new form of appropriation and enclosures of Indigenous lands through the state, driven by a marketization of forests for carbon to serve Global North interests in carbon offsetting and nurtured by an 'economic cost–benefit worldview' (Brockhaus et al, 2021). This results in the further reproduction of power asymmetries and various forms of injustices (Ramcilovic-Suominen et al, 2021). REDD+ has now become part of mainstream climate change negotiations and debates, its objectives have been incorporated in the Paris Agreement and many signatories address forests in their NDCs (Mills-Novoa and Liverman, 2019).

Spurred by the Paris Agreement, international climate governance now corresponds to the notion of a regime complex comprising a network of states, international organization and non-state actors (Kuyper et al, 2018). According to scholars, hierarchical forms of governance have shifted towards a more complex polycentric governance system, with states no longer at the centre and with non-state actors playing an increasingly important role (Kuyper et al, 2018). With all countries having their own NDCs, the Paris Agreement has overcome the previous differentiation between Annex 1 and non-Annex 1 countries. A more active role is also attributed to non-public actors of all sorts in (country-led) implementation and finance (Articles 6 and 9). However, pertaining power imbalances are well documented. For example, local communities, minorities and women are seen to be rarely involved in designing and implementing REDD+ (eg Bayrak and Marafa, 2016; Schroeder and González, 2019).

Forest biodiversity

Like the UNFCCC, the CBD has further developed protocols and amendments since the Rio Conference. The Nagoya Protocol on Access to

Genetic Resources and the Fair and Equitable Sharing of Benefits, adopted in October 2010, represents one of these developments. It further elaborates on power issues such as access to genetic resources, associated traditional knowledge and benefit sharing, which had always been a key focus of the CBD (Oberthür and Pożarowska, 2013). The Nagoya Protocol aims to help Indigenous peoples to empower themselves (Mile and Puno, 2021). The broader Nagoya outcome included the Strategic Plan for Biodiversity 2011–2020 with the so-called Aichi Biodiversity Targets, adopted in 2010 (Sotirov et al, 2020), strongly interlinked with the SDGs (particularly SDG 15); for example, Target 15 refers to the concepts of 'degraded ecosystems' and 'restoration' (CBD, 2010).

As in international climate governance, scholars have observed a shift from a primarily state-driven approach towards the inclusion of private actors (Pattberg et al, 2017). Aichi Target 4 acknowledges the diverse types of actors included in international biodiversity governance.

Forest landscape restoration

Forest (and) landscape restoration has gained increased international attention in the political sphere. FLR resonates with several SDGs, as restoring land is expected to help tackle climate change and desertification, support biodiversity and ensure human well-being (UNCCD, 2022). Already before the SDGs were agreed upon, numerous initiatives had begun to promote FLR as a solution for diverse environmental challenges including improving the livelihoods of poorer people. These initiatives have emerged at diverse levels from the local to the global, the latter prominently represented by the Bonn Challenge and the New York Declaration on Forests, both of which involve a broad range of actors, including public and private actors and civil society.

Current FLR concepts and initiatives seek to advance century-long experience with site-level forest restoration efforts to larger landscape and even global scales. In July 2000 WWF and the IUCN defined FLR as 'a planned process that aims to regain ecological integrity and enhance human well-being in deforested or degraded landscapes' (Mansourian, 2005: 10). Recognizing possible trade-offs, Mansourian concludes that it is 'possible to enhance the overall benefits to people and biodiversity at that scale' (Mansourian, 2005: 11). FLR is embedded in and linked to the wider concept of ecosystem restoration pursued by the Society for Ecological Restoration (SER) for more than three decades. FLR (re)gained international popularity with the UN Decade on Ecosystem Restoration (2021–30) (UNEP, 2019). The global FLR agenda is mirrored in a global proliferation of restoration investments and interventions nourishing a win–win narrative of FLR, which suggests that planting trees not only serves wood production and captures carbon but also fosters the well-being of humans and nature. The high level of attention builds on a range of powerful actors supporting the concept, well beyond SER. Laestadius et al (2015) affirm the key role of IUCN in mainstreaming the FLR concept, recommending that the World Bank takes up this approach in its investment (Maginnis et al, 2004). Most recently, with the seventh Global Environment Facility (GEF) replenishment, forest restoration has taken centre stage in international development finance as a solution reconciling food security, ecological integrity and economic prosperity (GEF, 2021). Additionally, the ITTO, together with WWF and the Center for International Forestry Research (CIFOR), investigated the concept of degraded forests (ITTO, 2002), and the UNFF focused on the restoration of forests (UNFF, 2005).

However, more recently critical voices have questioned the win–win narrative around FLR, pointing to losses and losers, and new conflicts of interest (Kleinschmit et al, 2023). Concerns are now growing over the social and economic implications of large-scale tree-planting schemes, including flaws in the governance of FLR (Brockhaus et al, 2021; Pritchard, 2021). Important concerns include unequal access to and control of land leading to the exclusion of local communities and minorities, specifically women (eg Chazdon et al, 2021).

3.2 Three logics of international forest governance and its relations to SDG 15

Section 3.1, while not fully exhaustive, provided a historical overview of the interests and institutions that shape and are shaped in the international forest governance complex, which have evolved over the years. The main aim of this chapter has been to identify three distinct logics underpinning and partly predominating the institutions and the interests of supportive actors over time. These logics, as with many analytical categorizations, simplify the complexity of international forest governance and thus do not reflect the blurred boundaries between them. In essence, the three key logics identified in international forest governance and encapsulated in SDG 15 may be understood as rationales each of which leans more towards and highlights mainly one of the pillars of sustainability (economic, ecologic, social): (1) production and market logic; (3) the ecological sustainability logic, addressing environmental challenges such as climate change and biodiversity loss; and (3) the community and empowerment logic. Their relevance, and predominance in international forest governance, vary over the years and between governance settings. At the same time they may be seen to compete with each other regarding interpretative dominance in international governance.

3.2.1 Logic of production and markets

The period preceding the Rio conference was dominated by interests devoted to production and international markets for tropical timber. This focus followed the historical heritage of colonial exploitation and was supported by powerful private actors, for example the ITTO supporting the commercial interests of private and state actors. The logic of production and in particular marketization is also reaffirmed by CITES. It reinforces the major role of markets despite its focus on the ecological aspects of endangered species and the restrictions imposed on imports and export. In contrast to the ITTO, interests of societal actors have gained some relevance in CITES. However, CITES has been criticized as representing an agenda that aligns mainly with Northern interests.

The powerful interests of private actors shaping the logic of production and markets have continued in the post-Rio development, where hybrid and market-based regimes are interlinked with public policies like FLEGT, private certification schemes and others. Hybrid governance has accelerated, with transnational public–private partnerships joining forces in international FLR governance, supported by dominant international organizations such as ITTO, World Bank and GEF, and diverse NGOs. State and private sector actors are pledging restoration action, thereby 'greening' their forest operations. SDG 15, and in particular Target 15.2, with its aim to end deforestation and restore degraded forests, builds directly on and is interlinked with the powerful institution and public and private sector interests pursuing FLR. In this sense, the SDG is also embracing the production and markets logic.

From a critical perspective, FLR movements presenting productive industrial and large-scale plantations as an opportunity for the restoration of multiple landscape functions (Sayer and Elliot, 2005) follow the same exploitative production and market logic as neocolonial trade relationships between world regions. However, creating regional value chains may as well directly benefit local people, and enabling local economies based on renewable materials also supports the logic of communities and empowerment (section 3.2.3).

3.2.2 Logic of ecological sustainability

With the Rio Conference, the logic of ecological sustainability gained increasing relevance in international forest governance, emphasizing environmental concerns and responding to challenges like climate change and the loss of biodiversity. Since then, this logic has been strongly supported by the powerful interests of governments and international and transnational actors, such as UN organizations including the UNFCCC, CBD and

UNCCD. The dominance of this logic has exacerbated with the FLR concept. Both FLR institutions and SDG 15 follow the assumption that planting more trees on 'abandoned land' is good in principle. However, it is assumed that targets of SDG 15, such as those devoted to biodiversity, might be missed as a result of the challenges facing forests (Krause and Tilker, 2022). One of these challenges results from the trade-off with interests driven by the logic of production and markets, for example the commercial use of monocultural plantations and its impact on biodiversity (Zhang et al, 2021). Trade-offs with particular local interests other than the ecological benefits, for example informal land use, have long been neglected. More recently, calls for a rights-based approach have become louder, questioning the overly optimistic assumptions about abandoned landscapes with little value (van Oosten and Merten, 2021).

3.2.3 Logic of communities and empowerment

The community and empowerment logic is devoted to local livelihood and empowerment. It originates from interests associated with development cooperation. Likewise, it may be seen as a reactive logic, responding to the trade-offs and conflicts arising from the practices pursued under the two aforementioned logics. This logic is mainly represented by the interests of rather marginalized actor groups, and is far less dominant than the others in international forest governance. Community and empowerment have been issues in UN-driven governmental institutions addressed sometimes more (eg CDB), sometimes less (eg CITES). However, the issues have gained increasing attention, as the tenure rights of marginalized groups, especially Indigenous peoples, youth and women, have been acknowledged in, for example, REDD+ safeguards as well as in the FLR. The need to involve, consider and empower communities has evolved with the evidence that the implementation of policies and programs (eg in climate or biodiversity governance) is weak without the involvement and acceptance of local people (Reinecke and Blum, 2018). The actors behind this logic and its respective stakes include not only local actors, but also international and local social movements, NGOs, development agencies, activists and social scientists, among others. However, local communities, in contrast to other actors (eg state, UN or international organizations, private sector), play a rather peripheral and partly symbolic role in international forest governance practice. This is despite the by now commonly accepted superiority of their techniques to coexist with, manage and protect the integrity of ecosystems and biodiversity, in contrast to western scientific and managerial mechanisms (Dawson et al, 2021). These actors are partly portrayed as 'losers' in REDD+ or FLR implementation, spurred by the shift of attention away from their territories and the natural resources they need to support their livelihoods with, to country-led 'international forest affairs' that puts nation-states back in the driving seat. The SDGs in general address the importance of communities and the demand of empowerment.

However, those taking a critical perspective argue that SDG 15 does partly fall back behind former frameworks in promoting these needs. Krauss (2021) criticizes the absence of either an explicit commitment in SDG 15 or an explicit connection to participatory governance. Instead, it is assumed that ecological goals might be prioritized in case of trade-offs with the logic of communities and empowerment (eg Baumgartner, 2019).

3.3 SDG 15 and the legacy of international forest governance

The present analysis shows the dominance of a persisting logic of production and markets combined with a Northern logic of ecological sustainability in international forest governance and in SDG 15. Both dominant logics are strongly supported by governance forms favouring the interest of governmental actors, UN-led initiatives and large-scale private sector interests. The way in which planting trees and FLR is presented as a win–win narrative, including in SDG 15, favours carbon capture and wood production over community and 'minor ecological' interests, while underestimating trade-offs between climate goals, local food production, biodiversity conservation and questions about which actors are benefiting in practice from planting trees. In this sense, the potential universality of the SDGs has not been fully tapped.

The interest of local (Southern) actors has gained increasing attention in recent years. The legacy of these logics, and their underlying interests and institutions as also encapsulated in SDG 15, have been a missed opportunity to develop a strong agenda for global development and for overcoming the binary view on and bridging the interests between the Global North and the Global South. The separating binary concept is rooted in international forest governance concepts and terminology. It is to be found in the differentiation of producer and consumer (countries) in ITTO, the legality verification regimes and the negotiations on a global forest convention and the international governance of biodiversity and climate change (Rosendahl, 2001), with annexes differentiating parties along those lines. The Global North-South divide has also been pursued in relation to the FLR concept, where 'global' solutions to environmental challenges are addressed with pledges and technical support from Northern to Southern countries like development interventions. These solutions are reaffirming international rather than global development (for a discussion about both concepts, see Horner, 2020). By fostering partnerships between all kinds of countries, multilateral organizations and other stakeholders, SDG 15 may be partly contributing to the overarching aim of overcoming power asymmetries. However, when unpacking who these 'other' stakeholders are, the private sector dominates.

Acknowledgements

We want to thank Symphorien Ongolo, our coauthor of a chapter on forest landscape restoration (Kleinschmit et al, 2023), on which this chapter draws. Sabaheta Ramcilovic-Suominen gratefully acknowledges funding from the Academy of Finland (grant number 332353), which supported her work on this chapter.

References

Acheampong, E. and Maryudi, A. (2020) 'Avoiding legality: timber producers' strategies and motivations under FLEGT in Ghana and Indonesia', *Forest Policy and Economics*, 111: 102047.

Angelsen, A., Brockhaus, M., Duchelle, A.E., Larson, A.M., Martius, C., Sunderlin, W.D. et al (2017) 'Learning from REDD+: a response to Fletcher et al.', *Conservation Biology*, 31(3): 718–20.

Arts, B. (1998) *The Political Influence of Global NGOs: Case Studies on the Climate and Biodiversity Conventions*, Utrecht: International Books.

Arts, B. (2006) 'Non-state actors in global environmental governance: new arrangements beyond the state', in M. Koenig-Archibugi and M. Zürn (eds) *New Modes of Governance in the Global System: Exploring Publicness, Delegation and Inclusiveness*, London: Springer, pp 177–200.

Auld, G., Balboa, C., Bernstein, S., Cashore, B., Delmas, M. and Young, O. (2009) 'The emergence of non-state market-driven (NSDM) global environmental governance', in M.A. Delmas and O.R. Young (eds) *Governance for the Environment: New Perspectives*, Cambridge: Cambridge University Press, pp 183–218.

Barnett, M. and Duvall, R. (2005) 'Power in global governance', in M. Barnett and R. Duvall (eds) *Power in Global Governance*, New York: Cambridge University Press, pp 1–32.

Baumgartner, R.J. (2019) 'Sustainable Development Goals and the forest sector – A complex relationship', *Forests*, 10(2): 152.

Bayrak M.M. and Marafa L.M. (2016) 'Ten years of REDD+: a critical review of the impact of REDD+ on forest-dependent communities', *Sustainability*, 8: 1–22.

Bernstein, S. and Cashore, B. (2004) 'Non-state global governance: is forest certification a legitimate alternative to a global forest convention', in J.J. Kirton and M.J. Trebilcock (eds) *Hard Choices, Soft Law: Voluntary Standards in Global Trade, Environment and Social Governance*, Aldershot: Ashgate, pp 33–63.

Brockhaus, M., Di Gregorio, M., Djoudi, H., Moeliono, M., Pham, T.T. and Wong, G.Y. (2021) 'The forest frontier in the Global South: climate change policies and the promise of development and equity', *Ambio*, 50(12): 2238–55.

Cashore, B. and Stone, M.W. (2012) 'Can legality verification rescue global forest governance? Analyzing the potential of public and private policy intersection to ameliorate forest challenges in Southeast Asia', *Forest Policy and Economics*, 18: 13–22.

Cashore, B.W., Auld, G. and Newsom, D. (2004) *Governing through Markets: Forest Certification and the Emergence of Non-State Authority*, New Haven, CT: Yale University Press.

CBD (Convention on Biological Diversity) (2010) 'Decision adopted by the Conference of the Parties to the Convention on Biological Diversity at its Tenth Meeting. X/2 The Strategic Plan for Biodiversity 2011–2020 and the Aichi Biodiversity Targets', UNEP/CBD/COP/DEC/X/2, Nagoya: UN/UNEP/CBD, Available from: https://www.cbd.int/doc/decisions/cop-10/cop-10-dec-02-en.pdf [Accessed 31 May 2023].

Challender, D.W., Harrop, S.R. and MacMillan, D.C. (2015) 'Towards informed and multi-faceted wildlife trade interventions', *Global Ecology and Conservation*, 3: 129–48.

Chazdon, R.L., Wilson, S.J., Brondizio, E., Guariguata, M.R. and Herbohn, J. (2021) 'Key challenges for governing forest and landscape restoration across different contexts', *Land Use Policy* 104: 104854.

Davenport, D.S. (2005) 'An alternative explanation for the failure of the UNCED forest negotiations', *Global Environmental Politics*, 5(1): 105–30.

Dawson, N., Coolsaet, B., Sterling, E., Loveridge, R., Gross-Camp, N.D., Wongbusarakum, S. et al (2021) 'The role of Indigenous peoples and local communities in effective and equitable conservation', *Ecology and Society*, 26(3): 19.

Di Gregorio, M., Fatorelli, L., Paavola, J., Locatelli, B., Pramova, E., Nurrochmat, D.R. et al (2019) 'Multi-level governance and power in climate change policy networks', *Global Environmental Change*, 54: 64–77.

Dimitrov, R.S. (2003) 'Knowledge, power, and interests in environmental regime formation', *International Studies Quarterly*, 47: 123–50.

Duffy, R. (2013) 'Global environmental governance and north–south dynamics: the case of the CITES', *Environment and Planning C: Government and Policy*, 31(2): 222–39.

Feintrenie, L. (2014) 'Agro-industrial plantations in Central Africa, risks and opportunties', *Biodiversity and Conservation*, 23(6): 1577–89.

Fernàndez-Blanco, C.R., Burns, S.L. and Giessen, L. (2019) 'Mapping the fragmentation of the international forest regime complex: institutional elements, conflicts and synergies', *International Environmental Agreements: Politics, Law and Economics*, 19: 187–205.

Fleischman, F., Basant, S., Fischer, H., Gupta, D., Lopez, G.G., Kashwan, P. et al (2021) 'How politics shapes the outcomes of forest carbon finance', *Current Opinion in Environmental Sustainability*, 51: 7–14.

GEF (Global Environment Facility) (2021) *Food Systems, Land Use and Restoration Impact Program*, Available from: https://www.thegef.org/sites/default/files/documents/2021-11/gef_food_systems_land_use_restoration_folur_impact_program_2021_11.pdf [Accessed 31 May 2023]

Glück, P., Carvalho Mendes, A. and Neven, I. (2003) *Making NFPs Work: Supporting Factors and Procedural Aspects. Report on COST Action. National Forest Programmes in a European Context*, Vienna: Institute for Forest Sector Policy and Economics, University of Natural Resources and Applied Life Sciences.

Gulbrandsen, L.H. (2006) 'Creating markets for eco-labelling: are consumers insignificant?', *International Journal of Consumer Studies*, 30(5): 477–89.

Gulbrandsen, L.H. and Humphreys, D. (2006) *International Initiatives to Address Tropical Timber Logging and Trade: A Report for the Norwegian Ministry of the Environment*, FNI Report 4/2006, Lysaker: Fridtjof Nansen Institute.

Horner, R. (2020) 'Towards a new paradigm of global development? Beyond the limits of international development', *Progress in Human Geography*, 44(3): 415–36.

Humphreys, D. (1996a) *Forest Politics: The Evolution of International Cooperation*, London: Earthscan.

Humphreys, D. (1996b) 'The global politics of forest conservation since the UNCED', *Environmental Politics*, 5(2): 231–57.

Humphreys, D. (2001) 'Forest negotiations at the United Nations: explaining cooperation and discord', *Forest Policy and Economics*, 3: 125–35.

Humphreys, D. (2004) 'Redefining the issues: NGO influence on international forest negotiations', *Global Environmental Politics*, 4(2): 51–74.

Humphreys, D. (2009) 'Discourse as ideology: neoliberalism and the limits of international forest policy', *Forest Policy and Economics*, 11: 319–25.

Humphreys, D. (2012) *Logjam: Deforestation and the Crisis of Global Governance*, London: Routledge.

International Tropical Timber Agreement (2006) Geneva, Available from: https://treaties.un.org/doc/Treaties/2006/02/20060215%2004-26%20PM/Ch_XIX_46p.pdf [Accessed 20 August 2022].

ITTO (International Timber Trade Organization) (2002) *ITTO Guidelines for the Restoration, Management and Rehabilitation of Degraded and Secondary Tropical Forests*, Yokohama, Available from: www.cbd.int/forest/doc/itto-guidelines-restoration-management-rehabilitation-degraded-forests-2002-en.pdf [Accessed 20 August 2022].

Kadam, P., Dwivedi, P. and Karnatz, C. (2021) 'Mapping convergence of sustainable forest management systems: comparing three protocols and two certification schemes for ascertaining the trends in global forest governance', *Forest Policy and Economics*, 133: 102614.

Katila, P., Colfer, C.J.P., De Jong, W., Galloway, G., Pacheco, P. and Winkel, G. (eds) (2019) *Sustainable Development Goals*, Cambridge: Cambridge University Press.

Karlsson, S. (2002) 'The North–South knowledge divide: consequences for global environmental governance', in D.C. Esty and M.H. Ivanova (eds) *Global Environmental Governance: Options and Opportunities*, New Haven, CT: Yale School of Forestry & Environmental Studies, pp 1–24.

Kleinschmit, D., Blum, M., Brockhaus, M., Karambiri, M., Kröger, M., Ramcilovik-Suominen, S. et al (2023) 'Forest (landscape) restoration governance: institutions, interests, ideas, and their interlinked logics', in P. Katila, C.J.P. Colfer, W. de Jong, G. Galloway, P. Pacheco and G. Winkel (eds) *Restoring Forests for Sustainable Development: Policies, Practices and Impacts*, Oxford: Oxford University Press.

Krause, T. and Tilker, A. (2022) 'How the loss of forest fauna undermines the achievement of the SDGs', *Ambio*, 51: 103–13.

Krauss, J.E. (2021) 'Decolonizing, conviviality and convivial conservation: towards a convivial SDG 15, life on land?' *Journal of Political Ecology*, 28(1).

Kunzmann, K. (2008) 'The non-legally binding instrument on sustainable management of all types of forests – Towards a legal regime for sustainable forest management?', *German Law Journal*, 9(8): 981–1006.

Kuyper, J.W., Linnér, B.-O. and Schroeder, H. (2018) 'Non-state actors in hybrid global climate governance: justice, legitimacy, and effectiveness in a post-Paris era', *WIREs Climate Change*, 9(1): e497.

Laestadius, L., Buckingham, K., Maginnis, S. and Saint-Laurent, C. (2015) 'Before Bonn and beyond: the history and future of forest landscape restoration', *Unasylva*, 66(245): 11.

Levin, K., McDermott, C. and Cashore, B. (2008) 'The climate regime as global forest governance: can reduced emissions from Deforestation and Forest Degradation (REDD) initiatives pass a dual effectiveness test?', *International Forestry Review*, 10(3): 538–49.

Linser, S., Wolfslehner, B., Bridge, S.R., Gritten, D., Johnson, S., Payn, T. et al (2018) '25 years of criteria and indicators for sustainable forest management: how intergovernmental C&I processes have made a difference', *Forests*, 9(9): 578.

Lipschutz, R.D. (2000) 'Why is there no international forestry law: an examination of international forestry regulation, both public and private', *UCLA Journal of Environmental Law and Policy*, 19(1): 153–79.

Locatelli, B., Kanninen, M., Brockhaus, M., Pierce Colfer, C.J., Murdiyarso, D. and Santoso, H. (2008) *Facing an Uncertain Future: How Forests and People Can Adapt to Climate Change*, Bogor Barat, Indonesia, Center for International Forestry Research (CIFOR).

Lomazzi, M., Borisch, B. and Laaser, U. (2014) 'The Millennium Development Goals: experiences, achievements and what's next', *Global Health Action*, 7(1): 23695.

Maginnis, S., Jackson, W. and Dudley, N. (2004) '14. Conservation landscapes: whose landscapes? whose trade-offs?', in *Getting Biodiversity Projects to Work*, New York: Columbia University Press, pp 321–39.

Mansourian, S. (2005) 'Overview of forest restoration strategies and terms', in S. Mansourian, D. Vallauri and N. Dudley (eds) *Forest Restoration in Landscapes: Beyond Planting Trees*, New York: Springer, pp 8–13.

McDermott, C., Acheampong, E. Arora-Jonsson, S. Asare, R., de Jong, W., Hirons, M. (2019) 'SDG 16: Peace, justice and strong Institutions – A political ecology perspective', in P. Katila, C.J.P. Colfer, W. De Jong, G. Galloway, P. Pacheco and G. Winkel (eds) *Sustainable Development Goals*, Cambridge: Cambridge University Press, pp 510–40.

Meidinger, E.E. (2003) 'Forest certification as a global civil society regulatory institution', in E. Meidinger, C. Elliott and G. Oesten (eds) *Social and Political Dimensions of Forest Certification*, Buffalo, NY: University at Buffalo School of Law, pp 265–89.

Mile, A. and Puno, R. (2021) 'Indigenous peoples, the SDGs, and international environmental law', in N. Kakar, V. Popovski and N.A. Robinson (eds) *Fulfilling the Sustainable Development Goals*, London: Routledge, pp 485–500.

Mills-Novoa, M. and Liverman, D.M. (2019) 'Nationally determined contributions: material climate commitments and discursive positioning in the NDCs', *Wiley Interdisciplinary Reviews: Climate Change*, 10(5): e589.

Menashy, F. (2019) *International Aid to Education: Power Dynamics in an Era of Partnership*, New York: Teachers College Press.

Oberthür, S. and Pożarowska, J. (2013) 'Managing institutional complexity and fragmentation: the Nagoya Protocol and the global governance of genetic resources', *Global Environmental Politics*, 13(3): 100–118.

Pattberg, P. (2006) 'Private governance and the South: lessons from global forest politics', *Third World Quarterly*, 27(4): 579–93.

Pattberg, P.H., Kristensen, K.E.G. and Widerberg, O.E. (2017) 'Beyond the CBD: environmental policy analysis, multi-layered governance in Europe and beyond (MLG)', IVM Report, Amsterdam: Institute for Environmental Studies/IVM.

Peluso, N.L. and Vandergeest, P. (2001) 'Genealogies of the political forest and customary rights in Indonesia, Malaysia, and Thailand', *Journal of Asian Studies*, 60(3): 761–812.

Pritchard, R. (2021) 'Politics, power and planting trees', *Nature Sustainability*, 4(11): 932.

Pülzl, H. (2005) 'Die Politik des Waldes: Governance natürlicher Ressourcen bei den Vereinten Nationen', Dissertation, Universität für Bodenkultur.

Ramcilovic-Suominen, S., Carodenuto, S., McDermott, C. and Hiedanpää, J. (2021) 'Environmental justice and REDD+ safeguards in Laos: lessons from an authoritarian political regime', *Ambio*, 50(12): 2256–71.

Reinecke, S. and Blum, M. (2018) 'Discourses across scales on forest landscape restoration', *Sustainability*, 10(3): 613.

Roe, D., Mulliken, T., Milledge, S., Mremi, J., Mosha, S. and Grieg-Gran, M. (2002) 'Making a killing or making a living: wildlife trade, trade controls, and rural livelihoods', Biodiversity and Livelihoods Issues No. 6, London: International Institute for Environment and Development (IIED), Available from: https://www.iied.org/sites/default/files/pdfs/migrate/9156IIED.pdf [Accessed 31 May 2023].

Rosendal, G.K. (2001) 'Overlapping international regimes: the case of the Intergovernmental Forum on Forests (IFF) between climate change and biodiversity', *International Environmental Agreements*, 1(4): 447–68.

Sayer, J. and Elliot, C. (2005) 'The role of commercial plantations in forest landscape restoration', in *Forest Restoration in Landscapes*, New York: Springer, pp 379–83.

Schally, H.M. (1994) 'Problems and opportunities in intergovernmental coordination: a Northern perspective', in B.I. Spector (ed) *Negotiating International Regimes: Lessons learned from the United Nations Conference on Environment and Development*, London: Graham &Trotman, pp 123–33.

Schmidt, V. (2008) 'Discursive institutionalism: the explanatory power of ideas and discourse', *Annual Review of Political Science*, 11: 303–26.

Schroeder, H. and González, N.C. (2019) 'Bridging knowledge divides: the case of Indigenous ontologies of territoriality and REDD+', *Forest Policy and Economics*, 100: 198–206.

Shaffer, G. and Bodansky, D. (2012) 'Transnationalism, unilateralism and international law', *Transnational Environmental Law*, 1(1): 31–41.

Slaughter, A-M., Tulumello, A. and Wood, S. (1998) 'International law and international relations theory: a new generation of interdisciplinary scholarship', *American Journal of International Law*, 92: 367–97.

Sotirov, M., Pokorny, B., Kleinschmit, D. and Kanowski, P. (2020) 'International forest governance and policy: institutional architecture and pathways of influence in global sustainability', *Sustainability*, 12(17): 7010.

UN (United Nations) (1992) *Convention on Biological Diversity*, Available from: www.cbd.int/doc/legal/cbd-en.pdf [Accessed 20 August 2022].

UNCCD (United Nations Convention to Combat Desertification) (1994) 'Elaboration of an International Convention to Combat Desertification in Countries Experiencing Serious Drought and/or Desertification, particularly in Africa', United Nations A/AC.241/27, Available from: http://www.un-documents.net/a-ac241-27.pdf [Accessed 20 August 2022].

UNCCD (2022) *Global Land Outlook 2*, Available from: www.unccd.int/ sites/default/files/2022-04/UNCCD_GLO2_low-res_2.pdf [Accessed 20 August 2022].

UNEP (United Nations Environmental Programme) (2019) 'UN Environment Programme Annual Report 2019', Available from: www. unep.org/annualreport/2019/index.php [Accessed 20 August 2022].

UNFCCC (United Nations Framework Convention on Climate Change) (1992) 'United Nations Framework Convention on Climate Change', FCCC/INFORMAL/84 GE.05-62220 (E) 200705, Available from: https://unfccc.int/resource/docs/convkp/conveng.pdf [Accessed 20 August 2022].

UNFF (United Nations Forum on Forests) (2005) 'Report of the fifth session', Economic and Social Council, Official Records, 2005, Supplement No. 22, Available from: https://documents-dds-ny.un.org/doc/UNDOC/ GEN/N05/396/66/PDF/N0539666.pdf?OpenElement [Accessed 20 August 2022].

UN System Task Team (2012) 'Building on the MDGs to bring sustainable development to the post-2015 development agenda', Thematic Think Piece, Available from: www.un.org/millenniumgoals/pdf/Think%20Pie ces/17_sustainable_development.pdf [Accessed 20 August 2022].

van Oosten, C. and Merten, K. (2021) 'Securing rights in landscapes: Towards a rights based landscape approach', Report WCDI-21-135, Wageningen Centre for Development Innovation, Wageningen: Wageningen University & Research, Available from: https://www.wwf.nl/globalassets/pdf/secur ing-rights-in-landscapes.pdf [accessed 31 May 2023].

Velázquez Gomar, J.O. and Stringer, L.C. (2011) 'Moving towards sustainability? An analysis of CITES' conservation policies', *Environmental Policy and Governance*, 21(4): 240–58.

Winkel, G. Leipold, S., Buhmann, K., Cashore, B., de Jong, W., Nathan, I. et al (2017) 'Narrating illegal logging across the globe: between green protectionism and sustainable resource use', *International Forestry Review*, 19(S1): 81–97.

Zentilli, B. (1995) 'What progress has been made on Agenda 21, chapter 1? A critical examination of developments since UNCED', *RECIEL* 4(3): 222–9.

Zhang, J., Fu, B., Stafford-Smith, M., Wang, S. and Zhao, W. (2021) 'Improve forest restoration initiatives to meet Sustainable Development Goal 15', *Nature Ecology & Evolution*, 5(1): 10–13.

Interview with Daniela Kleinschmit: The Value of Nature

Liam Gavin and Silvia Panini

We can infer from your chapter that, while in the past forests served only to generate money through the extraction of wood, value can now also be produced through its preservation (eg through REDD+). Would you agree with the suggestion that we are now 'selling nature to save it'?

Kleinschmit: I absolutely agree that this phenomenon is happening – we attach values and numbers to nature, which is well in line with the concept of payments for ecosystem services and which has gained increasing importance in the last decade. This is economic valuation of the diverse services offered by forests (and other ecosystems) to humanity. Hence, we can perceive it as a commodification – similar to the goods we exploit like wood. Just that we in this case accept payments to preserve the ecosystem. It's a very anthropocentric perspective on nature rather than understanding nature as something that has value just for being there, for its own sake. Instead, we have an act of economic valuation: we put a price tag on nature, and from this certain actors profit more likely than others.

We can increasingly see private actors buying up massive swathes of land and forests to establish carbon-offsetting projects to reduce their net emissions. However, this is often linked to mass evictions and human rights abuses, as these are often lands where Indigenous people usually live. What alternative visions can you put forward for sustainable reforestation in this light?

Kleinschmit: I think the local communities of the areas notified as 'degraded land', in particular, should have a large say in that: whether there is potential for such projects and whether that fits their context. In my opinion this is a precondition – in contrast to the idea of localizing areas by satellite images and from that pretending that 'we identified potential degraded land and we know exactly what we can plant and where from above'. The latter is a very technical way of approaching the issue and is neglecting the people in the area, in a top-down technical approach.

Many have drawn attention to the difference between natural forests and 'fake' – monocultural, plantation – forests. Do you notice much interest within the forestry governance regime in avoiding this?

Kleinschmit: Yes, certainly. For example, we do have many monoculture forests in Europe; however, we have also had a shift towards more natural and diverse forests. Because it has an impact on other parts of ecosystems, this debate plays a role in all the relevant conventions and agreements. However, there are different perspectives on this, for example the perspective of segregation versus the perspective of integration – so either fully protecting primary forests while also having separate timber plantations or having forests that integrate both conservation and production goals in forests. And this is a tricky conversation to have without a one-size solution for all.

4

Protecting Life below Water: Competing Normative, Economic and Epistemic Orders (SDG 14)

Alice B.M. Vadrot

The United Nations' Sustainable Development Goal 14, 'Life below water', captures the need to both protect the oceans and acknowledge the socio-economic dependence of humans – especially in coastal communities and Small Island Developing States (SIDS) – on marine resources (Ntona and Morgera, 2018). SDG 14 is embedded within a broad spectrum of recent international efforts to protect the ocean and to regulate sectoral aspects (including fisheries, navigation and shipping, whaling, deep-seabed mining and trade in endangered marine species). Although many scholars consider SDG 14 to be a landmark or, at least, a great step forward in advancing partnerships and synergies between various strands of contemporary ocean governance (eg Singh et al, 2017; Unger et al, 2017; Haas et al, 2019), there is also a broad recognition of its shortcomings when it comes to properly addressing pressing environmental problems and global inequality (eg Cormier and Elliott, 2017; Armstrong, 2020; Johansen and Vestvik, 2020; Haward and Haas, 2021).

SDG 14 is different from other environmental goals of the global sustainability agenda. The implementation of SDG 14, unlike SDG 13 (Climate action) and SDG 15 (Life on land), cannot rely on an existing international institutional landscape comparable to the web of multilateral agreements, subsidiary bodies and global expert organizations established to support international responses to tackle climate change (with the UNFCCC and IPCC) and biodiversity loss (with the CBD and the Intergovernmental Platform on Biodiversity and Ecosystem Services (IPBES)) at the global scale.

The Division for Ocean Affairs and the Law of the Sea (UNDOALOS), which could play a catalysing role, lacks environmental expertise and resources to fulfil the task of coordinating existing insertional efforts and figuring as a higher institutional arrangement overseeing the implementation of SDG 14. These dynamics underpin the need for critically assessing the potential of a global sustainability agenda for the oceans and whether SDG 14 is well enough positioned to support the transition towards an overarching and consistent political and legal framework for ocean protection.

Human societies have always depended on the ocean as a space for transport and trade and a source of nutrients. It has always provided a local livelihood, with approximately 40 per cent of human populations living along the coast. Dependence seems to be growing: today, over three billion people rely on the ocean for their livelihoods. In spite of this, the protection of the marine environment is more urgent than ever. The expansion of human activities into deeper regions of the ocean is accelerating the extraction of marine resources and the destruction of ocean ecosystems. Overfishing, ocean acidification, climate pressures (eg extreme events, sea level rise, the changing chemistry of the ocean) and diverse sources of pollution (eg shipping, land-based pollution, microplastics) have already left their marks on the marine environment, affecting the health of marine ecosystems and species. Emerging industrial activities such as deep-water fisheries, deep-seabed mining and deep-water oil and gas drilling may – in the near future – worsen the situation and cause irrecoverable damage to ecologically and biologically significant marine areas.

Several treaties and other agreements including ocean protection provisions (eg Article 192 of UNCLOS) have been signed, and international organizations and programmes established, such as the International Oceanographic Commission of UNESCO (IOC-UNESCO), the International Maritime Organization (IMO), the International Seabed Authority (ISA), and regional fisheries management organizations (RFMOs) and regional seas programmes established. This has led to a fragmented institutional landscape that perpetuates ineffective action. The UN Convention on the Law of the Sea (UNCLOS) divides the ocean into several maritime zones where different normative and legal principles apply, which complicates ownership issues and the accountability of both state and non-state actors. The ordering effects of this law and the way it has been shaping state interests since the 1970s calls for a political debate that goes beyond institutional fixes. Global environmental problems cannot be solved without questioning the normative, economic and epistemic orders within which they operate. Together with (geo)political factors and scientific uncertainties, current institutional and legal conditions – and the historical context in which they emerged – pose severe problems for the global ocean sustainability agenda and the implementation of SDG 14.

This chapter aims to critically assess the role of SDG 14 in contemporary ocean governance by bringing to the fore the normative, economic and epistemic orders that have shaped the institutional responses to an endangered ocean. It will argue that, while SDG 14 may symbolize a unifying imperative to put 'life below water' on the global sustainability agenda, it masks the deep-rooted tensions and inequalities underlying knowledge and governance of the ocean between the Global North and South as to how marine resources are exploited and explored. It will unpack these tensions by conceptualizing multilateral negotiations as order-making sites and examining ongoing international negotiations to establish a new legally binding agreement for the conservation and sustainable use of marine biodiversity beyond national jurisdiction. The BBNJ case illustrates the importance of negotiation sites and their related practices; it also enables discussion of the contest over normative, economic and epistemic orders that continues to hinder simple intergovernmental compromise and usher in an unsustainable ocean future. Finally, the chapter discusses alternative innovative approaches that deserve more research attention for a transition towards ocean sustainability.

4.1 Multilateral negotiations as order-making sites

The conceptual entry point of this chapter is the sites where global sustainability agendas and environmental agreements are negotiated. Rather than focusing on the outcome of multilateral negotiations – treaties, provisions, targets, goals – these negotiations are conceptualized as sites where order is made, contested and sustained (Vadrot, 2020; Hughes et al, 2021; Hughes and Vadrot, 2023). The negotiation site constitutes a political arena where different actors, interests and orders compete for recognition. Actors perceive their interests and develop their strategies within the selectively structured context in which they act and the insights that they have gained as to the likeliness that their strategies will be successful. This includes the knowledge they have to support their interests and the economic means to realize them. In other words, the formation and articulation of interests are always relational and depend on the position of actors in relation to the dominant order and the means they have – or are believed to have – to influence it. Negotiating practices and events are intelligible and can be explored empirically in various ways, including through collaborative event ethnography (CEE), which is used in the framework of the author's research project MARIPOLDATA to study ongoing marine biodiversity negotiations (Vadrot, 2020; Vadrot et al, 2021) and as a case to critically investigate the obstacles of global ocean sustainability agendas.

Negotiation practices are intended to impose meaning, especially on the various elements of the treaty text, the form of knowledge that is supposed to underpin these elements, and the exact language and terminology that are used to buttress negotiating positions and that governments consensually agree on in the end. Yet meaning making in the negotiation room always implies that a relatively stable normative and economic order is either reproduced or contested. Furthermore, different epistemic orders – or ways of knowing and using (scientific) knowledge to support negotiating positions and strategies – are mobilized and compete for recognition in the negotiation room. The strategic use of legal and scientific expertise is a negotiation practice used by state and non-state actors to legitimize and underpin either their support or their contestation of specific text elements, concepts and terminologies used in the negotiations (Gray et al, 2014; Vadrot, 2020).

4.2 SDG 14: a narrow understanding of ocean governance

A standalone Sustainable Development Goal for the ocean was a great step forward. In the preparation phase of the SDGs, several proposals were made that included the idea of a separate SDG for the oceans (Visbeck et al, 2014). During the eighth session of the United Nations' open working group for the SDGs, several proposals developed by civil society actors and academia were discussed (Houghton, 2015). While each of these proposals highlighted different facets of ocean sustainability, they all emphasized the coupling of human and environmental issues, including the benefits from ocean resources (eg 'blue wealth', 'blue growth', 'inclusive economic and human development', 'equitable access', 'prosperous and resilient people communities') and the life-sustaining functions of the ocean (Houghton, 2015). Framings such as 'blue growth' or 'blue wealth' imply a very specific norm,ative understanding of ocean protection in terms of economic growth, along with a utilitarian approach to nature that does not entail the necessity to rethink the ocean. In contrast, notions such as inclusivity and equity do imply that the structural conditions under which societies access, use and share marine resources need to be reflected on and applied if a mode of governance of the ocean that goes beyond the status quo is to be achieved.

All SDG proposals were negotiated between United Nations General Assembly member states. An agreement to include a standalone ocean SDG was reached, but it did not reflect the level of detail of previously discussed proposals and offered 'little to articulate a compelling sustainability narrative for the ocean that has not already been repeated in numerous other texts' (Houghton, 2015: 5). The official overarching goal of SDG 14 is to 'conserve and sustainably use the oceans, seas and marine resources for sustainable development'. SDG 14 is composed of ten targets targeting Marine Pollution (14.1), Healthy Oceans (14.2), Ocean Acidification (14.3), Sustainable Fisheries (14.4), Marine Protected Areas (14.5), Fisheries Subsidies (14.6), Economic benefits for Small Island

Developing States & Least Developed Countries (14.7), Knowledge and Technology (14.a), Small Scale Fisheries (14.b) and Law Development and Implementation (14.c) (SDG 14, Supplementary Appendix 1: see Table 4.1, UNSTATS, 2016).

Scholars from diverse disciplines view SDG 14 as a major step forwards in considering ocean protection as a key challenge that needs to be addressed by the international community (eg Unger et al, 2017; Ntona and Morgera, 2018; Armstrong, 2020). That governments adopted a standalone SDG for the ocean was perceived as a strong signal to environmentalists and ocean scientists who had raised awareness of the harmful effects of human activities on marine ecosystems. For instance through the publication of the *World Ocean Assessment* (United Nations, 2017), SDG 14 not only addresses key issues such as climate change, overfishing and pollution but also adds an explicit socio-economic dimension featuring the needs of the most vulnerable and ocean-dependent communities (see Target 14.7), such as small island developing states (SIDS) and least developed countries (LDCs). The needs of SIDS and LDCs are also mentioned in Target 14.6, which accepts that they should receive 'appropriate and effective special and differential treatment' regarding limitations to harmful fishery subsidies. This logic is carried further into Target 14.a, which acknowledges the need to increase scientific knowledge, develop research capacities and transfer marine technology, a topic that has been the crux of several discussions between countries from the Global North and the Global South, at least since the UNCLOS negotiations in the 1960s and 1970s.

Although the targets do incorporate some aspects reflecting the overall goal, the process has been criticized for leaving little room to discuss the link between the overarching goal, the specific targets and appropriate indicators (Houghton, 2015). While, for instance, the increase in scientific knowledge (as expressed in Target 14.a) may indeed be measured by the 'proportion of total research budget allocated to research in the field of marine technology', such an indicator completely neglects the progress made in past years – for instance, through activities related to the Census of Marine Life Project and the UN Decade of Ocean Science – towards addressing deep-seated inequalities between the Global North and the Global South beyond marine technology. The research and dialogue among scientists, and between scientists and policy makers, on how to ensure equitable access to research infrastructure and technology has been advancing rapidly; it includes a more reflexive approach to the conduct and role of ocean science more generally (Visbeck, 2018) and concrete suggestions on how to transform research and collaboration practices (Harden-Davies et al, 2022). Thus, SDG 14 neither reflects the complex intertwinements between ocean science and power, nor the scientific debate that has emerged on the need to broaden the scope of what is understood as ocean knowledge.

Table 4.1: SDG 14

Target	Description	Measured by
14.1	By 2025, prevent and significantly reduce marine pollution of all kinds	Index of coastal eutrophication and floating plastic debris density
14.2	By 2020, sustainably manage and protect marine and coastal ecosystems … including by strengthening their resilience, and take action for their restoration in order to achieve healthy and productive oceans	The proportion of countries' EEZs managed using ecosystem-based approaches
14.3	Minimize and address the impacts of ocean acidification	Average marine acidity measured at representative sampling stations
14.4	By 2020, effectively regulate harvesting and end overfishing; illegal, unreported and unregulated fishing; and destructive fishing practices; and implement science-based management plans to restore fish stocks in the shortest time feasible	Proportion of fish stocks within biologically sustainable levels
14.5	By 2020, conserve at least 10 per cent of coastal and marine areas	Geographical coverage of MPAs
14.6	By 2020, prohibit certain forms of fisheries subsidies that contribute to overcapacity and overfishing; eliminate subsidies that contribute to illegal, unreported and unregulated fishing; and refrain from introducing new such subsidies, recognizing that appropriate and effective special and differential treatment for developing and least developed countries should be an integral part of the World Trade Organization fisheries subsidies negotiation	Progress by countries in implementing international instruments aiming to combat illegal, unreported and unregulated fishing
14.7	By 2030, increase the economic benefits to SIDS and LDCs from the sustainable use of marine resources, including through sustainable management of fisheries, aquaculture and tourism	Sustainable fisheries as a percentage of GDP in SIDS and LDCs
14.a	Increase scientific knowledge, develop research capacities and transfer marine technology, taking into account the Intergovernmental Oceanographic Commission Criteria and Guidelines on the Transfer of Marine Technology to improve ocean health and to enhance the contribution of marine biodiversity to the development of developing countries, in particular SIDS and LDCs	Proportion of total research budget allocated to research in the field of marine technology

(continued)

Table 4.1: SDG 14 (continued)

Target	Description	Measured by
14.b	Provide access of small-scale artisanal fishers to marine resources and markets	Progress by countries in the degree of application of a legal/ regulatory/policy/institutional framework that recognizes and protects access rights for small-scale fisheries
14.c	Ensure the full implementation of international law, as reflected in UNCLOS for states party to it, including, where applicable, existing regional and international regimes for the conservation and sustainable use of oceans and their resources by their parties	Number of countries making progress in ratifying, accepting and implementing through legal, policy and institutional frameworks ocean-related instruments that implement international law, as reflected in UNCLOS, for the conservation and sustainable use of the oceans and their resources

Source: Based on UNSTATS (2016).

This 'narrow transformational vision' extends to many targets of SDG 14, which tends to either focus on single economic sectors (most notably the fishery sector) (Webster, 2022) or the needs of developing states, without adequately addressing participation (of diverse stakeholders and knowledge holders, including local communities and Indigenous peoples), gender equality (in ocean science, governance and the day-to-day practices of living with and from the ocean) or human rights (regarding the life-sustaining functions of the ocean, such as food, recreation, labour and the fight against climate change), even though these were repeatedly mentioned during the SDG negotiation process (Ntona and Morgera, 2018). By applying rather traditional and sector-specific approaches, SDG 14 tends to reproduce a narrow understanding of ocean governance and discourses in which oceans are framed as natural capital, good for business and integral to the economies of Pacific small island developing states and small-scale fishery livelihoods (Silver et al, 2015). Regrettably, this is not the only problem with SDG 14. Scholars have pointed to a lack of engagement with the financial cost of actually implementing SDG 14 (Johansen and Vestvik, 2020), social considerations such as labour conditions on vessels (Rudolph et al, 2020; Haward and Haas, 2021) and poverty eradication (Ntona and Morgera, 2018).

While these may all be valid points, they tend not to go far enough and fail to unscramble the deeper philosophical and theoretical implications of SDG 14's framing of the ocean. In particular, the focus on the ocean as a commodity to be indexed, traded and used, along with the narrow appreciation of ocean justice and global equality (Armstrong, 2020), prohibits a more ethical conversation on ocean sustainability. This calls for a political debate that goes beyond institutional fixes and takes competing normative, economic and epistemic order into account. While it is true that the fragmented ocean governance framework, including the legal order that has contributed to its formation, significantly obstructs a profound transformation of ocean policies and politics towards sustainability, this does not lead to ready-made explanations or conclusions. The institutional landscape that has been put in place in the past century serves specific interests and is the result of a contest between states on the normative and economic principles and knowledge base that should guide the exploration and exploitation of the ocean.

If SDG 14 – and this is what several critical voices seem to confirm – reproduces a narrow ocean governance approach, including all the pitfalls and failures of the past, it stands to reason that the sites where the future of the ocean is being negotiated need to be studied and intergovernmental negotiation practices to be unpacked. The ongoing discussion that seeks to address legal gaps in governing marine biodiversity beyond national jurisdiction offers such an opportunity.

4.3 The BBNJ case

The BBNJ negotiations and the empirical study thereof provide an opportunity par excellence to unpack the challenges of sustainable ocean governance (De Santo et al, 2019; Tiller et al, 2019). After seven Intergovernmental Conferences (IGCs) governments adopted the new legally binding international instrument on the conservation and sustainable use of marine biological diversity in areas beyond national jurisdiction on 19 June, at the United Nations headquarters in New York, United States. The BBNJ negotiations demonstrate that historically rooted conflicts revolving around competing normative, economic and epistemic orders may be carried into the present and potentially the future.

4.3.1 Extending sustainability agendas to the high seas

Areas beyond national jurisdiction (ABNJ) account for 64 per cent of the ocean's surface and more than 90 per cent of its volume. Although scientists consider that marine biodiversity is concentrated in coastal waters and exclusive economic zones (EEZs) – with a high degree of diversity in the euphotic zone (ie close to the surface) and 25 per cent of marine species living in coral reefs – research is progressively uncovering the diversity of life offshore, both in high and deep-sea areas. For instance, the Census of Marine Life Project, which ran from 2000 to 2010, found more than 6,000 species previously unknown to science, including in several samples from deep-sea hydrothermal vents, where several groups of animals and microbes live under extreme conditions.

Yet scientists are not the only ones to have developed an interest in exploring these mostly untapped areas in the high and deep seas: several industries with an economic interest in both living and non-living marine resources have emerged. Ecologists have been warning that the rate of marine biodiversity loss may increase owing to accelerated and expanded exploitation and a lack of international regulation of ABNJ. Although UNCLOS includes provisions requiring states to protect the marine environment, it does not mention marine biodiversity and the CBD, which has no mandate for ABNJ, thus leaving actors in legal limbo (Harrison, 2017).

As a result, state and non-state actors started consultations within the framework of ad hoc open-ended informal working groups and a preparatory committee (PrepCom) was established by the United Nations General Assembly Resolution 69/292, with support provided by UNDOALOS. This process can be traced back to 2006 and resulted in the adoption of a report on the elements of a new international legally binding instrument under UNCLOS; this was submitted to the UN General Assembly by PrepCom members at their last session (July 2017) (Wright et al, 2016). In December of the same year, the General Assembly decided to convene an intergovernmental conference under the auspices of the UN to consider the preparatory committee's recommendations and to mandate governments to prepare a new treaty.

Thus, the BBNJ negotiations on an implementing agreement under UNCLOS were launched in 2018 within the framework of seven intergovernmental conferences (2018–23). Four so-called package elements had been identified by the PrepCom as key issues where consensus should be reached: marine genetic resources (MGRs); area-based management tools (ABMTs), including marine protected areas (MPAs); environmental impact assessments (EIAs); and capacity building and marine technology transfer (CB&TT). ABMTs and MPAs, which deal with the conservation and sustainable use of marine biodiversity, may be considered the core elements of the treaty, while MGRs and CB&TT respond to demands made during PrepCom meetings by Global South governments, most notably the G77 and China, the African Group, Latin American countries and SIDS, aiming to narrow the economic and technological gaps that prevent many of them from accessing and using BBNJ in an equitable manner.

4.3.2 Competing normative orders

At the risk of oversimplifying, it could be said that the way the ocean is known and governed has always been shaped by tensions between those who possess the economic and technological means to explore and exploit the marine environment and those who do not. In his well-known book, *The Social Construction of the Ocean*, Philipp Steinberg (2001) pointed to important technological and economic developments that shaped how hegemonic powers – most notably Europe – and some non-European Oceanian societies accessed and used the ocean over the past centuries. From the fifteenth to the eighteenth centuries ocean navigation led to an asymmetry between different parts of the globe. Cannon-armed caravels and new methods for harvesting and shipping distant fish stocks significantly expanded the secure extraction and trade of marine resources (Steinberg, 2001: 8). For a long time, the ocean was governed by the notion of the 'freedom of the seas', codified by Hugo Grotius in the seventeenth century and translated into modern international law in 1982 when, after nine years of negotiation, over 160 participating nations signed the United Nations Convention on the Law of the Sea.

As a global constitution for the ocean, UNCLOS may be viewed as a response to the expanding sovereignty claims of coastal states after the Second World War, which partly preserve the core of the freedom of the seas (FOS) principle and grants all states specific freedoms such as the freedom of navigation and the freedom to conduct marine scientific research. But it also limits national rights and jurisdictions over the ocean by setting boundaries between areas within and beyond national jurisdiction. UNCLOS divides the ocean into several maritime zones: national waters (within 12 nautical miles); exclusive economic zones (up to 200 nm); and the high seas (beyond 200 nm) – where different legal principles apply (see Figure 4.1). This territorialization of the ocean has curbed sovereignty claims, most notably those of the US and several Latin American countries, but has also extended the areas that states are meant to control, for instance through surveillance or through the monitoring and management of fish stocks up to where international waters begin. While it seems that UNCLOS has ordered the ocean, boundaries continue to be disputed, negotiated and contested. This is apparent in cases of territorial disputes such as in the South China Sea or in the work of the Commission on the Limits of the Continental Shelf (CLCS) establishing the outer limits of the continental shelf beyond 200 nautical miles from the baselines from which the breadth of the territorial sea is measured (UN, 1958).

UNCLOS lists many rights and obligations of states, including the protection and preservation of the marine environment (UNCLOS, 1982: Article 192), which lawyers interpret as a 'common concern of humankind' (Harrison, 2017). However, at the time when UNCLOS was being negotiated, little was known about areas beyond national jurisdiction, especially their marine ecosystems and biological diversity. For this reason, provisions – including obligations to protect the environment – do not refer to biodiversity conservation, above all not in ABNJ. This legal gap has

Figure 4.1: Maritime zones and principles as determined by UNCLOS

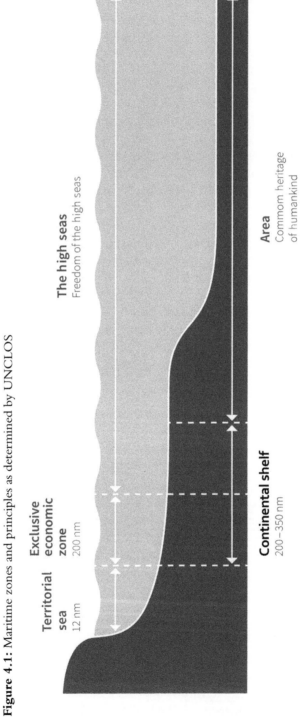

turned into a problem now that the destructive effects of expanding human activities in these areas has become apparent and well documented by science.

The BBNJ negotiations may thus be interpreted as a response to legal gaps in UNCLOS (and the CBD) in relation to environmental protection within ABNJ and an acknowledgement of momentous scientific advances which, in turn, have also delivered the knowledge and instruments needed to design a new treaty in such a specialized and technically intense policy area. Indeed, the BBNJ Treaty will be an 'implementing agreement' of UNCLOS, implying that UNCLOS is the dominant legal ocean governance framework, and will implement existing provisions such as the aforementioned Article 192. Yet, while most states involved in the negotiations would agree with the significance and pre-eminence of UNCLOS, palpable tensions illustrate the extent to which UNCLOS established – and sustains – a world order whose beneficiaries are a select few. Since its establishment, UNCLOS has tended to be either affirmed or contested by state and non-state actors in line with their interests and forms of knowledge; this has been visible during BBNJ negotiations.

One case in point are the states that have not signed or ratified the convention, but nonetheless participate in the conferences: Turkey, Colombia and Israel have not signed UNCLOS, while Egypt, Sudan and the US have not ratified it. Turkey and Colombia have constantly used environmental negotiations such as the Conferences of the Parties (COPs) of the CBD or the BBNJ intergovernmental conferences (IGCs), to contest UNCLOS and question its supremacy as a constitution and normative order for the ocean. For instance, during the CBD's COP 14 Turkey refused to accept a reference to UNCLOS in provisions on ecologically and biologically significant marine areas (EBSAs), arguing that the CBD should be the accepted legal and normative framework for the negotiation and international regulation of ocean protection matters. Similarly, during BBNJ IGCs, Colombia and Turkey used strong language to oppose UNCLOS, to ensure that the BBNJ Treaty would be open to non-UNCLOS members and to make certain that it would consider their interests and needs. Opposition to UNCLOS may be explained by dissatisfaction with territorial boundaries and the dispute settlement mechanism of UNCLOS, which these states interpret as a geopolitical disadvantage, as exemplified by disputes between Turkey and Greece.[1]

These dynamics have played out in BBNJ negotiations – and other environmental agreement-making settings such as the CBD. Territoriality and sovereignty issues keep thwarting agreement on ocean protection, especially regarding the rights and responsibilities of coastal states in adjacent marine areas; landlocked countries have also voiced reservations, trying to prevent coastal states from misusing environmental protection measures to expand their rights and duties beyond their EEZs. While it might make sense for a coastal state to be involved in the designation and management of a marine protected area or the conduct of an environmental impact assessment close to its EEZ because it might have relevant data, knowledge or management experience regarding neighbouring marine ecosystems, this would also imply an extension of rights that contradicts the 'freedom of the high seas' (FOS) principle enshrined in UNCLOS (Art. 87). Attempts by scientists to draw attention to the ecological connectivity of the ocean – exemplified by migratory species and ocean currents – tend to be instrumentalized rather than be used as essential starting points for rethinking the current normative order and whether it really is suitable for developing and implementing ocean sustainability policies, which might involve a global body to oversee ocean protection in ABNJ.

The existence of competing normative orders is also illustrated by tensions over the principles that should govern marine biodiversity conservation and sustainable use in ABNJ. For instance, the USA have signed but not ratified UNCLOS, although they do agree with most of its provisions and generally support the BBNJ Treaty. In a nutshell, the problem resides in the ownership and governance of marine resources in ABNJ – in other words, the so-called 'global commons'. This is due to competition between two opposing principles, both of which are part of UNCLOS: the freedom of the high seas and the common heritage of humankind principles (CHP).

The FOS principle grants states specific freedoms in ABNJ, except for activities in the 'Area' (ie the seabed and the ocean floor and subsoil thereof beyond the limits of national jurisdiction), where potential reservoirs of valuable mineral resources are expected to be discovered. The CHP was included in UNCLOS as the result of a compromise between two opposing positions: the G77 argued for the need for fair access and a benefit-sharing mechanism to avoid the overexploitation of mineral resources by technologically advanced countries of the Global North, whereas OECD members, including the US, disagreed with the CHP and its implications (Vadrot et al, 2021). After protracted negotiations, the CHP was included in UNCLOS, leading to the establishment of the International Seabed Authority, which now regulates the exploration and exploitation of the Area. The inclusion of the CHP and establishment of an access and benefit-sharing mechanism can be interpreted as a concession to governments of the Global South for signing an agreement that benefited mostly traditional maritime powers.

During BBNJ negotiations, similar tensions revolved around access to and use of marine genetic resources in ABNJ, which many Global South governments consider to be 'global commons' that need to be equally shared. If the term was included in the treaty in relation to MGR provisions, this would imply some access and benefit-sharing mechanism ensuring that states lacking the technological means to exploit MGRs nonetheless do share in their economic benefits. The CHP is thus viewed as a principle that could promote global equality while discouraging the privatization of the ocean and tackling longstanding asymmetries between the Global North and Global South as regards the extraction of marine resources (Vadrot et al, 2021). In contrast, many Global North governments do not wish to see the CHP in the treaty text, arguing that it may impede marine scientific research and disincentivize investment in ocean science.

The dispute over the CHP, which was resolved by adding both principles tothe treaty text, may be interpreted as a contestation by some parties of the dominant normative order favouring traditional maritime powers and a negotiation tool to put pressure on positions regarding ownership, equality and accountability. However, as many critics have pointed out, the predominance of economic interests tends to stifle the debate on the CHP and to exclude important elements such as intergenerational justice (the conservation of marine resources for future generations) and a less anthropocentric understanding of nature, that is, the protection of marine ecosystems and resources for their own sake – regardless of their imminent economic value (Vadrot et al, 2021).

4.3.3 Competing economic orders

As illustrated earlier, the issue of the prevailing normative order and its contestation is an essential aspect of negotiation practice. It is closely linked to the competing economic interests of the actors involved, which tend to impede sustainable ocean policies on an international scale – most notably as regards the Law of the Sea, how it has codified costs and benefits and incorporated forms of contestation on the conditions for exploring and exploiting marine resources. The diverse economic interests that have shaped ocean politics include the large-scale fishery sector, oil and gas drilling industries, biotech companies, public and private scientific institutions and laboratories, the military, small-scale fishery communities and Indigenous peoples.

In the BBNJ case, competing economic orders during the pre-negotiation phase significantly impacted the general design and definition of the four package elements of the treaty. While ABMTs, including MPAs and EIAs, deal with the issue of environmental sustainability, two other elements, MGRs and CB&TT, have a very pronounced socio-economic and equality dimension. In simplified terms, governments of the Global North, most notably the European Union, advocated an environmental protection treaty, including the elements of ABMTs, MPAs and EIAs, where marine scientific research and tech companies play important roles in expanding research and data infrastructures, product development and scientific advice. In turn, several governments of the Global South would agree to this only if issues of global equality and development were properly addressed and research capacity shared. As a result, the PrepCom recommended a package deal where the interests of country blocks would be properly represented and balanced.

The MGR element includes provisions on an access and benefit-sharing mechanism and is closely linked to provisions on CB&TT insofar as the latter may account for some of the inequalities resulting from the rapid privatization of MGRs in ABNJ by global biotechnology companies. If states agreed on mandatory monetary forms of CB&TT, whereby developed countries would contribute to the strengthening of marine scientific research in developing countries through regular payments and investment, this might ease some of the tensions regarding free access to MGRs in ABNJ that stand in contrast to the regulations for accessing MGRs in national waters, where the Nagoya Protocol requires scientists to obtain special permits. While many states of the Global North – particularly those that are active in ocean science, including the US, Canada, Australia, the EU, the UK and Japan – argue that access to MGRs is already regulated by the FOS and is thus free for everybody, state actors from the South tend to argue that, in reality, access is restricted owing to the simple fact that it depends on the availability of expensive, technologically intense ocean infrastructure.

Indeed, access to ocean science is a key issue for many governments of the Global South, who argue that they will not be able to implement MPA and EIA provisions without the necessary knowledge and monitoring programmes. MPAs on the high seas would involve state actors making proposals to a future body – most probably a conference of the parties – on the geographical locations of areas in need of protection on the basis of a specific set of targets and indicators that are subject to negotiation. These proposals would have to be reviewed and adopted by participating states, and would have to include a procedure to manage and monitor protected areas to ensure that conservation goals are met and to avoid a proliferation of so-called paper parks. But what does the establishment of an MPA imply for fishing, shipping or marine scientific research in that area? Are MPAs no-take zones or is there simply a limit on the quantity and scope of certain activities? Should an international authority be in charge of monitoring and managing MPAs, and if so how representative can such an institution be?

It is a common negotiation practice to refer to existing institutional arrangements that already regulate a given ocean activity in a way that suits specific state interests. In the MPA case, for instance, governments referred to conventions of the IMO concerning navigation and shipping that deal with safety, vessel-source pollution and maritime security. Under these, states may submit proposals regarding so-called particularly sensitive sea areas (PSSAs) if these areas fulfil certain criteria, including ecological criteria, such as unique or rare ecosystem, diversity of the ecosystem or vulnerability to degradation by natural events or human activities; social, cultural and economic criteria, such as significance of the area for recreation or tourism; and scientific and educational criteria, such as biological research or historical value. (IMO, 2019)

EBSAs are another often mentioned instrument, which has been agreed through an intergovernmental and interdisciplinary process under the CBD. However, the CBD lacks the legal mandate to designate and establish such areas in ABNJ.

Thus, negotiating the relationship between the new BBNJ instrument and existing regional and sectoral bodies and multilateral agreements on ocean protection, such as IMO, CBD, ISA, IOC-UNESCO and UNDOALOS is a key practice employed by states to avoid inconsistency and overlap between treaties. Correspondingly, states continue to argue that the new BBNJ instrument should 'not undermine' existing agreements, most notably UNCLOS (Langlet and Vadrot, 2023). However, this tends to rule out the establishment of a progressive legal framework for ocean sustainability that would go beyond existing measures. Scholars have pointed to undesirable effects such as forum shopping or strategic inconsistency in cases of regime complexity – where several governance regimes representing norms and values that are in line with specific state interests compete against one another. The attempt by many states to exclude fish from the new treaty, on the basis of the argument that RFMOs were already regulating diverse aspects of fishery, also illustrates the difficulty of contesting economic order and tackling ocean protection beyond the dominant sector-specific, fragmented governance approach.

4.3.4 Competing epistemic orders

As with many other environmental issues, negotiating ocean protection strongly relies on technical knowledge and expertise. However, as indicated earlier, scientific knowledge about the ocean is unevenly distributed between actors and regions. The most obvious discrepancy is between countries of the Global North and Global South in terms of research capacity and scientific output, including funding expenditure; the number of ocean scientists, institutions and scientific publications; and citation counts (see Figure 4.2). Although ocean knowledge is not only about scientific expertise, ocean science has significantly influenced how marine ecosystems are monitored, governed and protected. According to the *Global Ocean Science Report*, the latter includes 'physical, biological, chemical, geological, hydrographic, health and social sciences, as well as engineering, the humanities and multidisciplinary research on the relationship between humans and the ocean'; in addition, it 'seeks to understand complex, multiscale socio-ecological systems and

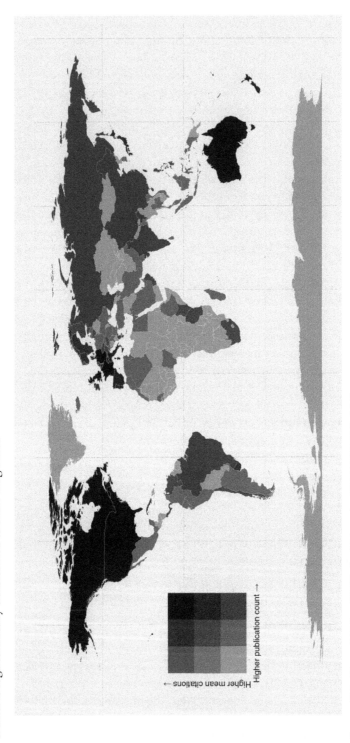

Figure 4.2: Geographic distribution of the total number of articles and average citation count by country, 1990–2018. The total number of articles is logarithmically transformed to show magnitude

Source: Tolochko and Vadrot (2021). Available under CC BY-NV-ND 4.0 license.

services, which requires observations and multidisciplinary and collaborative research' (IOC-UNESCO, 2017: 19).

Ocean science is a field where scientific, economic, military and political interests converge. UNCLOS negotiations in the 1970s unveiled the political-economic context of science and deep-rooted inequalities between the Global North and the Global South (Vanderpool, 1983). The lack of scientific infrastructure to monitor marine ecosystems poses a significant challenge to ocean protection, but at the same time this prevents marine resources from being exploited and extracted at a rapid rate. The private sector plays a central role in this regard. For instance, the demand for EIAs by the oil and gas industry (to obtain permits for the extraction of fossil fuels) has increased research funding opportunities. In many countries, public–private partnerships are an important driver of ocean science, more generally. Private companies, historically above all in the fishery sector but increasingly also in other sectors (such as oil and gas and deep-seabed mining), collect and own ocean data, including baseline data, to which public scientific institutions do not always enjoy access. Last but not least, ocean science depends on the industrial sector producing expensive, innovative research equipment and infrastructure, including research vessels, remotely operating vehicles (ROVs), underwater robots, environmental DNA technology, remote sensing and cameras. This blurs the demarcation between public and private.

Target 14.a of SGD 14 acknowledges the need to increase marine scientific research. However, since SDG 14 was negotiated, the debate has moved forwards (Visbeck, 2018), culminating in the UN Decade of Ocean Science for Sustainable Development (2021–30), which tackles issues that are at the forefront of current BBNJ negotiations. This includes the need for CB&TT to close gaps between developed and developing countries but it goes far beyond that. It also considers interdisciplinary and transdisciplinary approaches involving diverse stakeholders, who are a key factor in advancing ocean science agendas across the globe. In line with other processes such as assessments by IPBES, the UN Decade emphasizes the value of traditional knowledge held by local and Indigenous communities, who have been living with and from the ocean for centuries.

These dynamics – including rising tensions from global inequalities and the quest for a diversification of the knowledge base available to policy makers – are also at play in the BBNJ negotiations. The BBNJ case illustrates the heterogeneity of (scientific) knowledge on the ocean that is needed to monitor and manage unsustainable practices, including, for example, baseline monitoring data to inform EIAs. States, especially from the Global South, use scientific concepts such as *ecological connectivity* or *transboundary effects* to strengthen their lines of argumentation. Both terms run contrary to the idea that the ocean may be compartmentalized into maritime zones and acknowledge that marine ecosystems and species are connected in various ways across water areas. While this insight may be (and is) strategically used to argue for an expansion of coastal states' responsibilities beyond their EEZs, it can also potentially question the normative and epistemic order within which ocean sustainability policies are formulated and implemented, and provide a new narrative for ocean protection, where knowledge from local communities and Indigenous peoples becomes more important.

As argued earlier, the way the ocean is governed is linked to the way the ocean is known and represented. BBNJ negotiations, while leaving room for claims that recognize the value of non-scientific – most notably traditional, Indigenous and local – knowledge, have tended to reproduce 'surficial static ontologies typically associated with land' (Peters and Steinberg, 2019, 293).

4.4 Alternative approaches for ocean sustainability

As with the making of SDG 14, the BBNJ negotiations are threatened by geopolitical divides and a narrow understanding and representation of the ocean and its governance. The aforementioned tensions and how they are entrenched in the practices of negotiating ocean protection challenge the transition towards an overarching and consistent political and legal framework for ocean sustainability. However, and as argued earlier, the social world within which ocean sustainability is negotiated is ordered in a way that is both relatively enduring over time and open to contestation and change.

One alternative path would be to embrace the idea of an interconnected ocean in line with the concept of ecological connectivity and local and Indigenous traditional ecological knowledge about the ocean. The concept blurs the boundaries between national waters, EEZs and the high seas, and challenges the prevailing normative order and legal division of the ocean into different maritime zones. While enshrined in existing international law, the concept of managing the ocean in different spatial zones by different actors is increasingly criticized for not being effective for the conservation and sustainable use of marine biodiversity. Although some actors question the effectiveness of the fragmented ocean governance for ocean protection, claims to alter the legal order and renegotiate UNCLOS are, if at all, implicit. The power of the concept of ecological connectivity (and related knowledge forms) lies not only in its use as a negotiation tool but also in its capacity to question the normative order and to anticipate transformative change (Tessnow von Wysocki and Vadrot, 2022).

Transformative change as conceptualized by IPBES in line with the nature futures framework and the recognition of multiple values associated with nature (nature for nature; nature for culture; and nature for society) has not yet been associated with ocean sustainability agendas but can become a future avenue of both ocean governance (Mendenhall, 2019) and the study thereof (Brondizio et al, 2019). The recognition of multiple economic and non-economic values associated with ocean ecosystems and species is the first step towards an integrated and transdisciplinary framework for ocean protection that goes beyond the current fragmented landscape of ocean-related agreements and the narrow idea of 'blue growth' associated with SDG 14. Approaches recognizing the intrinsic value of nature and the *rights of nature* ('nature for nature') have already been articulated in relation to the governance of the global ocean commons (Harden-Davies et al, 2019) but have not yet been acknowledged within an international negotiation setting. As tools for contesting the dominant economic order, rights-based approaches question issues of ownership and responsibility, especially in view of the need to conserve nature for future generations.

An alternative to the idea of overarching goals, targets and indicators is provided by the concept of *polycentricity*, implying multiple centres of decision making operating with some degree of autonomy (Ostrom, 2008). The concept and idea of 'polycentricity' has been used to highlighting the need to strengthen regional and cross-sectoral cooperation (Carlisle and Gruby, 2019; Gjerde and Yadav, 2021). From this perspective, regional ocean agreements and decision making can increase ownership and support the formulation and implementation of multilateral global agreements; such an approach may also be more successful in accommodating area-based interests, tackling trade-offs between competing economic and non-economic values, and responding to the needs of different knowledge holders seeking to contribute to decision making at the regional level of policy making.

Last but not least, questioning the social construction and ontologies of the ocean, as outlined by Peters and Steinberg (2019), who contest land-oriented thinking on the basis of their concept of 'wet ontologies', may result in a fruitful debate about the (un)suitability of some of the spatial instruments put in place, such as MPAs, which have already been subjected to critique but continue to pass as key sustainable solutions.

4.5 Concluding remarks: beyond SDG 14

Ocean sustainability is an emerging and increasingly salient field of global environmental politics. While SDG 14 as a standalone goal for a sustainable ocean was in itself a breakthrough, the specific framing of its targets and how to achieve them contradict, or at least neglect, the essential factors that have made ocean governance – in particular, ocean protection – one of the hardest endeavours of contemporary environmental politics. It has been argued that a political debate was needed that would go beyond institutional fixes and recognize the role of multilateral negotiation settings as sites of order making and contestation. To this end, the BBNJ case has been used to illustrate how important it is to understand the sites and related practices where ocean sustainability is being negotiated, which involves a struggle over the dominant normative, economic and epistemic orders that have shaped ocean governance for decades.

The SDG 14 process most certainly would have profited from a formulation of ocean sustainability goals that was more open to alternative epistemologies and ontologies of the ocean. The same is the case for the ongoing BBNJ case. This is especially important as the political and epistemic landscape of global ocean sustainability politics is in the making and, compared to SDG 13 (Climate action) and SDG 15 (Life on land), less institutionalized and settled. Although ownership, equality and accountability are core issues in ocean governance, they tend to be overlain by issues championed by interests of a different nature, which view ocean resources as a commodity to be indexed, used and traded for the short-term benefit of today's societies, regardless of future generations' rights and needs. For SDG 14, it may be too late, but the BBNJ treaty is still under negotiation, and it may still be early enough to consider alternative approaches.

Acknowledgements

This work was supported by the by the European Research Council (804599, grant holder Alice B.M. Vadrot). I would like to thank the three anonymous reviewers who significantly increased the quality of this chapter with their constructive and knowledgeable feedback.

Note

[1] The US case is more complicated, with several US presidents since Bill Clinton arguing for ratification but failing to achieve acceptance domestically. One important reason for the US not joining is the inclusion in UNCLOS of the common heritage of humankind principle (Vadrot et al, 2021).

References

Armstrong, C. (2020) 'Ocean justice: SDG 14 and beyond', *Journal of Global Ethics*, 16(2): 239–55.

Brondizio, E.S., Settele, J., Díaz, S. and Ngo, H.T. (2019) *The Global Assessment Report on Biodiversity and Ecosystem Services of the Intergovernmental Science-Policy Platform on Biodiversity and Ecosystem Services*, Intergovernmental Platform on Biodiversity and Ecosystem Services (IPBES) Secretariat, Bonn.

Carlisle, K. and Gruby, R.L. (2019) 'Polycentric systems of governance: a theoretical model for the commons', *Policy Studies Journal*, 47(4): 927–52.

Cormier, R., and Elliott, M. (2017) 'SMART marine goals, targets and management – Is SDG 14 operational or aspirational, is 'Life below Water' sinking or swimming?' *Marine Pollution Bulletin*, 123: 28–33.

De Santo, E.M., Ásgeirsdóttir, Á., Barros-Platiau, A. and Biermann, F. (2019) 'Protecting biodiversity in areas beyond national jurisdiction: an earth system governance perspective', *Earth System Governance*, 2: 100029.

Gjerde, K.M. and Yadav, S.S. (2021) 'Polycentricity and regional ocean governance: implications for the emerging UN Agreement on Marine Biodiversity Beyond National Jurisdiction', *Frontiers in Marine Science*, 8: 704748.

Gray, N.J., Gruby, R.L. and Campbell, L.M. (2014) 'Boundary objects and global consensus: scalar narratives of marine conservation in the Convention on Biological Diversity', *Global Environmental Politics*, 14(3): 64–83.

Haas, B., Fleming, A., Haward, M. and McGee, J. (2019) 'Big fishing: the role of the large-scale commercial fishing industry in achieving Sustainable Development Goal 14', *Reviews in Fish Biology and Fisheries*, 29: 161–75.

Harden-Davies, H., Humphries, F. Maloney, M., Wright, G., Gjerde, K. and Vierros, M. (2019) 'Rights of nature: perspectives for global ocean stewardship', *Marine Policy*, 122: 104059.

Harden-Davies, H., Amon, D.J., Vierros, M., Bax, N.J., Hanich, Q., Hills, J.M. et al (2022) 'Capacity development in the Ocean Decade and beyond: key questions about meanings, motivations, pathways, and measurements', *Earth System Governance*, 12: 100138.

Harrison, J. (2017) *Saving the Oceans through Law: The International Legal Framework for the Protection of the Marine Environment*, Oxford: Oxford University Press.

Haward, M. and Haas, B. (2021) 'The need for social considerations in SDG 14', *Frontiers in Marine Science*, 8: 632282.

Houghton, K. (2015) 'A Sustainable Development Goal for the ocean: moving from goal framing towards targets and indicators for implementation', Institute for Advanced Sustainability Studies, Available from: www.iass-pots dam.de/sites/default/files/files/working_paper_ocean_sdg_houghton.pdf [Accessed 20 August 2016].

Hughes, H. and Vadrot, A. (eds) (2023) *Conducting Research on Global Environmental Agreement-Making*, Cambridge: Cambridge University Press.

Hughes, H., Vadrot, A.B.M., Allan, J.I., Bach, T., Bansard, J.S., Chasek, P. et al (2021) 'Global environmental agreement-making: upping the methodological and ethical stakes of studying negotiations' *Earth System Governance*, 10: 100121.

IOC-UNESCO (Intergovernmental Oceanographic Commission, UNESCO) (2017) *Global Ocean Science Report: The Current Status of Ocean Science around the World*, ed L. Valdés, UNESCO, Paris, Available from: https://unesdoc.unesco.org/ark:/48223/pf0000250428.locale=en [accessed 31 May 2023].

IMO (International Maritime Organization) (2019) 'Particularly Sensitive Sea Areas'. Available from: https://www.imo.org/en/OurWork/Environm ent/Pages/PSSAs.aspx [Accessed 19 June 2023].

Johansen, D.F. and Vestvik, R.A. (2020) 'The cost of saving our ocean – Estimating the funding gap of sustainable development goal 14', *Marine Policy*, 112(2): 103783.

Langlet, A. and Vadrot, A.B.M. (2023) 'Not 'underming' who? Unpacking the emerging BBNJ regime complex'. *Marine Policy*, 147(4): 105372.

Mendenhall, E. (2019) 'The Ocean Governance Regime: International Conventions and Institutions', in P. Harris (ed) *Climate Change and Ocean Governance: Politics and Policy for Threatened Seas*, Cambridge, NY: Cambridge University Press, pp 27–42.

Ntona, M. and Morgera, E. (2018) 'Connecting SDG 14 with the other Sustainable Development Goals through marine spatial planning', *Marine Policy*, 93(7): 214–22.

Ostrom, E. (2008) 'The challenge of common pool resources', *Environment*, 50(4): 8–21.

Peters, K. and Steinberg, P. (2019) 'The ocean in excess: towards a more-than-wet ontology', *Dialogues in Human Geography*, 9(3): 293–307.

Rudolph, T.B., Ruckelshaus, M., Swilling, M., Allison, E.H., Österblom, H., Gelcich, S. et al (2020) 'A transition to sustainable ocean governance', *Nature Communications*, 11(1): 1–14.

Silver, J.J., Gray, N.J., Campbell, L.M., Fairbanks, L.W. and Gruby, R.L. (2015) 'Blue economy and competing discourses in international oceans governance', *Journal of Environment & Development*, 24(2): 135–60.

Singh, G.G., Cisneros-Montemayor, A.M., Swartz, W., Cheung, W., Guy, J.A., Kenny, T.-A. et al (2017) 'A rapid assessment of co-benefits and trade-offs among sustainable development goals', *Marine Policy*, 93: 223–31.

Steinberg, P.E. (2001) *The Social Construction of the Ocean*, Cambridge: Cambridge University Press.

Tessnow-von Wysocki, I. and Vadrot, A.B.M. (2022) 'Governing a divided ocean: the transformative power of ecological connectivity in the BBNJ negotiations', *Politics and Governance*, 10(3): 14–28.

Tiller, R., De Santo, E., Mendenhall, E., and Nyman, E. (2019) 'The once and future treaty: towards a new regime for biodiversity in areas beyond national jurisdiction', *Marine Policy*, 99: 239–42.

Tolochko, P. and Vadrot, A.B.M. (2021) 'The usual suspects? Distribution of collaboration capital in marine biodiversity research', *Marine Policy*, 124: 104318.

UN (United Nations) (1958) 'The Convention on the Continental Shelf' November', Available from: https://treaties.un.org/Pages/ViewDetails. aspx?src=IND&mtdsg_no=XXI-4&chapter=21&clang=_en [Accessed 20 August 2022].

United Nations (ed.) (2017) *The First Global Integrated Marine Assessment: World Ocean Assessment I*, Cambridge: Cambridge University Press.

United Nations Convention on the Law of the Sea (UNCLOS) (1982). *Convention on the Law of the Sea*, Dec. 10, 1982, 1833 U.N.T.S.

United Nations Department of Economic and Social Affairs Statistics (UNSTATS) (2016) *Goal 14 Conserve and sustainably use the oceans, seas and marine resources for sustainable development1*. Available from: https://unstats.un.org/sdgs/files/metadata-compilation/metadata-goal-14.pdf [Accessed 20 August 2022].

Unger, S., Müller, A., Rochette, J., Schmidt, S., Shackeroff, J. and Wright, G. (2017) 'Achieving the Sustainable Development Goal for the oceans', IASS Policy Brief 1/2017, Available from: https://publications.iass-potsdam.de/rest/items/item_2041892_3/component/file_2041893/content

Vadrot, A.B.M. (2020) 'Multilateralism as a 'site' of struggle over environmental knowledge: the North–South divide', *Critical Policy Studies*, 14(2): 233–45.

Vadrot, A.B.M., Langlet, A. and Tessnow-von Wysocki, I. (2021) 'Who owns marine biodiversity? Contesting the world order through the "common heritage of humankind" principle', *Environmental Politics*, 31(2): 226–50.

Vanderpool, C. (1983) 'Marine science and the law of the sea', *Social Studies of Science*, 13(1): 107–29.

Visbeck, M. (2018) 'Ocean science research is key for a sustainable future', *Nature Communications*, 9(1): 690.

Visbeck, M., Kronfeld-Goharani, U., Neumann, B., Rickels, W., Schmidt, J., van Doorn, E. et al (2014) 'Securing blue wealth: the need for a special sustainable development goal for the ocean and coasts', *Marine Policy*, 48(9): 184–91.

Webster, D.G. (2022) *Beyond the Tragedy: Evolution and Sustainability in Global Fisheries Governance*, Cambridge, MA: MIT Press.

Wright, G., Rochette, J., Druel, E. and Gjerde, K. (2016) 'The long and winding road continues: towards a new agreement on high seas governance', Study No. 01/16, IDDRI, Paris, France.

Interview with Alice B.M. Vadrot: Ocean Governance: An Emerging Field for Political Science

Estefanía Lawrance Crespo and Sofie Jokerst

At first glance, the field of ocean governance appears to be technical and scientific. As a political scientist, how do you perceive this area of study?

Vadrot: Ocean governance, like many other environmental issues is based on very complex environmental findings. However, the further you go away from the coast, out of national territories, areas of the ocean known as the deep seas and the high seas, the less data is available, and it's not only a lack of scientific data but also a lack of legal and political frameworks to govern these areas. This is especially true for aspects related to ocean protection and marine biodiversity. Political science is not yet well represented in the field of marine biodiversity. There is some interesting work from political scientists on polar regions, on fisheries, and on regional ocean governance infrastructure, but only during the last five to ten years has there been an increasing awareness among political scientists that we need to turn our attention to the ocean.

By investigating these issues, you have applied an interesting methodological approach: collaborative event ethnography. How did you come to choose this approach?

Vadrot: The first time I experienced international negotiations was as a master's student, as I was extremely fascinated by the processes of on-site negotiations through which treaties and conventions are formed. I went to the negotiations that prepared the Intergovernmental Platform on Biodiversity and Ecosystem Services, and at the meetings and side events

I conducted interviews and tried to understand participants' narratives on why an IPCC for biodiversity was needed. Based on that material, I wrote my PhD thesis and published the book *The Politics of Knowledge and Global Biodiversity*, but I was dissatisfied with my methodological approach and the limits I experienced as a single researcher trying to capture everything that was happening at a global environmental conference. Within the MARIPOLDATA research project, funded by the European Research Council for the period of five years, we work collaboratively. With my team of PhD students, we have adopted a more coherent approach and developed a specific matrix that helps to collect quantitative and qualitative data at the same time and as a team. It enables us to reconstruct the negotiations, identify important struggles and quantify specific aspects, such as the number and time of interventions by a specific actor.

As knowledge plays a vital role in your research, how can we understand the concept of traditional knowledge and how does it differ from scientific knowledge?

Vadrot: First, I should remark that different terms for traditional knowledge can be used. For example, in the CBD context, they refer to 'knowledge from local communities and Indigenous peoples' since many sensitivities are implied. As Plato said, 'Knowledge is justified by true belief'. Scientific knowledge is what the community agrees on as the general mode of justification of knowledge and the method that makes it true, valid and objective, while traditional knowledge is different because it doesn't exist in codified forms. It is often oral and it's transmitted from generation to generation. There is a fascinating book from Philippe Descola, who lived with different Indigenous communities, and in it he says that local knowledge about the environment often includes their relationship with the environment. It is not the same modus operandus as science, which tries to measure the environment with facts and numbers. It's about the cosmological relationship between the people and the environment.

5

Sustainable Development and Water: Cross-sectoral, Transboundary and Multilevel Governance Arrangements in Bolivia, Ecuador and Switzerland

Manuel Fischer, Paúl Cisneros, Julie Duval,
Javier Gonzales-Iwanciw and Sofia Cordero Ponce

Water is a key resource for life, but it is also increasingly under pressure from urbanization, population growth, energy production, pollution and many other developments. Especially in mountain areas, rising temperatures, melting glaciers and changing precipitation patterns are disrupting water flows and affecting ecosystems, creating and worsening natural hazards, and threatening livelihoods and communities both within the mountains and downstream (Buytaert and De Bièvre, 2012; Wymann von Dach et al, 2018; IPCC, 2021). SDG 6 (Clean water and sanitation) aims at alleviating the increasing pressures on water resources and guiding society towards a sustainable use of the world's water resources and related ecosystems. Since 2015, when the SDGs were adopted as a non-binding international commitment, their implementation has been a challenge for national, subnational and local actors and existing institutions on all of these levels (eg Breuer et al, 2019; Rivera-Arriaga and Azuz-Adeath, 2019).

This chapter discusses the global water policies, SDG 6 and its implementation by actors and institutions at the subnational level, given three crucial challenges related to sustainable water governance (Mollinga et al, 2006; Lubell and Edelenbos, 2013; Fischer and Ingold, 2020). First,

SDG implementation in general, and water management specifically, should be coordinated across several sectors given the existence of important trade-offs but also potential synergies between different goals and targets (Weitz et al, 2018; Horan, 2020; Pham-Truffert et al, 2020). The specific focus of this chapter lies in the intersection of water (SDG 6) and climate (SDG 13). Second, tackling water governance often requires transboundary coordination, as the scales relevant for water resources do usually not correspond to jurisdictional boundaries. Third, multilevel governance, that is, coordination across different levels of governance, is relevant given that the SDGs have been established at the international level but must be implemented by national states that again often delegate these tasks to subnational levels of governance. This chapter contributes to the literature by highlighting the importance of cross-sectoral and transboundary challenges in situations (such as the implementation of SDGs) that require the coordination of international, national, regional and local policies. All three governance challenges are crucial to achieving more integrated and basin-wide water resources management.

The research question underpinning this chapter is: what governance arrangements tackle cross-sectoral, transboundary and multilevel challenges related to sustainable water management? The chapter thereby contributes to the three guiding questions of this book. For example, the changing perceptions of sustainable development (question 1) are discussed from the point of view of the development of global water policies, as well as by emphasizing the importance of the three challenges of cross-sectoral, transboundary and multilevel coordination. The chapter also discusses actors that are able to potentially address the three governance challenges, and therefore focuses on new types of actors (question 2) and governance arrangements (question 3).

The first part of the chapter discusses theoretical elements related to cross-sectoral, transboundary and multilevel governance challenges. It then presents the development of important global water-related agreements in the last 30 years, and how they considered these three challenges, and then discusses in more detail how the water SDG 6 interacts with the climate SDG 13. The third part of the chapter presents three case illustrations in mountain regions in Bolivia, Ecuador and Switzerland, and thereby complements the literature on water governance that has, for example, strongly focused on the cases of the Murray–Darling (Connell and Grafton, 2011) or the Mekong (Waibel et al, 2012) river-basins. The three case studies illustrate how actors and institutions on the subnational level address the cross-sectoral, transboundary and multilevel challenges to sustainable water governance. The subnational level is relevant as the United Nations' Agenda 2030 and the adaptation scholarship position local and subnational governments as key agenda setters and implementers for context-relevant adaptation measures (Cabrera-Barona and Cisneros, 2021; Lesnikowski et al, 2021). More specifically, the local case studies in this chapter provide concrete illustrations of the more general governance challenges, and highlight specific aspects that are crucial for understanding SDG implementation. These aspects, are, for example, local actors' perceptions of the SDGs, their importance contrasted with limited knowledge as to how they deal with the challenge of SDG implementation, and their capacity to provide governance innovations to deal with these challenges. The chapter relies on the qualitative analysis of documents and grey literature, and on expert interviews for the analysis of the three illustrative case studies, but has only very limited ambitions to systematically compare or explain findings from the three illustrative case studies.

5.1 The governance of water resources

5.1.1 Cross-sectoral interactions, transboundary arrangements and multilevel governance

A growing portion of the literature about the governance of water resources recognizes that the water crisis is mainly a crisis of governance (Gupta et al, 2013). Water scholars and practitioners recognize that governance arrangements are central to the effective management of water resources, in particular the requirements for sustainable and integrated water resource management (IWRM) capable of addressing the challenges related to climate change (eg Pahl-Wostl, 2019). In such an understanding, water governance is geared towards decentralizing institutions to improve fit with socio-ecological conditions but retaining a multiscale focus (Blomquist et al, 2005) such as watershed or water basin management councils and other decision-making or advisory venues for IWRM (see Margerum and Robinson, 2015). Integrated water governance tackles the inherent interdependency of different hydrological, ecological and socioeconomic aspects of water, and strives to incorporate a multitude of related stakeholder interests (Engle et al, 2011). Sustainable and integrated water management are challenging as they need to take three aspects into account: the cross-sectoral interactions related to different water-related sectors, transboundary arrangements to deal with waters flowing across jurisdictional boundaries, and multilevel governance to coordinate policies across different levels of governance (Mollinga et al, 2006; Fischer and Ingold, 2020).

In terms of cross-sectoral interactions, water management depends on and influences other sectors such as land, energy, agriculture, tourism, biodiversity and climate (Lubell and Edelenbos, 2013; Fischer and Ingold, 2020). Yet, on the level of public administrations, as well as in the logic of many other actors, these aspects are dealt with in different specialized units that interact only partially (eg Pahl-Wostl, 2009; Plummer and Armitage, 2010). Despite the adoption of cross-sectoral approaches such as IWRM in the past three decades, applying such concepts on the ground is still a great challenge for policy makers and practitioners (Hering and Ingold, 2012; Fischer and Ingold, 2020). Several studies show that adopting IWRM in many countries was limited to reorganizing existing policies to give an impression of conforming to the framework (Petit and Baron, 2009; Engle et al, 2011; Waibel et al, 2012). Lack of permanent funding, and deep-rooted sectoral priorities and state structures have also stymied IWRM implementation.

Cross-sectoral complexities interact with transboundary and multilevel governance challenges (Fischer and Ingold, 2020). The logic of political borders and boundaries has seldom followed the rules of nature: jurisdictions and legal units rarely match the physical, chemical or ecological area or the geological extent of a problem (Hering and Ingold, 2012; Bodin et al, 2019). The lack of alignment between jurisdictions and the territories where adaptation strategies are needed creates incentives for actors to form transboundary water governance arrangements. In this regard, countries and subnational governments make use of international treaties and forums to manage shared water bodies within basin-wide approaches (Sanchez and Roberts, 2014; Fischer and Jager, 2020). Mirumachi and Allan (2007) argue that basins differ in their international transboundary relations as these shift in intensities of both conflict and cooperation over time. For example, given power asymmetries, not all cooperation is equally appreciated by riparian states.

Finally, governance systems are organized across different levels, from local to international, not least where international agreements are relevant to subnational governance, and vice versa, such as with the SDG process. Competencies distributed across levels of governance create a need for multilevel coordination, particularly in federalist settings with important competences at the subnational level (Hooghe and Marks, 2001). Lundqvist (2011) points out that multilevel governance arrangements for sustainable water management often fail at handling the trilemma created by seeking effectiveness, participation and legitimacy.

5.1.2 Subnational government arrangements and SDG implementation

The SDG parlance uses the term 'SDG localization' to refer to the processes by which subnational and local actors translate globally defined goals and targets into policies, governance arrangements and practices shaped by their historically constructed realities (UCLG, 2015). Although there is increasing awareness of the need to set relevant priorities, determine effective means of implementation and adopt adequate indicators to measure progress (see Wymann von Dach et al, 2018), the roles of subnational governance arrangements in this process require further empirical analysis. Particularly in the context of mountainous regions, where issues of water governance and climate adaptation clearly converge, arrangements that tackle the three coordination challenges presented earlier may be key to promoting a more effective and just localization of the global agenda.

Tackling cross-sectoral, transboundary and multilevel challenges related to water governance on the subnational level requires multiple forms of collaboration. Collaboration can be voluntary or compulsory, more formal or informal, and it can happen bilaterally or within forums (eg Fischer and Leifeld, 2015). One of the most common reasons for the emergence of different forms of collaboration is the insufficient size and availability of resources of given actors, organizations or jurisdictions to deliver services, fulfil formal obligations or achieve other types of goals (Kersting et al, 2009; Teles, 2016). Specific policy instruments, such as 'cooperative instruments' asking actors to establish given forms of coordination and collaborations, might be formulated in both national and subnational policies, and transboundary treaties between different governments (Jordan et al, 2003; Fischer and Jager, 2020). Overall, governance arrangements to tackle the three challenges at the centre of this chapter involve some form of more or less institutionalized and rule-based coordination between actors.

5.1.3 The historical development of global water agreements

International agreements have been shaping the water sector for the last 30 years and have also aimed at tackling cross-sectoral, transboundary and multilevel challenges (see also Chapter 6 for a discussion on framings of water in the different agreements and reports over time). As a first important international water milestone, the Dublin Statement on Water and Sustainable Development highlighted the threat of scarcity and misuse of fresh water, and included four principles, among them the importance of a holistic approach for IWRM, insisting on intersectoral needs and planning (UN, 1992). Furthermore, the Dublin Statement considered transboundary governance for catchments crossing national boundaries and recognized the essential role of international cooperation in these cases, thus addressing the multilevel challenge. According to the second principle in the Dublin Statement, multilevel governance should be organized as follows: international governance should focus on water resource management coordination; the national level should focus on raising awareness by ensuring information circulation, and on coordination and planning activities; and IWRM 'should be delegated to those lowest appropriate levels' (UN, 1992: 15). This last statement lends importance to water governance at the subnational level.

In addition, the Rio Declaration on the Environment and Development in 1992 was accompanied by the adoption of the Agenda 21, a program regrouping strategies to achieve sustainable development in the 21st century

(UN, 1993). Agenda 21 insisted on cross-sectoral challenges by recognizing the multisectoral nature and multi-interest utilization of water resources (chapter 18), and by considering interlinkages between water resource management and other issues, such as agricultural and rural development. The Rio Declaration further emphasizes the transboundary challenge by mentioning the specific needs of riparian countries to 'formulate water resources strategies, prepare water resources action programs and consider, where appropriate, the harmonization of those strategies and action programs' (UN, 1993: 277). Lastly, the multilevel challenge appears in the Rio Declaration through a chapter dedicated to 'local authorities' (chapter 28) and further detailed IWRM that 'should be carried out at the level of the catchment basin or sub-basin' (UN, 1993: 277).

The Millennium Declaration adopted in 2000 included eight Millennium Development Goals to be achieved by 2015. The goal related to water considered the multilevel challenge by recommending stopping 'the unsustainable exploitation of water resources by developing water management strategies at the regional, national and local levels, which promote both equitable access and adequate supplies' (UN, 2000: 6).

At the Rio+20 UN Conference on Sustainable Development in 2012, the SDG formulation process was launched and three years later, in September 2015, all UN members adopted Agenda 2030, including 17 SDGs with 169 targets. Water is included as an individual goal (SDG 6), emphasizing aspects of the entire water cycle including drinking water and sanitation, hygiene, water quality and the integrated management of the water resource along with logics of IWRM. The entire Agenda 2030 and further related documents, for example about SDG implementation, emphasized the cross-sectoral challenge by identifying interlinkages between water SDG 6 and other SDGs.

5.1.4 Example of cross-sectoral interactions in SDG implementation: water and climate

The Agenda 2030 and related documents illustrate how cross-sectoral interactions – across different SDGs – are important for the implementation of international goals across multiple levels. In 2015 the International Council of Science (ICS) examined the related challenges and stated that the holistic nature of the goals and targets must be taken into account (ICSU and ISSC, 2015). These interactions between goals are of crucial importance as the synergies and trade-offs across them constitute an efficient way to achieve them (Weitz et al, 2018; Independent Group of Scientists appointed by the Secretary-General, 2019; Pham-Truffert et al, 2020). For example, Chapter 6 in this volume (Mehta et al) discusses the interactions between SDGs 6 and 2, and thus between water and human rights. Other important interactions are, for example, between water and energy (SDG 7), or between water and life on land (SDG 15) (Pham-Truffert et al, 2020).

This chapter focuses on the relations between the water SDG 6 and SDG 13 (Climate action). The focus is on these two areas, as they are described as two of the core SDGs for a transition to environmental sustainability (see Partzsch in the Introduction to this volume), and pursuing both goals creates particularly strong synergies (Alcamo, 2019; UN-Water, 2016). Furthermore, water management has been emphasized as being 'at the heart of adaptation to climate change' (ESCAP, 2016: 20) and that 'most of the impacts of climate change are felt through the water cycle, so the linkages are critical' (UN-Water, 2016: 39). A coordinated implementation of SDGs 6 and 13 is an aspect of sustainable water management, and, vice versa, sustainable (and integrated) water management is one aspect of (a successful implementation of) the SDGs. The strong potential synergies between SDGs 6 and 13 contrast with other pairs of SDGs (eg SDGs 6 and 7 on energy) that display strong trade-offs and potential conflicts (Pham-Truffert et al, 2020).

Pursuing the achievement of universal and equitable access to safe and affordable drinking water for all (target 6.1) – sometimes through increased water efficiency (target 6.4) – can lead to climate change measures being integrated into national policies, strategies and planning (target 13.2) (ESCAP, 2017) to achieve universal and equitable access to safe and affordable drinking water (target 6.1). In addition, protecting and restoring water-related ecosystems (target 6.6) is a direct driver for strengthening resilience to climate-related hazards (target 13.1) (ESCAP, 2016; UN-Water, 2016; UNCCD, 2017).

The influences of SDG 13 on SDG 6 are much more prominent than those of SDG 6 on SDG 13. Most targets under the climate change mitigation goal are linked to several targets under the water-related goal and therefore reinforce each other (Le Blanc, 2016; Pham-Truffert et al, 2020). Indeed, as increasing temperatures due to climate change imply heavier stress on water resources worldwide (Future Earth and Earth League, 2017), taking urgent action to combat climate change and its impacts (SDG 13) will lead to greater achievement regarding water management and sanitation for all (SDG 6) (Pham-Truffert et al, 2020). On the one hand, efforts at strengthening resilience and adaptive capacity to climate change-related issues (target 13.1) will reinforce access to drinkable and affordable water (target 6.1) and limit saltwater intrusion into surface water (target 6.3) (IPCC, 2014a). On the other hand, national climate policies might help achieve universal and equitable access to safe and affordable drinking water for all (target 6.1), reduced water pollution (target 6.3) (Haines et al, 2017), reduced water scarcity (UN-WATER, 2016), increased water use efficiency (target 6.4) (van Vuuren et al, 2015) and achieving IWRM (target 6.5). Indeed, including climate change measures in water agreements is of particular importance in countries where pressure on water threatens agreements on shared water (Swain, 2016), especially for most dry regions where water competition between sectors is intensifying (IPCC, 2014a) and water security is threatened, for example in the Andes Mountains (IPCC, 2021). This is also true for rural areas that 'are expected to experience major impacts on water availability and supply' (IPCC, 2014b: 16).

5.2 Illustrative case studies of subnational governance arrangements in mountain countries

This section, which is based on the analysis of relevant policy documents and on explorative expert interviews,[1] presents three illustrative examples of subnational governance arrangements in mountainous areas that aim at tackling cross-sectoral, transboundary and multilevel challenges. The three cases provide insights into a range of different governance arrangements and highlight how the challenge of SDG implementation and the more general governance challenges of cross-sectoral, transboundary and multilevel coordination happen at the local level. The institutional context conditions differ across the three countries, providing different types of incentives, resources and degrees of autonomy to subnational governments, which demonstrate how SDG implementation and tackling the three governance challenges happen at the local level in diverse contexts. First, while the economy of Switzerland is highly diversified, Bolivia and Ecuador rely on the primary sector, mainly in mineral and hydrocarbon extraction (IMF, 2017). The countries also have very different levels of GDP, or income. Second, according to data provided by the World Health Organization and the United Nations' Joint Monitoring Programme for Water Supply and Sanitation (JMP) (2022), Bolivia and Ecuador have narrowed the gap on improved drinking water and sanitation in relation to Switzerland. However, the gap is still considerable regarding safely managed drinking water. Finally, the countries are different in terms of decentralization. Bolivia is characterized by a model of decentralization that defines the functions of national authorities and local governments, which adoption is far from complete. Meanwhile, Ecuador adopted a decentralization model that promotes associations of municipalities to tackle common problems, but prescribes how these subnational arrangements must operate, allowing almost no room for adaptive governance (Cisneros, 2022). Switzerland traditionally corresponds to an ideal type of a federalist state with strong decision-making capacities at the subnational level and some tendencies to centralization over time (Sciarini et al, 2015).

The Lac Léman region in Switzerland and the Lake Titicaca region in Bolivia include sizeable urban areas while the Bosque Seco in Ecuador is composed of small urban centres. Accordingly, the cases also involve different sets of actors, although a systematic overview or comparison of actors involved cannot be provided in this chapter. However, all three regions have to deal with the three challenges within the context of mountainous regions that require securing ecosystemic services to meet growing demand under conditions of climatic change. Indeed, mountains are among the regions most affected by climate change because higher elevations suffer from amplified warming.

5.2.1 Bolivia: Lake Titicaca region

Lake Titicaca is the largest water body shared by Bolivia and Peru, located 3,800 metres above sea level in the northern part of the South American Altiplano. The lake is part of an endorheic basin that comprises other water bodies (Uru Uru, Poopo and the Coipasa Salt Marsh) and river basins defined as the TDPS System.[2] The main management issue in this region relates to water availability, since the maximum usable flow is dramatically less than the estimated demand. The region has attracted the interest of glaciologists and hydrologists concerned about the environmental changes produced by temperature increase and glacier melting (Francou et al, 1995; Ramírez et al, 2001). This has also raised concerns about the potential impacts of glacier melting on water resources and the reliability of water provision systems in major cities (Soruco et al, 2015; Kinouchi et al, 2019) such as the city region of La Paz–El Alto, home to more than two million people. Water pollution in certain regions of the lake has also become a major issue in recent decades, in particular the contamination of Puno Bay on the Peruvian side and Cohana Bay in the Bolivian side, from industrial and urban discharges (Archundia et al, 2016; Molina et al, 2017).

The transboundary nature of Titicaca Lake and its contributing watersheds has encouraged the governments of Bolivia and Peru to establish, through public international law in 1992, a binational authority, Autoridad Autónoma Binacional del Lago Titicaca (ALT), responsible for managing water resources in the Titicaca Lake region (Revollo, 2001). The ALT has been working with local communities to build capacity for water management through the promotion of networks of leaders and technology transfer (ALT, 2011). In 2017 both the Ministry of Environment and Water of Bolivia and the Ministry of Environment of Peru registered an SDG partnership[3] to address SDG 6, SDG 13 and others to promote IWRM in the binational TDPS System. This initiative also involves governments and civil society groups, and serves to guide the work towards SDG implementation and to track related progress (Int2).

In addition to this binational agreement, as part of its National Watershed Planning Plan (Plan Nacional de Cuencas (PNC)), the Bolivian government prioritized the management of the Katari River Basin (KRB). This basin is a principal tributary of the Menor Lake (a part of Titicaca Lake that constitutes 16 per cent of its surface), where more than one million people live in nine municipalities. The KRB receives untreated water from the city of El Alto and acid mine drainage from the area of Miyuni, which lead to the pollution of crops and livestock, endangering human health (Redextractivos, 2021). The KRB Director Plan identified five major challenges relating to SGDs 6 and 13, including water governance, regulation of different water uses, reducing water pollution, restoring relevant environmental functions for the hydrological cycle and climate change adaptation, and enhanced understanding and management capacities through dialogue between science and Indigenous knowledge. With support from international cooperation, this plan has led to agreements with local populations to implement actions and identify ways to increase social participation (IAGUA, 2018). The KRB Director Plan itself led to the municipalities involved agreeing a common plan for the management of water resources, with a defined budget, which can be seen as a form of municipal association (Int1).

Municipal governments are also mandated through the National Planning Law (Law 777 of 2016) to work together to restore key ecosystem features, adapt to climate change and reduce disaster risk in territorial plans (Plan Territorial de Desarrollo Integral) in line with SDG 13, that is, it demands cross-sectoral coordination at different governance levels (Int4). These plans, however, lack the necessary instruments to effectively implement SDG 13 as well as effective instruments for cross-sectoral coordination (Int3). Also, despite important adjustments in the governance of water resources with the implementation of the PNC (Ruíz and Gentes, 2008; Llavona, 2020) and climate change adaptation being mainstreamed at the level of the PNC (Gonzales-Iwanciw et al, 2021), local populations are still not participating at all levels of decision making that recognize their contribution to improving water management (IUCN and BRIDGE, 2019).

5.2.2 Ecuador: Mancomunidad of Municipalities in the Southwest Loja Province 'Bosque Seco' (MBS)

The Mancomunidad of Municipalities in the Southwest Loja Province 'Bosque Seco' was launched in 2005 by five small municipalities (Celica, Zapotillo, Puyango, Pindal and Macará) to tackle the social and economic challenges caused by extended droughts, desertification, water pollution, soil erosion and deforestation (MBS, 2012).

The tropical dry forest that makes most of the MBS's jurisdiction is one of the most threatened and under-researched land uses in the world (Blackie et al, 2014). The MBS sits in the larger area of the Puyango–Tumbes and Catamayo–Chira watersheds shared by Ecuador and Peru. Aquifers and watersheds in the area are an important, yet highly variable, water supply essential for the region's socio-economic development and the integrity of its ecosystems. Currently, inappropriate land use, overexploitation, pollution, inefficient management and climate change threaten these water resources. Climate change in the region is exacerbating dry and wet seasons, increasing the need for improved infrastructure, ecosystem restoration and alternative economic opportunities for the local population (Project GIRHT, 2020).

The most salient example of cross-sectoral collaboration dates from 2014 and includes MBS, three neighbouring municipalities and an NGO. With support from national sectoral authorities (the Ministry of Tourism, Ministry of the Environment, and Ministry of Foreign Affairs), they presented a proposal to UNESCO for the creation of the Bosque Seco Biosphere Reserve (BSBR). The project aims to promote tourism, create a regional identity around a site recognized worldwide for its natural and cultural richness, and increase the flow of international cooperation to the area (Int5).

The BSBR allowed the MBS to access funding from international and national programs to institutionalize its governance arrangement and strengthen local capacities to implement decisions related to SGDs 6 and 13. In late 2015 the MBS launched the Program for the Restoration of Forests for Environmental Conservation and the Protection of Watersheds, financed by the Ministry of Environment and Water (MAAE) under the National Program for Forest Restoration (part of the national Program of Incentives for Conservation, or SocioBosque Program). The SocioBosque Program supports restoration activities that intend to join the scheme for carbon capture through avoiding deforestation in the future. These activities further promote multilevel collaborations between municipalities and national agencies.

Transboundary interactions for MBS members have been sustained since 2001, with the implementation of the Binational Plan for the Development of the Ecuador–Peru Border. This plan is the main mechanism for enforcing the peace agreements of 1998, which ended territorial disputes between the two countries. The Puyango–Tumbes and Catamayo–Chira basins are part of a binational strategy for IWRM that includes the nine watersheds shared by Ecuador and Peru. With support from several European aid agencies, the binational plan has supported MBS members in developing watershed management plans and capacities to implement IWRM at all government levels (Int6). One relevant outcome, among others, of transboundary coordination was the creation in 2017 of the Transboundary Biosphere Reserve Bosques de Paz, which encompasses the BRBS and the Amotapes–Manglares Biosphere Reserve in Peru.

Multilevel interactions in the MBS territory are multiple and involve all levels of governments and international actors as well as local organizations. Most have been supported by the Ecuador–Peru binational plan or have interacted with it in a sustained manner. Since 2006, the rapid conversion of forest to grasslands and the subsequent reduction of water availability have motivated watershed restoration and management programs, and the creation of 75 areas where local governments and landowners invest in improved technology for water catchment, reforestation and improved cattle ranching. In addition, the regional water conservation fund (FORAGUA) was created with support from aid agencies. FORAGUA uses a trust fund, with special contributions made by its members, to buy land in critical areas to secure its restoration or conservation (Int7). In 2010 the municipalities of the MBS and FORAGUA signed an agreement to jointly implement a program for watershed restoration.

The prevalence of poverty in the rural areas of its jurisdiction (with poverty levels of 90 per cent in some areas) has led the MBS to prioritize the creation of economic opportunities through the involvement of the local population in the restoration of degraded ecosystems to increase water availability. Economic opportunities are offered through building capacity for using local species adapted to the dry conditions of the area to develop agribusinesses and promote ecological tourism. These species are grown locally and used for reforestation in areas managed by associations of local farmers and municipalities (Int5).

5.2.3 Switzerland: the Lac Léman region

Switzerland has long been considered the 'water castle of Europe' because of its abundant natural water reservoir, which represents 6 per cent of Europe's freshwater (Pflieger, 2009) and is mainly powered by precipitation (Blanc and Schädler, 2013). Lac Léman – also known as Lake Geneva – is the largest alpine lake in western Europe and provides drinking water for more than 900,000 people across Switzerland and France (CIPEL, 2021). It spreads across both territories: 88 per cent of the total lake area is composed of the north shore crossing over the three Swiss cantons of Geneva, Vaud and Valais, while the southern shore is located in the French department of Haute-Savoie (OFEV, 2016).

The unique abundance of water from high mountain sources in Switzerland was long seen as a guarantee of water quality (Pflieger, 2009). However, micropollutants now constitute one of the major ecological challenges for the region, despite efforts to remove them (CIPEL, 2021). Indeed, the pollution – mainly from industrial emissions upstream from the lake, in the Rhone River of the Valais canton (CIPEL, 2021) – is a significant threat to the lake's fauna (Faure et al, 2012). The use of pesticides is an additional threat to the fauna and to the agricultural soils, which are vulnerable to pesticide transfers to surface waters (CIPEL, 2021). Water quality is thus seen as the main water-related problem in the Lac Léman region.

In terms of cross-sectoral governance, the forum Grand Genève Agglomeration is a space for discussion about various usages that brings together authorities from France and from Geneva and Vaud cantons, as well as civil society. Climate change is an important issue in water governance arrangements in the region, and the International Commission for the Protection of the Waters of Lake Geneva (CIPEL) was set up as a binational body, with representatives from the agricultural sector and other interest groups, to tackle cross-sectoral issues, including sanitation and coordination (Int11). CIPEL conducts scientific evaluations of the consequences of climate change and produces action plans and recommendations. Experts at the Geneva Cantonal Office of Water (OCEau) also oversee projects including on climate change issues (Int10). Agricultural water issues, mostly irrigation related, are dealt with in informal collaborations between the cantonal offices for water and agriculture. Similarly, water needs related to territory development are handled in informal collaborations between the cantonal offices for water and tourism (Int10).

In terms of transboundary governance, elected officials from water or environment offices represent their respective cantons – Genève, Vaud and Valais – within CIPEL. Set up in 1963 via a constitutive convention, CIPEL remains one of the central actors in the transboundary water governance of the region, focusing mainly on water quality (Int11). CIPEL focuses on controlling the water quality of the lake and related water streams, studying and preventing water pollution, coordinating water management policies and cooperating with local partners on quality issues. Additionally, given the transboundary characteristics of and increasing demographic pressures on water resources in the region, the Transboundary Water Community (CTEau) was created in 2012. As a part of the environmental commission of the Regional Committee Franco-Genevois (CRFG), created in 1974, CTEau offers a space for discussion between French departments and other organizations, and Geneva and Vaud cantons, and collaborates with several actors including CIPEL and Geneva Industrial Services (SIG).

There are several aspects to multilevel governance on this issue. While at the national level SDG 6 is not identified as one of Switzerland's priorities for the implementation of Agenda 2030, the cantonal climate plans of Geneva and Vaud at least integrate important points from SDG 6; for example, on the issue of water management, 'for an equitable and sustainable sharing of the resource on the transboundary scale of the Geneva watershed' (DT, 2021). The implementation of these measures has been delegated to water- and environment-related offices (Int10; Conseil d'État vaudois, 2020). Municipalities have recently been encouraged to work on communal climate plans (Int18) that, again, implicitly include aspects of the implementation of SDG 6. Some cities have also created specific departments, such as the Durability Department in the city of Vevey. The transboundary forms of coordination discussed earlier do have a multilevel character, given that transboundary interactions take place at the cantonal, or departmental, and national levels in both France and Switzerland.

5.3 Discussion and conclusion

Subnational and local governments are called on to spearhead the implementation of the SDGs so that local conditions can be taken into account effectively. This chapter shows that subnational governance arrangements in Bolivia, Ecuador and Switzerland are addressing the three coordination challenges to different degrees. The three case studies provide examples demonstrating such governance challenges and potential solutions, but a more detailed comparison of the three cases is beyond the scope of this chapter. The focus is on SDGs 6 and 13, while many other areas such as, for example, human rights, life on land and energy, are crucially related to water. Table 5.1 summarizes how governance in the different regions address cross-sectoral, transboundary and multilevel challenges when dealing with water and climate change issues.

All the cases show interactions between SDG 6 and SDG 13 in different arrangements. In the case of Lac Léman (Switzerland), the challenge of improving water quality in a context of climate change adds to previously existing interactions between the industrial sector, agriculture and the recreation sector. It is complemented by interactions across levels of government and the capability to sustain transboundary commitments through formal and informal mechanisms. In the MBS (Ecuador), cross-sectoral interactions are driven by the need to create economic opportunities through new forms of agricultural production and sustainable tourism that depend on regional and transboundary adaptation projects financed by international cooperation. Given their availability of resources, transboundary interactions serve as an umbrella for cross-sectoral and multilevel coordination. In the Lake Titicaca region (Bolivia), rapid urbanization, urban–rural interactions and water pollution challenge the provision of water and sanitation in a context of transboundary governance arrangements. Centralism in water policies and limited local capacities preclude multisectoral coordination on competing water uses for drinking water, mining and agriculture. Moreover, climate change does not figure prominently in cross-sectoral interactions.

The extent to which these initiatives localize the global development agenda in the specific context of mountainous regions is hard to measure. The illustrative case studies provide some insights into a range of different governance arrangements and their tackling of the three governance challenges; however, initiatives related to the core of the water SDGs 6 and 13 lack a vocabulary consistent with the SDGs and are thus at best implicit. For example, despite clear recognition that sustainable development

Table 5.1: Governance arrangements and coordination challenges

Cases	Coordination challenges		
	Cross-sectoral	Transboundary	Multilevel
Lake Titicaca, Bolivia	*Municipal plans and watershed management plans* define cross-sectoral coordination goals for water and climate change adaptation.	*Binational authority* addresses water and climate change adaptation issues.	Water and climate change adaptation figure in *local development plans.* The National Watershed Plan (PNC) also addresses water and climate. Both instruments include stakeholder participation.
MBS, Ecuador	A *cross-sectoral committee* promotes the regionalization of development activities to secure water provision and promote climate change adaptation.	A *binational development program* focused on water governance supports cross-sectoral and multilevel initiatives with climate change adaptation components.	Water and climate change figure in *local development plans,* which are implemented with extended social participation.
Lac Léman, Switzerland	*Multisectoral forums and working groups* in subnational government units tackle cross-sectoral issues.	*Binational commissions and committees* focus on water quality and evaluate the impact of climate change on water resources.	Water and climate change figure in *national, subnational and local climate plans and transboundary initiatives.*

should be a concern when dealing with water resources and climate change adaptation, neither the KRB Director Plan nor the various Lake Titicaca region municipal government PTDIs (Plan Territorial de Desarrollo Integral) for the period 2016–20 explicitly mentions the SDGs, nor do any of these initiatives use the indicators of the SDG framework.

This situation makes the interactions between SDGs less tractable, especially in contexts where access to public information and records is scarce, as in Bolivia and Ecuador. According to Swiss interview partners, this absence of SDG vocabulary – for their case – is due to at least three reasons: water abundance and advancement in terms of sanitation, water quality and water access (Int10); the existence of other policy or legal instruments relating to the sustainable development of water and climate change concerns (Int14, Int15, Int19); and that this vocabulary does not speak to the different actors' realities (Int14). In Ecuador, informants argue that the SDGs are important only to maintain collaboration with international donors and national programmes but are not necessary for day-to-day interactions at the local level (Int5).

Overall, this chapter has emphasized the importance of three governance challenges that concern all types of natural resource management, and for which the water sector is a prime example: the coordination of actors across different thematic sectors, different administrative entities across borders and different levels of governance. By discussing how these three challenges – and potential solutions – appear in the global discussion on water policies as well as in subnational governance arrangements, this chapter contributes to the three guiding questions of this book. In relation to the first question of how perceptions of sustainable development have changed in politics and research since the 1992 Earth Summit, this chapter reviews the development of global water policies and emphasizes the importance of the three challenges of cross-sectoral, transboundary and multilevel coordination as important aspects of sustainable development at the global and the subnational levels. The second question – of which actors and institutions have most mattered for governance efforts over the past three decades and who should be held accountable for success and failure – the chapter has discussed actors and instruments that are able, or have been specifically designed, to potentially address the three governance challenges. However, actors working on these three challenges might be more difficult to hold accountable since, by definition, they are not clearly related to one specific sector, administrative unit or governance level. The third question – of which alternative and innovative forms of governance exist and deserve more research attention for a transition to environmentally salient sustainability – again points towards potentially new and innovative forms of governance that are able to tackle the three governance challenges.

This chapter has been able only to scratch the surface of these issues, and there are still many important research questions to be addressed. For example, it could be asked whether the types of governance arrangements, as well as the more or less explicit use of SDG language on the subnational level, depend on the types of actors involved (eg private companies, regional governments, international donors) and their resources, the institutional context or related challenges. A more explicit comparative approach than this chapter has taken should include a fine-grained discussion of case selection and of context conditions (institutions, economic capacity and so on) that may explain the existence and performance of given types of subnational governance arrangements. Finally, and related to the second guiding question of this book, questions of accountability of given actors were not at the centre of this chapter and will require more in-depth case analyses. These could consist of elaborating on the accountability challenges raised by new governance arrangements that address the cross-sectoral, transboundary and multilevel challenges relating to water and climate governance.

Notes

[1] The illustrative case studies are based on the qualitative and interpretative analysis of policy documents, as well as on expert interviews conducted between 2018 and 2022. The relevant policy documents are referenced in the text and appear in the list of references. Some of the following interviews are directly referenced in the text given that specific text elements rely on a specific interview, while others are not directly referenced in the text but provide context elements for the three cases. For Bolivia: Interview 1 (Int1): director of the Pilot Project for Climate Resilience (June 2018); Interview 2 (Int2): president of the Women Association for the Defence of Titicaca Lake (June 2018); Interview 3 (Int3): water consultant (March 2022); Interview 4 (Int4) natural resources specialist, Swiss Cooperation Integrated Water Project (March 2022). For Ecuador: Interview 5 (Int5): coordinator of the MBS (July 2021, online); Interview 6 (Int6): director of Articulación del Recurso Hídrico del Ministerio del Ambiente, Agua y Transición Ecológica (January 2022, online); Interview 7 (Int7): representative, Naturaleza y Cultura Internacional (February 2022, online); Interview 8 (Int8): representative, Consejo Nacional de Competencias (June 2021, online). For Switzerland: Interview 9 (Int9): researcher, University of Geneva (September 2021, online); Interview 10 (Int10): representative, Geneva Cantonal Office for Water (September 2021, online); Interview 11 (Int11): representative, International Commission for the Protection of the Waters of Lake Geneva (CIPEL) (September 2021, online); Interview 12 (Int12): associate professor, University of Geneva (October 2021, online); Interview 13 (Int13): representative, Vaud Climat Plan Unit (November 2021, online); Interview 14 (Int14): representative, Vaud Groundwater Section (January 2022, online); Interview 15 (Int15): representative, Department of Consumer and Veterinary Affairs, and member, Water Resources Management Commission (GRE), Vaud Canton (January 2022, online); Interview 16 (Int16): representative, Geneva Sustainable Development Department (January 2022, online); Interview 17 (Int17): representative, Vaud Water Protection Division (January 2022, online); Interview 18 (Int18): representative, Valais's Foundation for Sustainable Development of Mountain Regions (FDDM) (January 2022, online); Interview 19 (Int19): representative, Valais Surface Water and Waste Division (March 2022, online).

[2] T: Lake Titicaca basin; D: Desaguadero River basin; P: Lake Poopó basin; S: Salar de Coipasa basin.

[3] The Partnership Platform is a global registry of voluntary commitments and multistakeholder partnerships by stakeholders in support of the implementation of the SDGs (see Chapter 13).

References

Alcamo, J. (2019) 'Water quality and its interlinkages with the Sustainable Development Goals', *Current Opinion in Environmental Sustainability*, 36: 126–40.

ALT (Autoridad Autónoma Binacional del Lago Titicaca) (2011) *Memoria Anual 2011*, Available from: http://alt-perubolivia.org/sitio/pdf/transp arencia/memoria-anual-2011.pdf [Accessed 10 August 2022].

Archundia, D., Duwig, C., Spadini, L., Uzu, G., Guédron, S., Morel, M.C. et al (2016) 'How uncontrolled urban expansion increases the contamination of the Titicaca Lake Basin (El Alto, La Paz, Bolivia)', *Water, Air, & Soil Pollution*, 228(1): 1–17.

Blackie R., Baldauf C., Gautier D., Gumbo D., Kassa H., Parthasarathy N. et al (2014) *Tropical Dry Forests: The State of Global Knowledge and Recommendations for Future Research*, Discussion paper, Bogor: Centre for International Forestry Research.

Blanc, P. and Schädler, B. (2013) *L'eau en Suisse – un aperçu*, Bern: Commission suisse d'hydrologie, Available from: www.unibe.ch/unibe/portal/content/ e809/e810/e812/e832/e834/e488430/e114276/files114281/Leau_en_ Suisse_fra.pdf [Accessed 10 August 2022].

Blomquist, W., Dinar, A. and Kemper, K. (2005) *Comparison of Institutional Arrangements for River Basin Management in Eight Basins*, Washington, DC: World Bank.

Bodin, Ö., Alexander, S.M., Baggio, J., Barnes, M.L., Berardo, R., Cumming, G.S., et al (2019) 'Improving network approaches to the study of complex social–ecological interdependencies', *Nature Sustainability*, 2(7): 551–9.

Breuer, A., Leininger, J. and Tosun, J. (2019) *Integrated Policymaking: Choosing an Institutional Design for Implementing the Sustainable Development Goals (SDGs)*, Discussion paper, Bonn: Deutsches Institut für Entwicklungspolitik (DIE).

Buytaert, W. and De Bièvre, B. (2012) 'Water for cities: the impact of climate change and demographic growth in the tropical Andes', *Water Resources Research*, 48(8): 1–13.

Cabrera-Barona, P. and Cisneros, P. (2021) 'Explaining the effectiveness of forest and water management and its spatial distribution in the metropolitan district of Quito', *Geography, Environment, Sustainability*, 14(1): 53–62.

CIPEL (International Commission for the Protection of Lake Léman's Waters) (2021) *Plan d'action 2021–2030 Cap sur le Léman 2030*, Nyon: CIPEL, Available from: https://www.cipel.org/wp-content/uploads/2021/06/ plan-action-cipel-format-a4-vf.pdf [Accessed 10 August 2022].

Cisneros, P. (2022) 'Intergovernmental relations and the effective implementation of Agenda 2030 in South America', in R. Baikady, S.M. Sajid, J. Przeperski, V. Nadesan, I. Rezaul and J. Gao (eds) *Palgrave Handbook of Global Social Problems*, Cham: Palgrave Macmillan, pp 1–23 (living reference work entry).

Connell, D. and Grafton, R.Q. (2011) 'Water reform in the Murray–Darling Basin', *Water Resources Research*, 47: 1–9.

Conseil d'État vaudois (2020) *Stratégie du Conseil d'État vaudois pour la protection du climat. Plan climat vaudois – 1ère génération*, Available from: https://www.vd.ch/fileadmin/user_upload/themes/environnement/climat/fichiers_pdf/202006_Plan_climat.pdf [Accessed 10 August 2022].

DT (Département du Territoire) (2021) *Plan climat cantonal 2030. 2ème Génération*, Available from: www.ge.ch/document/plan-climat-cantonal-2030-2e-generation-0 [Accessed 10 August 2022].

Engle, N.L., Johns, O.R. and Lemos, M.C. (2011) 'Integrated and adaptive management of water resources: tensions, legacies, and the next best thing', *Ecology and Society*, 16(1): 1–11.

ESCAP (United Nations Economic and Social Commission for Asia and the Pacific) (2016) *Analytical Framework for Integration of Water and Sanitation SDGs and Targets Using Systems Thinking Approach*, Working paper, Bangkok: ESCAP, Available from: https://sdghelpdesk.unescap.org/sites/default/files/2018-02/integration%20sdg6.pdf [Accessed 10 August 2022].

ESCAP (2017) *Integrated Approaches for Sustainable Development Goals Planning: The Case of Goal 6 on Water and Sanitation,* Bangkok: ESCAP, Available from: www.unescap.org/sites/default/d8files/knowledge-products/Integrated%20Approaches%20for%20SDG%20Planning3.pdf [Accessed 10 August 2022].

Faure, F., Corbaz, M., Baecher, H. and De Alencastro, L.F. (2012) 'Pollution due to plastics and microplastics in Lake Geneva and in the Mediterranean Sea', *Archives des Science*, 65: 157–164.

Fischer, M. and Leifeld, P. (2015) 'Policy forums: why do they exist and what are they used for?', *Policy Sciences*, 48(3): 363–82.

Fischer, M. and Ingold, K.M. (eds) (2020) *Networks in Water Governance*, Cham: Palgrave Macmillan.

Fischer, M. and Jager, N.W. (2020) 'How policy-specific factors influence horizontal cooperation among subnational governments: evidence from the Swiss water sector', *Publius: The Journal of Federalism*, 50(4): 645–71.

Future Earth and Earth League (2017) *The 10 Science 'Must Knows' on Climate Change*, Prepared by the Earth League and Future Earth for the UNFCCC 23rd Conference of the Parties, Available from: https://futureearth.org/2017/11/12/the-10-science-must-knows-on-climate-change [Accessed 10 August 2022].

Francou, B., Ribstein, P., Saravia, R. and Tiriau, E. (1995) 'Monthly balance and water discharge of an inter-tropical glacier: Zongo Glacier, Cordillera Real, Bolivia, 16° S', *Journal of Glaciology*, 41(137): 61–7.

Gonzales-Iwanciw, J., Karlsson-Vinkhuyzen, S. and Dewulf, A. (2021) 'Multi-level learning in the governance of adaptation to climate change: the case of Bolivia's water sector', *Climate and Development*, 13(5): 399–413.

Gupta, J., Pahl-Wostl, C. and Zondervan, R. (2013) ' "Glocal" water governance: a multi-level challenge in the anthropocene', *Current Opinion in Environmental Sustainability*, 5(6): 573–80.

Haines, A., Amann, M., Borgford-Parnell, N., Leonard, S., Kuylenstierna, J., and Shindell, D. (2017) 'Short-lived climate pollutant mitigation and the Sustainable Development Goals', *Nature Climate Change*, 7(12): 863–9.

Hering, J.G. and Ingold, K.M. (2012) 'Water resources management: what should be integrated?', *Science*, 336(6086): 1234–5.

Hooghe, L. and Marks, G. (2001) 'Types of multi-level governance', *European Integration Online Papers (EIoP)*, 5(11): 1–32.

Horan, D. (2020) 'National baselines for integrated implementation of an environmental Sustainable Development Goal assessed in a new integrated SDG index', *Sustainability*, 12(17): 1–22.

iAgua (2018) 'Bolivia accrued acciones para descontaminar la Cuenca Katari y el lago Menor del Titicaca', Available from: https://www.iagua.es/notic ias/mmaya/bolivia-acuerda-acciones-descontaminar-cuenca-katari-y-lago-menor-titicaca [Accessed 10 August 2022].

ICSU (International Council for Science) and ISSC (and International Social Science Council) (2015) *Review of Targets for the Sustainable Development Goals: The Science Perspective*, Paris: ICSU, Available from: https://coun cil.science/wp-content/uploads/2017/05/SDG-Report.pdf [Accessed 10 August 2022].

IMF (International Monetary Fund) (2017) *Export Diversification and Quality*, Available from: https://data.imf.org/?sk=388dfa60-1d26-4ade-b505-a05a5 58d9a42 [Accessed 10 August 2022].

Independent Group of Scientists appointed by the Secretary-General (2019) *The Future is Now: Science for Achieving Sustainable Development. Global Sustainable Development Report 2019.* Available from: http://pure.iiasa. ac.at/id/eprint/16067/1/24797GSDR_report_2019.pdf [Accessed 10 August 2022].

IPCC (Intergovernmental Panel on Climate Change) (2014a) *Climate Change 2014: Impacts, Adaptation, and Vulnerability*, part A: *Global and Sectoral Aspects*, Contribution of Working Group II to the Fifth Assessment Report of the Intergovernmental Panel on Climate Change, New York: Cambridge University Press, Available from: https://www.ipcc.ch/site/assets/uplo ads/2018/02/WGIIAR5-PartA_FINAL.pdf [Accessed 10 August 2022].

IPCC (2014b) *Climate Change 2014: Synthesis Report*, Contribution of Working Groups I, II and III to the Fifth Assessment Report of the Intergovernmental Panel on Climate Change, Geneva: IPCC, Available from: https://www.ipcc.ch/site/assets/uploads/2018/02/SYR_AR5_FIN AL_full.pdf [Accessed 10 August 2022].

IPCC (2021) Summary for Policymakers. In V. Masson-Delmotte, P. Zhai, A. Pirani, S. L. Connors, C. Péan, S. Berger, et al (eds), *Climate Change 2021: The Physical Science Basis*, Contribution of Working Group I to the Sixth Assessment Report of the Intergovernmental Panel on Climate Change, Cambridge: Cambridge University Press, Available from: https://www.ipcc.ch/report/ar6/wg1/downloads/report/IPCC_AR6_WGI_Full_Report_smaller.pdf [Accessed 10 August 2022].

IUCN (International Union for Conservation of Nature) and BRIDGE (Building River Governance and Dialogue) (2019) 'Lago Titicaca: empoderando a las mujeres y mejorando la gobernanza del agua', Available at https://digital.iucn.org/agua/lago-titicaca [Accessed 10 August 2022].

JMP (Joint Monitoring Programme) (2022) Household data, Available from: https://washdata.org/data/info/household [Accessed 10 August 2022].

Jordan, A., Wurzel, R., Zito, A.R. and Brückner, L. (2003) 'European governance and the transfer of "new" environmental policy instruments (NEPIs) in the European Union', *Public Administration*, 81(3): 555–74.

Kersting, N., Caulfield, J., Nickson, R.A., Olowu, D. and Wollmann, H. (2009) *Local Governance Reform in Global Perspective*, Wiesbaden: VS Verlag Für Sozialwissenschaften.

Kinouchi, T., Nakajima, T., Mendoza, J., Fuchs, P. and Asaoka, Y. (2019) 'Water security in high mountain cities of the Andes under a growing population and climate change: a case study of La Paz and El Alto, Bolivia', *Water Security*, 6: 1–11.

Le Blanc, D. (2016) 'Sustainable Development Goals and policy integration in the nexus', in F. Dodds and J. Bartram (eds) *The Water, Food, Energy and Climate Nexus: Challenges and an Agenda for Action*, London: Routledge, pp 47–56.

Lesnikowski, A., Biesbroek, R., Ford, J. and Berrang-Ford, L. (2021) 'Policy implementation styles and local governments: the case of climate change adaptation', *Environmental Politics*, 30(5): 753–90.

Llavona, A. (2020) *Lecciones del Estado Plurinacional de Bolivia para la adopción del enfoque del Nexo: análisis del Plan Nacional de Cuencas, el Sistema Múltiple Misicuni y las políticas de riego*, Available from: https://repositorio.cepal.org/handle/11362/46546 [Accessed 10 August 2022].

Lubell, M. and Edelenbos, J. (2013) 'Integrated water resources management: a comparative laboratory for water governance', *International Journal of Water Governance*, 1(3–4): 177–96.

Lundqvist, L. (2011) 'Integrating Swedish water resource management: a multi-level governance trilemma', *Local Environment*, 9(5): 413–24.

MBS (Mancomunidad Bosque Seco) (2012) *Plan de Fortalecimiento y Desarrollo de la Mancomunidad 'Bosque Seco'*. https://www.mancomunidadbosqueseco.gob.ec/fortalecimiento-y-desarrollo-institucional/

Margerum, R.D. and Robinson, C.J. (2015) 'Collaborative partnerships and the challenges for sustainable water management', *Current Opinion in Environmental Sustainability*, 12: 53–8.

Mirumachi, N. and Allan, J.A. (2007) 'Revisiting transboundary water governance: power, conflict, cooperation and the political economy', in *Proceedings from CAIWA International Conference on Adaptive and Integrated Water Management: Coping with Scarcity*, Available from: http://citeseerx. ist.psu.edu/viewdoc/download?doi=10.1.1.490.9078&rep=rep1&type= pdf [Accessed 10 August 2022].

Molina, C.I., Lazzaro, X., Guédron, S. and Achá, D. (2017) 'Contaminación de la Bahía de Cohana, Lago Titicaca (Bolivia): desafíos y oportunidades para promover su recuperación', *Ecología en Bolivia*, 52(2): 65–76, Available from: www.scielo.org.bo/scielo.php?script=sci_arttext&pid=S1605-25282017000200002&lng=es&nrm=iso [Accessed 10 August 2022].

Mollinga, P.P., Dixit, A. and Athukorala, K. (2006) *Integrated Water Resources Management: Global Theory, Emerging Practice and Local Needs*, New Delhi: SAGE.

OFEV (Federal Office for the Environment) (2016) *Le Léman: qualité de l'eau du lac*, Available from: www.bafu.admin.ch/dam/bafu/fr/dokume nte/wasser/fachinfo-daten/Wasserqualität_Seen_Le_Léman.pdf.downl oad.pdf/Wasserqualität_Seen_Le_Léman.pdf [Accessed 10 August 2022].

Pahl-Wostl, C. (2009) 'A conceptual framework for analysing adaptive capacity and multi-level learning processes in resource governance regimes', *Global Environmental Change*, 19(3): 354–65.

Pahl-Wostl, C. (2019) 'The role of governance modes and meta-governance in the transformation towards sustainable water governance', *Environmental Science & Policy*, 91: 6–16.

Petit, O. and Baron, C. (2009) 'Integrated water resources management: from general principles to its implementation by the state: the case of Burkina Faso', *Natural Resources Forum*, 33(1): 49–59.

Pflieger, G. (2009) *L'eau des villes: aux sources des empires municipaux*, Lausanne: Presses Polytechniques et Universitaires Romandes.

Pham-Truffert, M., Metz, F., Fischer, M., Rueff, H. and Messerli, P. (2020) 'Interactions among Sustainable Development Goals: knowledge for identifying multipliers and virtuous cycles', *Sustainable Development*, 28(5): 1236–50.

Plummer, R. and Armitage, D. (2010) 'Integrating perspectives on adaptive capacity and environmental governance', in R. Plummer and D. Armitage (eds) *Adaptive Capacity and Environmental Governance*, Heidelberg: Springer, pp 1–19.

Project GIRHT (Integrated Water Resources Management in the Puyango-Tumbes, Catamayo-Chira and Zarumilla Transboundary Aquifers and River Basins) (2020) *Strategic Action Programme (SAP) for the Puyango-Tumbes, Catamayo-Chira and Zarumilla Transboundary Watersheds and Aquifers*, Available from: https://iwlearn.net/resolveuid/f99ac232-35fc-424d-bdae-4f5593f7e8b5 [Accessed 10 August 2022].

Ramírez, E., Francou, B., Ribstein, P., Descloitres, M., Guérin, R., Mendoza, J., et al (2001) 'Small glaciers disappearing in the tropical Andes: a case-study in Bolivia: Glaciar Chacaltaya (16 S)', *Journal of Glaciology*, 47(157): 187–94.

Redextractivos (Red Latinoamericana sobre las Industrias Extractivas) (2021) *La cuenca Katari en diálogo: impactos del extractivismo minero y el desarrollo urbano*, Available at: https://redextractivas.org/la-cuenca-katari-en-dialogo-impactos-del-extractivismo-minero-y-el-desarrollo-urbano/ [Accessed 10 August 2022].

Revollo, M.M. (2001) 'Management issues in the Lake Titicaca and Lake Poopo system: importance of developing a water budget', *Lakes & Reservoirs: Research and Management*, 6(3): 225–9.

Rivera-Arriaga, E. and Azuz-Adeath, I. (2019) 'Implementing the SDG14 in Mexico: diagnosis and ways forward', *Revista Costas*, 1(1): 219–42.

Ruíz, S.A. and Gentes, I.G. (2008) 'Retos y perspectivas de la gobernanza del agua y gestión integral de recursos hídricos en Bolivia', *European Review of Latin American and Caribbean Studies/Revista Europea de Estudios Latinoamericanos y Del Caribe*, 85: 41–59.

Sanchez, J.C. and Roberts, J. (2014) *Transboundary Water Governance: Adaptation to Climate Change*, Gland: IUCN.

Sciarini, P., Fischer, M. and Traber, D. (2015) *Political Decision-making in Switzerland: The Consensus Model under Pressure*, Basingstoke: Palgrave Macmillan.

Soruco, A., Vincent, C., Rabatel, A., Francou, B., Thibert, E., Sicart, J.E. and Condom, T. (2015) 'Contribution of glacier runoff to water resources of La Paz city, Bolivia (16 S)', *Annals of Glaciology*, 56(70): 147–54.

Swain, A. (2016) 'Water and post-conflict peacebuilding', *Hydrological Sciences Journal*, 61(7): 1313–22.

Teles, F. (2016) *Local Governance and Inter-municipal Cooperation*, Basingstoke: Palgrave Macmillan.

UCLG (United Cities and Local Governments) (2015) *The Sustainable Development Goals: What Local Governments Need to Know*, Available from: https://www.uclg.org/sites/default/files/the_sdgs_what_localgov_need_to_know_0.pdf [Accessed 10 August 2022].

UN (United Nations) (1992) *The Dublin Statement on Water and Sustainable Development*, Available from: http://www.un-documents.net/h2o-dub.htm [Accessed 10 August 2022].

UN (1993) *Report of the United Nations Conference on Environment and Development, Rio de Janeiro, 3–14 June 1992*, vol. 1: *Resolutions Adopted by the Conference*, Available from: https://www.un.org/esa/dsd/agenda21/Agenda%2021.pdf [Accessed 10 August 2022].

UN (2000) *Millennium Declaration*, Available from: www.ohchr.org/en/instruments-mechanisms/instruments/united-nations-millennium-declaration [Accessed 10 August 2022].

UNCCD (United Nations Convention to Combat Desertification) (2017) *Global Land Outlook*, Bonn, Available from: www.unccd.int/sites/default/files/documents/2017-09/GLO_Full_Report_low_res.pdf [Accessed 10 August 2022].

UN-Water (United Nations Water) (2016) *Water and Sanitation Interlinkages across the 2030 Agenda for Sustainable Development*, Available from: www.unwater.org/publications/water-sanitation-interlinkages-across-2030-agenda-sustainable-development [Accessed 10 August 2022].

van Vuuren, D.P., Kok, M., Lucas, P., L., Prins, A.G., Alkemade, R., van den Berg, M., et al (2015) 'Pathways to achieve a set of ambitious global sustainability objectives by 2050: explorations using the IMAGE integrated assessment model', *Technological Forecasting and Social Change*, 98: 303–23.

Waibel, G., Benedikter, S., Reis, N., Genschick, S., Nguyen, L., Cong Huu, P., and Thanh Be, T. (2012) 'Water governance under renovation? Concepts and practices of IWRM in the Mekong Delta, Vietnam', in F.G. Renaud and C. Kuenzer (eds) *The Mekong Delta System: Interdisciplinary Analyses of a River Delta*, Dordrecht: Springer, pp 167–98.

Weitz, N., Carlsen, H., Nilson, M. and Skanberg, K. (2018) 'Towards systemic and contextual priority setting for implementing the 2030 Agenda', *Sustainability Science*, 13(2): 531–48.

Wymann von Dach, S., Bracher, C., Peralvo, M., Perez, K., Adler, C. et al (2018) *Leaving No One in Mountains Behind: Localizing the SDGs for Resilience of Mountain People and Ecosystems*, Bern: Centre for Development and Environment and Mountain Research Initiative, with Bern Open Publishing (BOP).

Interview with Manuel Fischer and Paúl Cisneros: A Conflict of Priorities, Not of Knowledge

Felix Nütz and Elizaveta Kapinos

The SDGs were created as a framework at the international level. How successful is the implementation when it comes to freshwater governance at the local level?

Cisneros: In principle, the SDGs could have had a bigger impact. The timing for the adoption of the SDGs was very unfortunate. Most of the countries in South America started to enter an economic recession when the SDGs were adopted. Most of the government's funding before 2014 wasn't there anymore to keep expanding the water provision systems or to form water management councils. Specialized groups of consultants and public officials who are very well connected to the international community got resources for specific projects regarding the SDGs, but the wider community has not seen its benefits.

Fischer: Some of our interview partners said that they had a hard time even talking about the SDGs. Many people on the ground involved in water management working in the municipalities do not talk the SDG language. It's not used on the local level but in academic circles and on the international level. However, if you want to get money for projects, you have to mention the SDGs. This can be criticized because then you just put some of these 'nice' labels on top and continue to do whatever you've been doing already. However, the SDGs do also create a joint language and a framework for different types of research.

Do you see in your work in South America a conflict between western scientific knowledge and local knowledge of freshwater management?

Cisneros: In the areas where we have been doing research, the local communities are not Indigenous communities – they've been incorporated in international markets for centuries. It is thus very hard to see a conflict of knowledge – there is more of a conflict of priorities. Whereas communities would rather have clean drinking water and sanitation, private companies doing business in the area mostly prioritize increasing their production, and so use water to grow and export fruits.

How can these groups with similar interests but different priorities be brought together to collaborate?

Cisneros: In these complex fields, facilitators like NGOs or government institutions can play a crucial role by bridging the different lists of priorities of the local communities and agricultural exporters. They can see the similarities and possibilities for collaboration that the other actors do not always see. But there is one thing I've seen time and time again: it is rare that anyone actually takes the role of facilitator or even invests in the resources that are required to facilitate discussion. But there are also positive examples. I've seen some of these water councils being formed around what are called water funds. These are funding pots that are contributed to by private companies and the government. Interest from these funds then finances local development projects. So we can see real tangible action taken by private actors beyond the responsibility to water as dictated by regulations. But, again, it is always difficult to balance different priorities.

PART II

The Goals with Environmental Trade-offs and Synergies

6

Water for Life and Food: Synergies between SDGs 2 and 6 and Human Rights

Lyla Mehta, Claudia Ringler and Shiney Varghese

As the Qur'an states, 'By means of water, we give life to everything' (21: 3).[1] Water is a fundamental element on which human beings depend for their lives and livelihoods. It has multiple values, faces and meanings in the everyday contexts in which people live their lives, as this chapter will show. Safe drinking water and sanitation are fundamental to the health, nutrition, and dignity of all (UNDP, 2006; UNSCN, 2020). Despite the progress made in achieving past global targets, including the Millennium Development Goals and efforts in previous Water Decades, and now SDG 6 on clean water and sanitation, in 2017 about 2.2 billion people lacked access to safe, readily available water at home, and 4.2 billion lacked safely managed sanitation (UNICEF and WHO, 2019). Accessing water can be particularly challenging for smallholders, vulnerable and marginalized populations, and women, and thus also affects SDG 10 (Reduced inequalities). Even though the human right to water and sanitation was globally endorsed by the United Nations in 2010, it is violated every day across the globe. This situation undermines health, nutrition, human well-being and dignity, and is a global and moral outrage.

Water is also the lifeblood of most ecosystems, including forests, lakes and wetlands, and their functions. These ecosystem functions, directly or indirectly, are also essential for the survival of people and the nurturing of flora, fauna and the environment. They are particularly important for poor people, providing them with nutrition and livelihoods. Water is also fundamental for all other sectors of importance to human well-being, including energy and manufacturing. Finally, water has important cultural and aesthetic values (HLPE, 2015).

Global inequality in access to water and sanitation is unacceptable and one of the largest inequities of the 21st century. The 2006 Human Development Report (UNDP, 2006: 3) noted that 'no act of terrorism generates economic devastation on the scale of the crisis in water and sanitation'. However, this crisis occurs largely in silence. Unlike wars and natural disasters, it remains invisible and has been quasi-naturalized – that is, accepted as part of life – by both those who enjoy access to safe water and the millions who do not. The water crisis is largely caused and legitimized by different forms of unequal gender and social relations and power relations, which prevent universal access (Mehta et al, 2020). In the case of millions of women and girls who spend hours collecting water, this naturalized gendered nature of water collection has undermined their health, education and chances in life. Similarly, poor water quality affects human health and also the functioning of ecosystems, with adverse impacts for poor and vulnerable groups that directly depend on this resource base for their livelihoods. Climate change, including growing climate variability, affects everyone on the globe and adds irregularity and uncertainty to the availability of and demand for water, with known effects on the vulnerability of the poorest and their food security (see Bates et al, 2008).

This chapter focuses on SDG 6 and its linkages with realizing other SDGs, especially SDG 2 (Zero hunger) in the spirit of integrating development and environmental agendas, and makes a case for more joined-up thinking on water and food, given the central role of water in ensuring food security, nutrition and human survival. Other linkages across SDGs are described in Nilsson et al (2017).

6.1 The multiple framings of water

Water is a contested and multifaceted resource. People across the globe value water for both its economic and its non-economic values, including its deep spiritual significance in many cultures. The Dublin Principles developed at the 1992 International Conference on Water and the Environment, for the first time, focused on the economic values of water (see also Fischer, Chapter 5 in this volume). While many parallel water governance and management systems coexist (see Pahl-Wostl, 2015), there is still a tendency for global efforts to focus on the economic values that can undermine water's embeddedness in people's lives. Growing water scarcity may well further squeeze water's symbolic, cultural and social contexts, within which people live their lives (see Mehta, 2005).

More than most resources, water is highly variable across time and space. Its state and availability depend on temperature, rainfall, soil moisture and overall ecosystem health, as well as on human-built infrastructure such as wells and irrigation canals. But water availability does not necessarily translate to water access, which is mediated through institutions, gender, social and power relations, property rights, identity and culture. Water has symbolic as well as material dimensions and multiple ontologies (ie ways of being and existing). It is subjected to contests rooted in relations of power in both the discursive and the material realms (Mehta, 2005; Wilson and Inkster, 2018). Because of the fluid nature of water and its linkages with land, water rights are usually competing and overlapping, and entail a mixture of formal and informal arrangements (van Koppen and Schreiner, 2014). Customary law and practices, kinship networks, gender, caste and patronage tend to dominate in practice despite the existence of formal institutional arrangements (van Koppen and Schreiner, 2014).

Dominant modes of water management have conventionally been characterized by sectoral approaches that separate water and sanitation from water for food, energy, domestic supply, irrigation and floodwater management (Mehta et al, 2020). The International Decade on Water and Sanitation (the Water Decade) was launched in 1981. It was mainly a supply-led and government-focused initiative, in line with the thinking at the time that governments should be in the driving seat (see Nicol et al, 2012). The primary focus was on expanding and universalizing the coverage of drinking water supply and sanitation. In the 1990s supply-oriented paradigms gave way to demand-led approaches in the water sector. In parallel, there was increased emphasis on scaling back public sector investment, slimming down public services and deregulation and liberalization in line with World Bank and International Monetary Fund-led structural adjustment programmes. Consequently, the early 1990s saw substantial attention and output by World Bank economists on the necessity of treating water as an economic good and of using economic incentives to increase so-called water use efficiency (see World Bank, 1994; Briscoe et al, 1998). This period also saw the promotion of a holistic approach to water resources management, the birth of the Dublin Principles and the paradigm of integrated water resources management (Pahl-Wostl, 2015).

The four Dublin Principles, which have, over time, been considered integral to IWRM, recognize: (1) the finite nature of water and its key role in sustaining life, development and the environment (this principle has often been translated into a principle of managing water according to its resource-based boundary, ie the river basin); (2) the importance of participatory approaches in water development and management; (3) the central role played by women in the provision, management and safeguarding of water, thus making direct linkages to SDG 5 (Gender equality) and the empowerment of all women and girls (see Derman and Prabhakaran, 2016); and (4) the economic and competing values of water and the need to recognize water as an economic good. Since Dublin, IWRM has gradually emerged as the sanctioned discourse on water resources management in both the global water domain and national water policies and legislations (see Mehta et al, 2016 for a special issue on the politics of IWRM in southern Africa).

6.2 Unpacking SDG 6

The UN General Assembly adopted the SDGs in September 2015. A wide process of consultation and engagement had been implemented to develop the SDGs, which are successors to the Millennium Development Goals adopted in 2000, but with a crucial additional focus on sustainability. Even though the MDG period (2000–2015) coincided with the United Nations Human Rights Council elaborating on the General Comment on Right to Water (2002), and recognizing the right to water and sanitation (2010), resulting in the endorsement of the human right to water and sanitation by the United Nations (2010), the MDGs failed to directly address issues of equity and discrimination in water access (especially in challenging but rapidly growing areas such as peri-urban areas and slums). While SDG 6 on clean water and sanitation fails to take a rights-based approach, it is still a significant improvement on the MDGs. It seeks to achieve, by 2030, universal and equitable access to safe and affordable drinking water for all, and access to adequate and equitable sanitation and hygiene for all, and to end open defecation, paying special attention to the needs of women and girls and those in vulnerable situations.[2] For the SDGs, addressing inequality, which is also reflected in SDG 10, is key. In the case of SDG this is done by focusing on universality (ie achieving universal access to water and sanitation), monitoring and eliminating inequalities by improving service levels, and going beyond the household to focus on access in schools, health centres, and so on (see Cumming and Slaymaker, 2018). In addition, water quality concerns that were missing from the MDGs are considered, including a commitment to reduce the number of people suffering from water scarcity and to support and strengthen the participation of local communities in improving water and sanitation management. These trends are to be welcomed, alongside a commitment to improving data collection, monitoring and use at the country level. Given that water is used for various purposes, institutions and governance systems are needed to ensure coherence across these uses and to support access by disadvantaged users.

The greater breadth of SDG 6, however, led to a larger number of indicators and hence problems with monitoring and tracking, and to the risk of an SDG industry in every country (Swain, 2018). As with the MDGs, there is also a lack of clear mechanisms of accountability, and the challenge to define locally – where it matters – the meaning of each goal, target and indicator. Generalized, globalized arguments that underpin policy debates tend to remain disconnected from the everyday experiences of local people. For example, while SDG 6 is far more nuanced than the MDG in stating what constitutes an 'improved' water source by creating a 'service ladder' from 'safely managed' to 'basic', 'unimproved' and surface water sources, there is a large gap, as Welle's (2013) research in Ethiopia has demonstrated, between how global agencies, national agencies and local people understand, define and measure water access and inequality (see also Cumming and Slaymaker, 2018).

Moreover, while there is greater awareness of gender inequalities, sex-disaggregated data are seldom collected, making it impossible to monitor progress or to devise gender sensitive policies. It is, furthermore, alarming that a source that can be accessed within 30 minutes (round-trip collection time) is considered an 'improved water source' in accordance with the SDG indicators tracking progress on drinking water, sanitation and hygiene. No doubt, poor women in many parts of the Global South may need to make multiple trips of 30 minutes or more with heavy water pots. That such a situation is considered acceptable reflects how critical it is to ensure that more women are part of the decision-making processes when such standards and targets are being set. Finally, as argued by Mitra and Rao (2019), in the enthusiasm to meet the global targets, governments tend to ignore the interconnectedness between water services and food production, and their wider embeddedness in ecology (ie land use and water flows), gendered labour patterns and how these are determined by power relations.

6.3 The centrality of SDG 6 for multiple human well-being outcomes and ecosystem health

Water is not only essential to achieving drinking water and sanitation goals but is also integral to food security and nutrition (HLPE, 2015). As illustrated in Figure 6.1, the key dimensions of water that are of importance for human survival and well-being are its availability, access, stability and quality (see also Ringler et al, 2021). These have close linkages with key dimensions of food systems that include the production, processing, distribution, preparation and consumption of food within a wider socioeconomic, political and environmental context.

According to UN World Water Development Report (WWAP 2014), 70 per cent of human water withdrawals are used for agriculture, 20 per cent for industrial uses and 10 per cent for domestic uses. However, these large withdrawals for agriculture only support 40 per cent of agricultural crops, while the remainder rely entirely on rain or 'green' water (Ringler, 2017). Despite rapid increases in non-irrigation uses of water, water for food production remains the largest user of human freshwater withdrawals, making farmers, in many ways, the main stewards of the world's water resources. Thus, without water there is no food security, which makes it important to join up action and efforts on water with those on food. Similarly, without better food systems, including the incorporation of environmental concerns into both crop production practices and diets, there will be no water security, given the large water use for food production and the considerable pollution of water bodies from agricultural pollution (fertilizers, pesticides, food processing and waste disposal).

As argued by Ringler et al (2021), for hunger and malnutrition to end, SDG 2 and SDG 6 need to be achieved together. This is because access to

Figure 6.1: The linkages between water and food systems

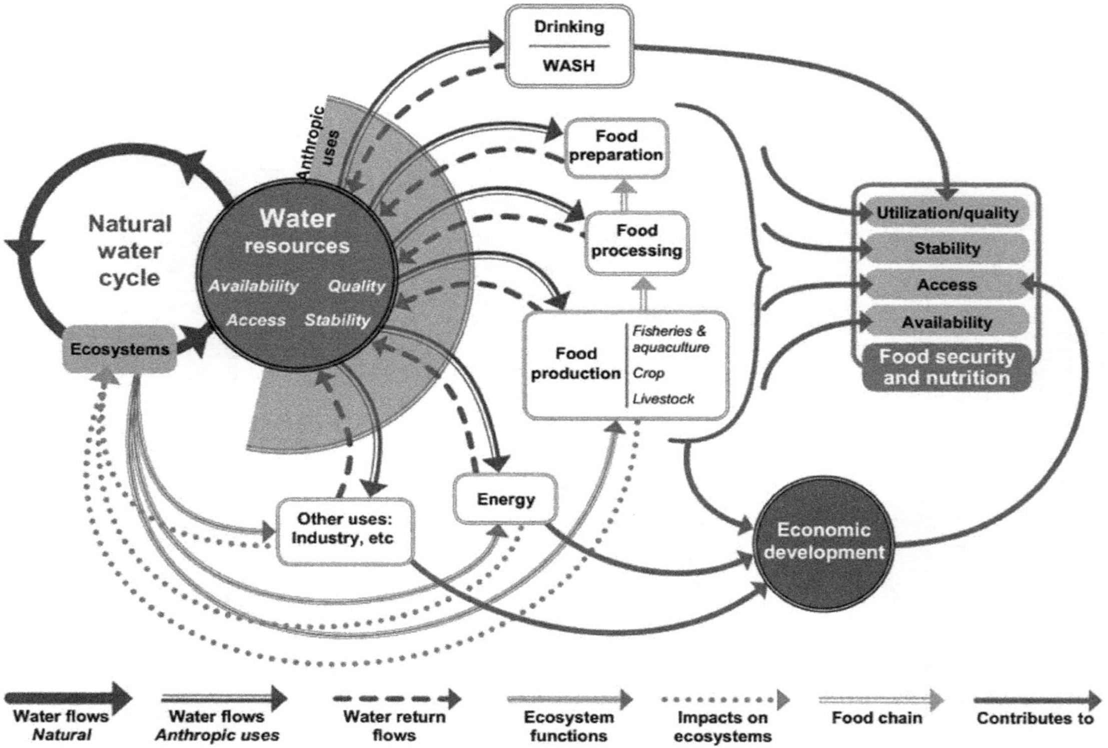

Source: HLPE (2015). Reproduced with permission.

safe drinking water (SDG 6.1) and equitable sanitation and hygiene (SDG 6.2) are necessary to ensure good nutrition and bodily health and well-being. Furthermore, good-quality water ensures the productivity and sustainability of agricultural systems (including irrigation systems), landscapes and other ecosystem services. Finally, access to water and food are also linked to access to land and land rights. Inequality in ownership and control over land (SDG 1.4) can seriously undermine overall food and water security, especially for women, smallholders and landless people.

Thus, SDG 2 (Zero hunger) and SDG 6 are closely linked and, as argued by Ringler et al (2021), can be achieved only if the water and food communities work closely together, something that is sadly lacking, especially at the national and global levels.

6.4 Competing demands on water and the politics of scarcity

Inequality within and between countries, communities and households means that many people continue to have inadequate access to water; this is also directly linked to unacceptably high and growing rates of undernutrition (FAO et al, 2021) and lack of access to clean drinking water and sanitation, which also affects food and nutrition outcomes.

While accessible water resources are adequate at global levels to meet the water needs of the world (HLPE, 2015), these resources are unevenly distributed across the globe, with per capita resources particularly low in the Middle East, North African and southern Asia regions (see Mehta et al, 2020). There is also a lot of variation in water availability within regions and countries. Moreover, in parts of the globe, historical rainfall patterns are changing, adding significant uncertainty to the reliable availability of water in many regions in the future (IPCC, 2021).

While future water demand estimates vary, there is agreement that domestic, municipal and industrial demands are growing faster than irrigation demands and that rapid urbanization will increase pressure on water and food. These various trends highlight the dilemma of competing demands on a very limited natural resource that is crucial to all life and particularly to the food security and nutrition of all humanity. There are competing pathways and discourses regarding water security and food security. According to the European Commission (2012), pressures on water availability will continue to grow – not only through the need to feed and hydrate a growing global population, but also through changes in consumption patterns. In the context of the Organisation for Economic Co-operation and Development's (OECD) 2050 projections, global water demand is projected to increase by 55 per cent as a consequence of increases in manufacturing, electricity and domestic use, leaving little scope for increasing water use for irrigation (OECD, 2012).

This supply-side vision is based in part on neo–Malthusian visions of scarcity and crises. The authors follow the UNDP (2006) in rejecting this 'gloomy arithmetic' vision and acknowledge the massive water injustices that poor women and men around the world encounter daily in accessing water for their survival, including food security.

Similarly, as argued by Amartya Sen, the fixation with per capita food availability decline (FAD) is a misleading way to approach hunger and famine, since hunger is more about people not having access to food owing to wider social and political arrangements rather than to there not being enough food to eat (Sen, 1981; 1983). The per capita availability of a resource lacks relevant discrimination and is even more questionable when it is applied to the population of the world as a whole (Sen, 1981). Water scarcity is also often misleadingly perceived as per capita water availability rather than as inequality in access to water. Instead, as argued earlier, water access is determined by social and political institutions, cultural and gender norms, and property rights. Some groups may suffer from lack of water even when there is no decline in water availability in the region. Thus, water shortages such as in famines are best understood as entitlement failures requiring effective and democratic governance solutions that can be accepted as legitimate by all (see also Anand, 2007; Mehta, 2014).

It is also important to look at unsustainable consumption and production practices that are fuelling land and water crises. For example, rising meat production and consumption through the increased use of food crops (and underlying water) as animal feed, and the increased use of biofuels as transportation fuel, are placing growing pressures on land and water resources. Growing middle classes in both high- and low-income countries are switching to water-intensive diets. But these diets are both environmentally unsustainable and often not nutritious. Changing diets containing excessive levels of meat, sugar and a handful of refined grains to more diverse and sustainable diets could improve both nutritional and health outcomes and put less pressure on the land and water resources of marginal resource users (Mehta et al, 2020).

This requires a critical approach to water and food that is concerned with social justice, human rights and the politics of framing. It is also important to do away with silo-driven discourses (ie between water, food and land, and between water supply and water for food production) that are highly problematic from the perspective of local users, for whom there is little sense in separating out these dimensions, which are crucial for survival.

All this makes questions of governance and decision making with regard to water an urgent imperative, both within countries and at the regional and transboundary levels. While it is often observed that 'water flows uphill to money and power', it is also clear that water is a resource that ignores national boundaries, thus complicating the challenge of water governance even further (Zhang et al, 2020).

6.5 Access to water as a sociopolitical construct

The persistence of water inequalities globally can be attributed in part to various power imbalances that prevent universal access. Mehta (2014) has argued that invisible power (ie how powerless groups lack awareness of their rights and interests, and internalize dominant values and norms) allows structural violence (ie prevailing political, social and cultural arrangements) to persist in the water domain that disadvantage and cause harm to marginalized social groups. This structural violence, in particular, disadvantages powerless groups such as migrants, poor women, ethnic minorities, Indigenous peoples and lower castes.

Gender and other markers of identities largely determine water allocation and access among users. Cultural norms in much of the Global South dictate that women and girls are responsible for water collection, and they may spend several hours a day collecting water. Unequal power relations within the household, and women's minimal control over household finances or spending, can force them into a daily trudge (taking precious time) to fetch cheaper or free untreated water, which may result in health problems and increased poverty and destitution. This time could instead be used to focus on livelihood and agricultural activities, to attend school and to improve maternal and infant health. This situation is worsened by women rarely playing key roles in decision-making processes regarding water from the household to the community levels and to more formal realms. This means that their interests and needs are often ignored and remain unarticulated, and this has implications for the realization of SDG 6.

According to the 2006 Human Development Report, which focuses in depth on water scarcity from a human development perspective, the global water crisis is overwhelmingly a crisis of the poor. The distribution of water access in many countries mirrors the distribution of wealth, and vast inequalities exist in both. The UNDP (2006) reports that those who lack access to clean water and adequate sanitation tend to live on less than $2 a day. Furthermore, not only do the poorest people get less access to water, and even less to clean and safe water, but they also pay some of the world's highest prices for water (see below).

Elite biases, democratic deficits (and distortions) and market-based mechanisms compound the structural violence that lead to such groups largely bearing the brunt of water-related injustices.

6.6 The role of the private sector and growing corporate involvement in water management

Over recent decades there has been significant and heated discussion on the role of the private sector, particularly in relation to the provision of drinking water. In the early 1990s the solution to the failure in the universal delivery of water services by the public sector was to increase the role of private water providers (World Bank, 1994). A number of Europe-based transnational water corporations played a significant role in the attempted privatization of water services provision and management, with variable results. The promotion of private sector involvement often accompanied structural adjustment programmes imposed on debt-ridden countries by the Bretton Woods institutions (Varghese, 2013). Privatization proponents have emphasized poor people's willingness to pay for water (Altaf et al, 1992) and, relatively speaking, poor people pay far more than the rich for water. The UNDP's Human Development Report for 2006 notes that:

> the poorest 20% of households in Argentina, El Salvador, Jamaica and Nicaragua allocate more than 10% of their spending to water. In Uganda water payments represent as much as 22% of the average income of urban households in the poorest 20% of the income distribution. (UNDP, 2006: 51)

With failures in regulation, in some privatization cases prices rose beyond contractually agreed levels, resulting in cut-offs for those unable to pay. Popular resistance to water privatization has been widespread. Several transnational water corporation contracts have failed, leading to a retreat of the private sector in some areas, with a re-municipalization of water services (Pigeon et al, 2012; Lobina et al, 2014), primarily in the Global North but also in the Global South (Lobina et al, 2014), or to public–public partnerships.

While there is a role for the private sector, in many countries of the Global South there is insufficient regulatory capacity to ensure that private sector management and provision of water is pro-poor without compromising on basic rights to water and food. Effective regulation is required to control the drive of the private sector to make profit out of what is a human right and a social good, to counteract the monopolistic nature of the water provision sector and to ensure that the private sector provides adequate services in poor urban and rural areas (see Bayliss, 2014). These debates echo the controversies around the 1992 Dublin Declaration, which tended to highlight water as an economic good over and above its cultural and symbolic characteristics. In many countries of the Global South, there is a lack of effective legal and institutional frameworks to protect the rights and interests of poor and marginalized communities, particularly in rural areas where they often access water under unprotected customary practices. In such cases, the introduction of the private sector in the management of water may reduce local control over water sources and undermine the access of local communities to realizing the SDGs and the human right to water (Cullet, 2014), and to sufficient water to meet their own food needs.

6.6.1 Corporatization of land and water resources

The food price crisis of 2007–8 contributed to increased investments by some companies and countries in land resources and associated water resources for the production of biofuels and in some cases food (see Franco and Borras, 2009; Cotula, 2012). This phenomenon was popularly known as 'land grabbing'. Many of the planned investments did not go forward or were later cancelled as the complexity of large-scale land transactions had not been clear to all involved parties. However, all of these negotiations over land also included a water dimension as the planned agricultural production processes required the concomitant development of water resources for production (Mehta et al, 2012; Franco et al, 2013). As argued by several authors (see Mehta et al, 2012), 'water grabbing' associated with land investments has led to a significant reappropriation of water resources by elites, affecting water tenure relations, with implications for basic human rights and local water and food security. Land that should have been used by local communities to grow food to ensure their own nutritional and food security was instead diverted to global land investments and commercial agriculture.

Since 2011, global corporations have spent more than $84 billion on how they manage, conserve or obtain water (Clark, 2014). The reasons range from having to deal with physical water shortages and the need to appear concerned about water scarcity and water crises. While some argue that the growing corporate involvement in water management is negligible and is also to be welcomed because it will lead to new technological innovation (Clark, 2014), others argue that this engagement has risks and implications for current and future water and food security (see Sojamo and Larson, 2012). These include the potential reallocation of water to the 'highest economic value', with potential detrimental impacts on local lives, livelihoods and water and food security (see for example Franco et al, 2013). Importantly, risks are often unequally shared between companies and other local water uses, and new water stresses may be created. Furthermore, as with the private sector described earlier, companies are often more legally bound to be accountable to distant shareholders than to local stakeholders, who are often voiceless and powerless. Their structural and bargaining power and influence over global and national policies and processes allow them to shape and frame powerful discourses, subjecting water governance institutions to processes of capture (see Sojamo and Larson, 2012). Because of these inconsistencies and discrepancies, an interlinked human rights approach to water and water for food is required. This will help both to achieve the SDGs and to integrate the norms, standards and principles of the international human rights system into the plans, policies and developmental processes related to resource security at the international, national and subnational levels. The norms and principles include accountability, transparency, empowerment, participation, non-discrimination (equality and equity) and attention to vulnerable groups (see HLPE, 2015).

6.7 The right to water and synergies with the right to food

Water is integral to human food security and nutrition, and safe water is fundamental to the nutrition, health and dignity of all (see UNDP, 2006; HLPE, 2015). This notwithstanding, the right to water (RTW) was not explicitly acknowledged in the 1948 Universal Declaration of Human Rights. Until the turn of the 21st century, there remained considerable resistance to the RTW on the part of some nations and corporations (see Sultana and Loftus, 2 011; Mehta, 2014). The human right to safe drinking water and sanitation was recognized by the UN General Assembly in 2010 (United Nations, 2010),[3] following a protracted struggle.

It is telling that 41 nations, including Australia, Canada and the US, abstained from recognizing water as a human right in 2010 (and these three countries have abstained up to the present). The long road in explicitly recognizing water as a human right has been attributed to a lack of political will and resources in this area compared to investment in other sectors (UNDP, 2006).

> Since the poor – who suffer the most from a lack of access to improved water and sanitation services – tend to have a limited voice in political arenas, as is often argued, their claims for these services can be more easily ignored if the human RTW and sanitation is not explicit. (Hall et al, 2014: 852)

Through the establishment of water as a human right, states have been obliged 'as duty bearers to ensure that every citizen has affordable access to water infrastructure services for drinking, personal and other domestic uses and sanitation' (van Koppen et al, 2017: 130). The human rights emphasis also highlights the weaknesses of mainstream notions of efficiency that promote profit maximization and disregard associated social costs (Lobina, 2017), as noted earlier in the discussion on privatization.

Despite the long overdue global recognition, the RTW remains conceptually ambiguous. For example, there have been heated debates about whether or not the RTW is compatible with parallel global trends of water commodification and privatization (see Sultana and Loftus, 2011). It is also still unclear what constitutes the RTW, that is, in terms of the actual amount but also whether its narrow scope should be expanded to also look at wider livelihood and survival needs beyond domestic issues.

The water domain has been traditionally divided into two sectors: water supply/services and water resources management or, as the 2006 Human

Development Report puts it, 'water for life' and 'water for production' (UNDP, 2006). Water for life refers to water for drinking and domestic purposes and is considered key to human survival. Water for production refers to water in irrigation, industry and small-scale entrepreneurial activities, as well as water used in producing food for subsistence. This distinction, however, is highly problematic from the perspective of local users whose daily activities encompass both domestic and productive elements of water and for whom there is little sense in separating water for drinking and washing from water for small-scale productive activities that are also crucial for survival. Empirical research demonstrates the critical role played by water for productive purposes and livelihoods, especially for poor women (van Houweling et al, 2012; van Koppen et al, 2017). Hall et al (2014) argue that water plays a key role in livelihood activities both in rural areas, for example crop irrigation, livestock watering or brick making, and in peri-urban areas. Furthermore, the distinction also results in a narrow scope for the RTW, especially compared to the right to food (RTF) (see HLPE, 2015).

This chapter builds on Franco et al (2013), Hellum et al (2015), HLPE (2015), van Koppen et al (2017) and Mehta et al (2020), who call for an elaboration of a human rights perspective to water and food that encompasses the productive uses of water while being more interconnected than the current RTW. This broader conceptualization of the RTW is truer to how water is understood and embedded in the daily lives of local women and men around the world. Local communities rarely distinguish water for domestic and subsistence purposes. It is thus important for the RTW to go beyond the current domestic focus to embrace a more holistic definition of well-being and human survival.

Mehta (2014) and Anand (2007) draw on Amartya Sen's capabilities approach to promote a holistic view regarding the RTW and its links to wider survival issues and livelihoods to highlight that the one cannot be guaranteed without the other. Capabilities refer to the 'actual living that people manage to achieve' (Sen, 1999: 730). This approach focuses on 'substantive freedoms – the capabilities – to choose a life one has reason to value' (Sen, 1981; 1983; 1999: 74) and the freedoms that an individual can enjoy. In this capabilities' approach the focus is not on the quantity of the bundles of entitlements but instead on the principle of equality and capability to do and to be. Such an approach translated to water would mean that a basic amount of water is required for basic human functioning (drinking, washing and being free of disease), and it has therefore been argued that this minimum requirement for human functioning should also capture livelihood and subsistence purposes (see Mehta, 2014). This strongly resonates with work by Jepson et al (2019) who argue that water and access to it should be understood as a hydro-social and cultural process to include water flows, water quality and water services to capture socionatural dynamics.

The capabilities' approach refrains from outlining what exactly this minimum threshold should be, as 'conventional, established measures and metrics do not fully reflect the unique hydrosocial conditions or historical marginalization that produce water insecurity' (Wutich et al, 2017: 7). Evidence from the water sector in setting up standards around what constitutes a 'basic water requirement' highlights the variations by country and by institution. Basic water requirements have been suggested by various donor agencies, ranging from 20 to 50 litres a day, regardless of culture, climate or technology. Nevertheless, culture, climate, livelihoods, and urban and rural contexts clearly do matter. The WHO definition – again seemingly blind to context – prescribes between 20 and 100 litres a day (WHO, 2003), but this amount excludes water for productive or survival activities such as growing food (Mehta, 2014). Clearly, in not considering livelihood and subsistence needs, low-end provision takes a very narrow view of the water needs of the poor. Rarely is there any discussion of the maximum amount of water people can consume per day; for example, in the US it is about 300 litres without productive activities (EPA, 2022).

In terms of capability, people ultimately need different basic amounts of water to enjoy the same standard. Take the case of South Africa, one of the first countries to explicitly recognize the RTW. Its Free Basic Water policy provided a minimum of 25 litres per capita per day based on a household size of eight people, initially free to all citizens (see McDonald and Ruiters, 2005; Mehta, 2006). But implementing the RTW in South Africa has been fraught with difficulties, and there are huge debates as to whether the right has had a significant impact on improving the well-being of poor South African citizens (McDonald and Ruiters, 2005; Mehta, 2006; see also Flynn and Chirwa, 2005). There are further heated debates about whether the right to water is compatible with parallel trends of water privatization or rather runs contradictory to citizens' basic right to water while also creating new forms of poverty and ill-being (see Flynn and Chirwa, 2005; Loftus, 2005; McDonald and Ruiters, 2005). Last but not least, it has also led to the aforementioned debates concerning the sufficiency of 25 litres per day per person, especially if the household number is large. All these issues further emphasize the need for an expansion of the narrow scope of the RTW and for the need to link it to the RTF.

The RTW as recognized by the UN General Assembly in 2010 largely focuses on drinking water and sanitation services, and has not been deployed to look at the productive use of water, despite earlier broader interpretations in the General Comment No. 15, reinforced in later reports such as UNHRC (Hall et al, 2014). The RTW and the RTF have close ties because water and sanitation are crucial for health and nutrition, and because access to water is indispensable for food producers and the RTF of producers.

On 27 November 2002, the UNCESCR adopted the General Comment No. 15 on the RTW (UN CESCR, 2002). The committee defined water as a social and cultural good and not solely as an economic commodity, and stressed the state's legal responsibility to fulfil the right. Interestingly the CESCR's General Comment No. 15 (GC 15) on the RTW identified other aspects of the RTW that have remained underexplored and underdeveloped. GC 15 recognized that:

> Water is required for a range of different purposes, besides personal and domestic uses, to realize many of the Covenant rights. For instance, water is necessary to produce food (right to adequate food) and ensure environmental hygiene (right to health). Water is essential for securing livelihoods (right to gain a living by work) and enjoying certain cultural practices (right to take part in cultural life). (UN CESR, 2002: para 6)

Despite this broader framing, the UN General Assembly and the UN Human Rights Commission decided on a rather narrow focus on safe drinking water, personal and other domestic uses, and sanitation in their recognition of the human right to water. This represents a political prioritization that does not pay adequate attention to other uses of water, for example by subsistence farmers.

By contrast, the RTF was recognized as part of the right to an adequate standard of living in the 1948 Universal Declaration of Human Rights, and has been part of the International Covenant on Economic, Social and Cultural Rights (ICESCR) from 1976. It has a far broader framing than the RTW. In 2014 the UN special rapporteur on the right to food wrote about the 'transformative potential of the Right to Food', defining it as the right of every individual, 'alone or in community with others, to have physical and economic access at all times to sufficient, adequate and culturally acceptable food, that is produced and consumed sustainably, preserving access to food for future generations' (UNGA, 2014: 4). The mandate of the special rapporteur on the right to food has historically been broader, and many special rapporteurs on food have played an active role in tracking violations. Such tracking has served as a powerful tool to counteract food-related injustices, allowing the special rapporteur to respond to allegations with respect to violations and also enabling them to write to relevant governments to ask them to take action to ensure redress and accountability. Initially at least the special rapporteur on the right to water lacked this explicit mandate. Catarina de Albuquerque was the first UN special rapporteur on the right to safe drinking water and sanitation, and had initially been appointed as an independent expert in 2008 before the right to water had been recognized by the UN General Assembly. In the early years, she focused mainly on best practices and took a very narrow view of the RTW, perhaps because of its controversial nature and the initial resistance to its existence on the part of many powerful (corporate) players. This has changed over time and her successors have been more vocal and critical about water privatization and other wider issues beyond the domestic scope of the right to water. Until very recently, the RTW had not been deployed to focus explicitly on water management issues or the water implications of so-called land and water grabs because of the limiting of its scope to domestic uses of water. This had been in sharp contrast to different special rapporteurs on the right to food who have frequently commented on land acquisitions and grabs and their impacts on local people's food security (see Franco et al, 2013).

There is thus much scope to strengthen the interpretation and understanding of different aspects of the RTW, and of its interlinkages with the RTF. A positive step in this direction was taken in 2018 when the UN General Assembly passed the resolution on the rights of Indigenous peasants and other people working in rural areas (United Nations Human Rights Council, 2018). It recognized the rights 'of peasants and other people working in rural areas' to water for personal and domestic use, farming, fishing and livestock keeping and to securing other water-related livelihoods, ensuring the conservation, restoration and sustainable use of water.

6.8 Conclusion

This chapter has focused on SDG 6 on water and the challenges of achieving it, such as poor governance and a lack of gender-disaggregated data collection. It has also made a case for linking SDG 6 with SDG 2 since both are necessary to ensure food security and human well-being, survival as well as ecosystem integrity. These linkages are also key to supporting the integration of sustainability and human development goals, as envisioned in Agenda 21. Finally, the chapter has also made a case for linking the realization of the SDGs to human rights and for expanding the RTW beyond its narrow domestic focus to embrace wider productive and livelihood issues. This broader conceptualization of the RTW is more true to how water is understood and embedded in the daily lives of local women and men around the world, and will also help realize both SDGs 6 and 2 as well as improve gender equity.

All this calls for joining up governance and human rights processes around water, food and land. Water, food and land governance regimes tend to be highly disconnected, often doubly disadvantaging marginal land and water users. While approaches such as integrated water resources management are intended to break down existing silos and physical boundaries, they are not centred around the human rights of the communities concerned, are often executed in a top-down manner and are difficult to implement, leaving the poor still marginalized. In addition, as discussed, large-scale land acquisitions that have been taking place in recent years have often tended to exclude local populations from their lands and water resources, and have increased local-level conflicts. It is thus important to look at the culpability of large-scale users and owners of land and water and to hold them accountable for the growing water and food insecurities of poor, vulnerable and marginalized people.

The lack of integration in major global and national initiatives around water, land and food governance is also true of the SDGs as well as of the human rights to water and food. This chapter has thus called for their integration to support the achievement of both sustainable and equitable development – a need that is greater today than ever before. Breaking down the silos between SDG 6 and SDG 2 requires better policies, institutions and investments that consider the joint achievement of their targets with a much stronger focus on providing access to marginalized populations.

Notes

[1] This chapter builds on arguments articulated in our coauthored book (Mehta et al, 2020).

[2] By contrast, the sanitation target was not part of the MDG on water, but was added two years after it was launched, in 2002, at the World Summit on Sustainable Development (WSSD), in response to sustained campaigning by CSOs.

[3] The RTW and to sanitation were jointly recognized by the 2010 UN General Assembly. They are two different rights in effect, following the position of the Special Rapporteur on the Human Right to Safe Drinking Water and Sanitation (HRC, 2014: 27). In this chapter, we focus largely on the human RTW and not on sanitation issues.

References

Altaf, A., Jamal, H. and Whittington, D. (1992) *Willingness to Pay for Water in Rural Punjab, Pakistan*, Washington, DC: UNDP-World Bank Water and Sanitation Program.

Anand, P.B. (2007) *Scarcity, Entitlements and the Economics of Water in Developing Countries*, Cheltenham: Edward Elgar.

Bates, B.C., Kundzewicz, Z.W., Wu, S. and Palutikof, J.P. (eds) (2008) *Climate Change and Water*, Intergovernmental Panel on Climate Change Technical Paper VI, Geneva: IPCC Secretariat, Available from: https://www.ipcc.ch/publication/climate-change-and-water-2 [Accessed 20 June 2023].

Bayliss, K. (2014) 'The financialization of water', *Review of Radical Political Economics*, 46(3): 292–307.

Briscoe, J., Anguita Salas, P. and Peña T.H. (1998) *Managing Water as an Economic Resource: Reflections on the Chilean Experience*, Environment Department Working Paper No. 62, Environmental Economic Series, Washington, DC: World Bank.

Clark, P. (2014) 'A world without water', *Financial Times*, 14 July, Available from: http://ig-legacy.ft.com/content/8e42bdc8-0838-11e4-9afc-00144feab7de#slide0 [Accessed 20 August 2022].

Cotula, L. (2012) 'The international political economy of the global land rush: a critical appraisal of trends, scale, geography and drivers', *Journal of Peasant Studies*, 39(3–4): 649–80.

Cullet, P. (2014) 'Groundwater law in India: towards a framework ensuring equitable access and aquifer protection', *Journal of Environmental Law*, 26(1): 55–81.

Cumming, O. and Slaymaker, T. (eds) (2018) *Equality in Water and Sanitation Services*, London: Routledge.

Derman, B. and Prabhakaran, P. (2016) 'Reflections on the formulation and implementation of IWRM in Southern Africa from a gender perspective', *Water Alternatives*, 9: 644–61.

EPA (United States Environmental Protection Agency) (2022) 'Statistics and facts', Available from: https://www.epa.gov/watersense/statistics-and-facts [Accessed 22 August 2022].

European Commission (2012) *European Report on Development 2011/ 2012: Confronting Scarcity: Managing Water, Energy and Land for Inclusive and Sustainable Growth*, Available from: https://ecdpm.org/publications/ european-report-development-2011-2012-confronting-scarcity-water-ene rgy-land-inclusive-and-sustainable-growth [Accessed 20 August 2022].

FAO (Food and Agriculture Organization), IFAD (International Fund for Agricultural Development), UNICEF, WFP (World Food Programme) and WHO (World Health Organization) (2021) *The State of Food Security and Nutrition in the World. 2021: Transforming Food Systems for Food Security, Improved Nutrition and Affordable Healthy Diets for All*, Rome: FAO.

Flynn, S. and Chirwa, D.M. (2005) 'The constitutional implications of commercialising water in South Africa', in D. McDonald and G. Ruiters (eds) *The Age of Commodity: Water Privatization in Southern Africa*, London: Earthscan, pp 59–77.

Franco, J.C. and Borras, S.M. Jr (2009) 'Paradigm shift: the 'September thesis' and rebirth of the 'open' peasant mass movement in the era of neoliberal globalization in the Philippines' in D. Caouette and S. Turner (eds) *Agrarian Angst and Rural Resistance in Contemporary Southeast Asia*, London: Routledge, pp 226–46.

Franco, J., Mehta, L. and Veldwisch, G.J. (2013) 'The global politics of water grabbing', *Third World Quarterly*, 34(9): 1651–75.

Hall, R., van Koppen, B. and van Houweling, E. (2014) 'The human right to water: the importance of domestic and productive water rights', *Science Engineering Ethics*, 20(4): 849–66.

Hellum, A., Kameri-Mbote, P. and van Koppen, B. (2015) The human right to water and sanitation in a legal pluralist landscape: perspectives of southern and eastern African women', in A. Hellum, P. Kameri-Mbote and B. van Koppen (eds) *Water is Life: Women's Human Rights in National and Local Water Governance in Southern and Eastern Africa*, Harare, Zimbabwe: Weaver Press, pp 1–31.

HLPE (High Level Panel of Experts on Food Security and Nutrition) (2015) *Water for Food Security and Nutrition*, Report by the High Level Panel of Experts on Food Security and Nutrition of the Committee on World Food Security, Report No. 9, Rome: FAO.

HRC (Human Rights Council) (2014) 'Report of the Special Rapporteur on the human right to safe drinking water and sanitation, Catarina de Albuquerque. Common violations of the human rights to water and sanitation', Available from: https://digitallibrary.un.org/record/731356 [Accessed 22 June 2023].

IPCC (Intergovernmental Panel on Climate Change) (2021) *Climate Change 2021: The Physical Science Basis*, Contribution of Working Group I to the Sixth Assessment Report of the Intergovernmental Panel on Climate Change, Available from: https://www.ipcc.ch/report/ar6/wg1 [Accessed 20 June 2023].

Jepson, W., Wutich, A. and Harris, L.M. (2019) 'Water-security capabilities and the human right to water', in F. Sultana and A. Loftus (eds) *Water Politics*, London: Routledge, pp 84–98.

Lobina, E. (2017) 'Water remunicipalisation: between pendulum swings and paradigm advocacy', in S. Bell, A. Allen, P. Hofmann and T.H. Teh (eds), *Urban Water Trajectories*, Cham: Springer, pp 149–61.

Lobina, E., Kishimoto, S. and Petitjean, O. (2014) *Here to Stay: Water Remunicipalisation as a Global Trend*, Report by PSIRU, Transnational Institute, and the Multinationals Observatory, November, Available from: https://www.tni.org/en/publication/here-to-stay-water-remunic ipalisation-as-a-global-trend [Accessed 20 June 2023]

Loftus, A. (2005) ' "Free water" as a commodity: the paradoxes of Durban's water service transformations', in D. McDonald and G. Ruiters (eds) *The Age of Commodity: Water Privatization in Southern Africa*, London: Earthscan, pp 189–203.

McDonald, D. and Ruiters, G. (eds) (2005) *The Age of Commodity: Water Privatization in Southern Africa*, London: Earthscan.

Mehta, L. (2005) *The Politics and Poetics of Water: Naturalising Scarcity in Western India*, New Delhi: Orient Longman.

Mehta, L. (2006) 'Do human rights make a difference to poor and vulnerable people? Accountability for the right to water in South Africa', in P. Newell and J. Wheeler (eds) *Rights, Resources and the Politics of Accountability*, London: Zed Books.

Mehta, L. (2014) 'Water and human development', *World Development*, 59: 59–69.

Mehta, L., Veldwisch, G.J. and Franco, J. (eds) (2012) *Water Grabbing? Focus on the (Re)Appropriation of Finite Water Resources*, special Issue, *Water Alternatives*, 5(2): 193–468.

Mehta, L., Movik, S., Bolding, A., Derman, A. and Manzungu, E. (2016) 'Introduction to the Special Issue', in *Flows and Practices: The Politics of Integrated Water Resources Management (IWRM) in Southern Africa*, special issue, *Water Alternatives*, 9(3): 389–411.

Mehta, L., Oweis, T., Ringler, C., Schreiner, B. and Varghese, S. (2020) *Water for Food Security, Nutrition and Social Justice*, London: Routledge.

Mitra, A. and Rao, N. (2019) 'Gender, water and nutrition in India: an intersectional analysis', *Water Alternatives*, 12(1): 169–91.

Nicol, A., Mehta, L. and Allouche, J. (2012) "Some for all rather than more for some'? Contested pathways and politics since the 1990 New Delhi Statement', *IDS Bulletin*, 43: 1–9.

Nilsson, M., Griggs, D., Visbeck, M., Ringler, C. and McCollum, D. (2017) 'Introduction: A framework for understanding Sustainable Development Goal interactions', in D.J. Griggs, M. Nilsson, A. Stevance and D. McCollum (eds) *A Guide to SDG Interactions: From Science to Implementation*, Paris: ICSU, pp 18–30.

OECD (Organisation for Economic Co-operation and Development) (2012) *OECD Environmental Outlook 2050: The Consequences of Inaction*, Paris: OECD.

Pahl-Wostl, C. (2015) *Water Governance in the Face of Global Change: From Undestanding to Transformation*, Berlin: Springer.

Pigeon, M., McDonald, D.A., Hoedeman, O. and Kishimoto, S. (eds) (2012) *Remunicipalisation: Putting Water Back into Public Hands*, Amsterdam: TNI.

Ringler, C. (2017) *Investment in Irrigation for Global Food Security*, IFPRI Policy Note, Washington, DC: IFPRI

Ringler, C., Agonlahor, M., Baye, K., Barron, J., Hafeez, M., Agbonlahor, M. et al (2021) *Water for Food Systems and Nutrition: Food Systems Summit Brief*, prepared by Research Partners of the Scientific Group for the Food Systems Summit, May, Available from: https://opendocs.ids.ac.uk/opend ocs/handle/20.500.12413/16604 [Accessed 20 June 2023].

Sen, A. (1981) *Poverty and Famines: An Essay on Entitlement and Deprivation*, Oxford: Clarendon Press.

Sen, A. (1983) 'Development: which way now?', *The Economic Journal*, 93(372): 745–62.

Sen, A. (1999) *Development as Freedom*, Oxford: Oxford University Press.

Sojamo, S. and Larson, E. (2012) 'Investigating food and agribusiness corporations as global water security, management and governance agents: the case of Nestlé, Bunge and Cargill', *Water Alternatives*, 5(3): 619–35.

Sultana, F. and Loftus, A. (eds) (2011) *The Right to Water: Politics, Governance and Social Struggles*, London: Routledge.

Swain, R. B. (2018) 'A critical analysis of the Sustainable Development Goals', in W. Leal Filho (ed) *Handbook of Sustainability Science and Research*, Cham: Springer, pp 341–55.

United Nations (2010) *Resolution adopted by the General Assembly on 28 July 2010 64/292. The human right to water and sanitation-* Available from: https://digitallibrary.un.org/nanna/record/687002/files/A_RES_64_292-EN.pdf?withWatermark=0&withMetadata=0&version=1&registerDownload=1 [Accessed 20 June 2023].

UN CESCR (United Nations Committee on Economic, Social and Cultural Rights) (2002) 'General Comment No. 15 (2002): The right to water (arts. 11 and 12 of the International Covenant on Economic, Social and Cultural Rights)', E/C.12/2002/11, Available from: https://digitallibrary.un.org/record/486454?ln=en [Accessed 20 June 2023].

UNDP (United Nations Development Programme) (2006) *Beyond Scarcity: Power, Poverty and the Global Crises*, Human Development Report 2006, New York: UNDP.

UNGA (United Nations General Assembly) (2014) 'Report of the Special Rapporteur on the right to food, Olivier De Schutter. Final report: The transformative potential of the right to food', A/HRC/25/57, Available from: https://digitallibrary.un.org/record/766914?ln=en [Accessed 20 June 2023].

UNICEF (United Nations Children's Fund) and WHO (2019) *Progress on Household Drinking Water, Sanitation and Hygiene 2000-2017. Special Focus on Inequalities*, New York: UNICEF and WHO.

United Nations Human Rights Council (2018) 'Draft United Nations declaration on the rights of peasants and other people working in rural areas', A/HRC/WG.15/5/3, Available from: https://www.ohchr.org/sites/default/files/Documents/HRBodies/HRCouncil/WGPleasants/Session5/A-HRC-WG.15-5-3.pdf [Accessed 20 June 2023].

UNSCN (United Nations Standing Committee on Nutrition) (2020) *Water and Nutrition. Harmonizing Actions for the United Nations Decade of Action on Nutrition and the United Nations Water Action Decade*, Rome: UNSCN.

UN WWAP (2014) *UN World Water Development Report 2014: Water and Energy*, Paris: UNESCO.

van Houweling, E., Hall, R., Diop, A.S., Davis, J. and Seiss, M. (2012) 'The role of productive water use in women's livelihoods: evidence from rural Senegal', *Water Alternatives,* 5(3): 658–77.

van Koppen, B. and Schreiner. B. (2014) 'Priority General Authorizations in rights-based water use authorization in South Africa', *Water Policy*, 16: 59–77.

van Koppen, B., Hellum, A., Mehta, L., Derman, B. and Schreiner, B. (2017) 'Rights-based freshwater governance for the twenty-first century: beyond an exclusionary focus on domestic water uses', in E. Karar (ed) *Freshwater Governance for the 21st Century*, Cham: Springer.

Varghese, S. (2013) *Water Governance in the 21st Century: Lessons from Water Trading in the U.S. and Australia*, Minneapolis, MN: Institute for Agriculture and Trade Policy, Available from: www.iatp.org/files/2013_03_27_WaterTrading_SV_0.pdf [Accessed 20 August 2022].

Welle, K. (2013) 'Monitoring performance or performing monitoring. The case of rural water access in Ethiopia', unpublished PhD thesis, Brighton: University of Sussex.

WHO (World Health Organization) (2003) *The Right to Water*, Health and Human Rights Publication Series 3, Geneva: WHO.

Wilson, N.J. and Inkster, J. (2018) 'Respecting water: Indigenous water governance, ontologies, and the politics of kinship on the ground', *Environment and Planning E: Nature and Space*, 1(4): 516–38.

World Bank (1994) *World Development Report 1994: Infrastructure for Development*, New York: Oxford University Press.

Wutich, A., Budds, J., Eichelberger, L., Geere, J., Harris, L.M., Horney, J.A., et al (2017) 'Advancing methods for research on household water insecurity: studying entitlements and capabilities, socio-cultural dynamics, and political processes, institutions and governance', *Water Security*, 2: 1–10.

WWAP (World Water Assessment Programme) (2014) *UN World Water Development Report 2014: Water and Energy*, Paris: UNESCO.

Zhang, W., El Didi, H., Swallow, K., Meinzen-Dick, R., Ringler, C., Masuda Y. and Aldous, A. (2020) *Community-Based Management of Freshwater Resources: A Practitioners' Guide to Applying TNC's Voice, Choice, and Action Framework*, Arlington, VA: The Nature Conservancy.

Interview with Lyla Mehta and Claudia Ringler: Push Them, Name Them, and Shame Them

Ruth Krötz and Maren Lorenzen-Fischer

Universal approaches to water governance address inequality very generally, but the area becomes much more nuanced when we consider local specifics. How can we bring these levels of discourse together?

Mehta: Universal approaches certainly help give validity to national and local struggles but they can also neglect local and regional specificities. However, the universal human rights framework is essential. One must work at different levels and be strategic about what to use and when. If it helps in your struggle, then universal rights can be evoked, but sometimes there's also push-back against them. You need to be flexible and iterative in drawing on the universal. The best way to do it is to bring these multiple perspectives together.

You adopt Amartya Sen's approach saying that hunger is about people not having access to food due to social and political arrangements as opposed to there not being enough food. Why does the narrative that there is a lack of food persist so tenaciously?

Ringler: The poor distribution of food is analogous to the poor distribution of money. There is enough money in the world. If it were distributed equally, everyone could buy food to sustain themselves. But, in reality, in 2021, 1 per cent of US citizens owned a record 32 per cent of US wealth. So why do extremely poor populations need to survive with less than a dollar a day? Because humans are not willing to share. The same applies to food: there would be enough food, *if* people would be willing to share, but

149

they don't and we cannot wait for human behaviour to change while watching people starving or dying from hunger. The food access problem is directly linked to income inequality. The first-best solution would be to distribute money to make richer people poorer and poorer people more well-off. The second-best solution (currently being pursued) is to produce so much food that it becomes cheap enough for poor people to afford, though it has raised questions of sustainability.

How do the concepts of water as an economic good and as a human right contrast with each other? What role does highlighting the voices of voiceless people play in this context?

Mehta: Water has multiple meanings and is everything; it is an economic resource, a human right, and a social and cultural good. However, the economic values and perspectives tend to dominate. At the turn of the century, there was a huge reluctance on the part of powerful players to recognize the human right to water. Now all big corporations, even Nestlé, claim they are supporting the human right to water, even though they make huge profits from this life-giving resource and also at times violate poor people's basic rights to water, land and food through some of their operations. In our work we have also been trying to bring together water, food and land issues. We have been exploring whether it makes sense to have a human right to water for food or a right to water that also captures livelihood perspectives. This is important because local people (especially in rural areas) do not separate these issues in their daily lives and also because rights to water, land and food are routinely violated through water and land grabs and extractivism. We thus need to integrate water security, food security and human rights since they're very siloed – we need to push for looking at these interlinked issues together. We have to repeat the same messages, even if we sound like a broken record, highlight the injustices, hold the powerful accountable, push them, name them and shame them.

How is the impact of market-based mechanisms linked to gender discrimination in the sector of food and water governance?

Ringler: Women are not the primary breadwinners but are responsible for care, sanitation and domestic water use. They have all these essential roles but they do not have the voice to ensure that decisions on water allocation guarantee domestic water security. At the same time, they are often excluded from formal and informal water and land markets. To change this requires their involvement in water management committees and water-user associations. Also, women have to put the food on the table – if women don't have a voice in crop production and livestock management, decisions regarding the use of household income for water and food security are suboptimal. Many approaches involving women are time consuming, and women have a lot of skills and resources, but time is not among them. The challenge is to involve them in a way to ensure that they can actually save time by affecting these decisions.

7

SDG 2 and the Dominance of Food Security in the Global Agri-food Norm Cluster

Sandra Schwindenhammer and Lena Partzsch

The world is facing severe levels of food insecurity and environmental degradation related to agri-food practices. World hunger is increasingly facilitated by the negative impacts of the COVID-19-pandemic and the consequences of the Russian invasion of Ukraine. Between 702 and 828 million people are affected by hunger (FAO et al, 2022: 10). Since 2015, the UN's Agenda 2030 for Sustainable Development has aimed to end world hunger and achieve food security, improved nutrition and sustainable agriculture through the SDG 2 (Zero hunger) (UN, 2015a). However, according to the High Level Panel of Experts on Food Security and Nutrition to the Committee on World Food Security (CFS), it is impossible to accomplish this goal by 2030 without radical governance transformation (HLPE, 2020).

Global agri-food governance typifies current sustainability challenges and is well suited to inform our knowledge on global norm stability and change. Even though SDG 2 as a key component of Agenda 2030 serves as the commonly accepted global normative reference for agri-food governance, different agri-food norms compete with each other, and ongoing norm debates hamper the fight against hunger (Breitmeier et al, 2021a; 2021b). Moreover, agri-food norms are linked to a range of environmental issues, and there are several synergies and trade-offs between SDG 2 and the 'green goals' (SDGs 6, 13–15) of Agenda 2030 (Griggs et al, 2017). While, on the one hand, the agri-food sector is a major driver of environmental pollution, on the other hand, agri-food production systems are prone to negative environmental changes with effects on global food security (Rockström et al, 2009).

Building on research on global norms and norm clusters in international relations (eg Lantis and Wunderlich, 2018), and on agri-food governance and norm development (Clapp and Scott, 2018; Breitmeier et al, 2021a; 2021b; Clapp et al, 2022), this chapter reflects on the normative foundations of SDG 2. After laying out the theoretical framework, the chapter traces different historical phases of norm development in global agri-food governance since the Second World War (see Table 7.1).

Food security is at the core of a norm cluster in addition to improved nutrition and sustainable agriculture. As a core norm, food security perpetuates approaches that are primarily designed to increase agri-food production and technological innovations and do not inherently acknowledge environmental considerations of sustainability. A subsequent section assesses how SDG 2 reflects the historically grown cluster of global agri-food norms. The goal formally integrates different agri-food norms, while the dominance of food security continues to hinder environmentally salient governance approaches. Alternative policy actors, such as organic and food sovereignty movements, have not yet succeeded in promoting sustainable agriculture.

7.1 Norm clusters and global agri-food governance

Norms define what can be considered appropriate behaviour. The content of a global norm can be shaped and localized in political discourses and thus varies between different institutional settings, cultural practices and over time (Acharya, 2004). A *norm cluster* is understood as a collection of 'aligned, but distinct, norms or principles that relate to a common, overarching issue area; they address different aspects and contain specific normative obligations' (Lantis and Wunderlich, 2018: 571). Such a norm cluster can consist of a mix of old norms that have received social weight over time, for example by their application in practice or their formalization in international law (Fehl and Rosert, 2020), and new norms that emanate from new and politicized norm debates (Sandholtz, 2023). Change in norm clusters can occur either when the substantial content of an already established norm is modified, for example by including new normative elements, or when new norms are institutionalized and constitutionalized next to established ones (Lantis and Wunderlich, 2018).

Whether a norm is being strengthened, maintained, or weakened depends on the degree to which policy actors advance or support pro-norm arguments in norm contestation (Sandholtz, 2023). Public and private *norm entrepreneurs* can raise awareness of new norms, establish new ways of talking about and understanding issues, reframe a formerly unproblematic phenomenon to become problematic, and attempt to convince a critical mass to embrace newly established norms (Finnemore and Sikkink, 1998: 895). Norm entrepreneurship in agri-food governance relates to the agency of consumers and producers (individuals or groups), that is, their capacity to make self-determined decisions, for example, about what foods they eat or produce, how that food is distributed within food systems and their ability to engage in processes that shape agri-food system policies (HLPE, 2020: 9). Norm entrepreneurs politicize or depoliticize agri-food issues by constituting or removing an issue area from the public agenda, respectively (Breitmeier et al, 2021a). Such politicization renders ongoing norm contestation visible (Wiener, 2014).

The politicization of a norm cluster triggers normative clarification and refinement processes on both procedural and substantive grounds (Lantis and Wunderlich, 2018). It also provokes the development of alternative norms and goals that not only facilitate the replacement of old norms with new ones but can also foster contextual sense making (Zimmermann, 2016) and the discursive reframing and development of established norms (Breitmeier et al, 2021a).

Norm contestation in the agri-food sector is accompanied by asymmetric power structures that enable donor countries (such as the US and the UK), international institutions (such as the FAO), and transnational business actors (such as Bayer or Nestlé) to shape the agenda in accordance with their interests and norm understandings (Breitmeier et al, 2021b). The advancing globalization of food markets over the past three decades has put business actors in a position to influence the broad lines of agri-food research and development and to make governance decisions themselves, for example, by setting private labelling standards (Clapp and Scott, 2018). Even though policy actors from civil society have succeeded in raising alternative normative frames in norm contestation, such as La Via Campesina promoting *food sovereignty*, they usually have fewer resources compared to corporate actors and therefore tend to have less agency (Clapp and Scott, 2018; McKeon, 2021).

Breitmeier et al (2021a: 626) have applied the norm cluster concept to the agri-food sector and highlight that policy actors display different interpretations of sustainability, prioritizing certain elements of a norm cluster and neglecting others. In this vein, food movements politicize old norms, promote alternative normative frames and thereby redirect policy foci (Lang and Barling, 2012: 317; McKeon, 2021: 49). Their emergence leads to changes in norm prioritization and fuels new norm dynamics. In this vein, as the global agri-food norm cluster is characterized by ongoing contestation, there is also an ongoing debate about appropriate governance approaches (Breitmeier et al, 2021a; Schwindenhammer, 2023). However, as outlined in the following sections, despite ongoing norm contestation and new links between old and new norms and principles, the food security norm continues to be the most robust one in the global agri-food norm cluster. Environmental aspects have been left behind.

7.2 Historical norm development in global agri-food governance

This section traces different and cumulating historical phases of norm development in global agri-food governance (see also Table 7.1). While, in addition to food security, norms for nutrition security and sustainable agriculture have emerged, food security continues to be the primary norm and it perpetuates production- and technology-oriented agri-food approaches. As these approaches neglect detrimental environmental and social impacts, they have given rise to alternative policy actors, such as organic and food sovereignty movements, that promote normative counter-frames in global agri-food governance (see Figure 7.1).

7.2.1 The emergence of the food security norm

The global agri-food norm cluster is essentially characterized by the fight against hunger and dominated by the food security norm. After the Second World War, the right to food was embodied in the 1948 UN Universal Declaration of Human Rights. It was then codified in international human rights law in the International Covenant on Economic, Social and Cultural Rights in 1966 (Khoo, 2010).

Figure 7.1: Normative counter-frames

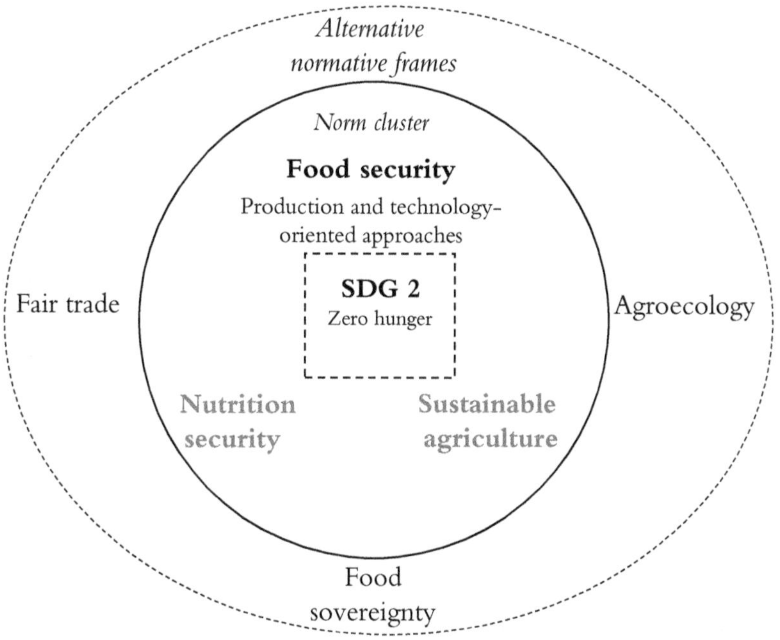

Table 7.1: Historical milestones in global agri-food governance

1945	Foundation of FAO
1948	Universal Declaration of Human Rights embodies the right to food
Since 1950s	Green Revolution (industrialization of agriculture)
1966	International Covenant on Economic, Social and Cultural Rights (ICESCR) codifies the RTF in international human rights law (Article 11)
Since 1970s	Fair trade movements (eg Worldshops)
1972	Foundation of International Federation of Organic Agriculture Movements (IFOAM)
1974	World Food Conference: foundation of Committee on World Food Security (CFS)
1993	Foundation of La Via Campesina (food sovereignty movement)
1996	World Food Summit establishes 'right to adequate food'; Rome Declaration establishes four pillars concept of food security and envisages halving the proportion of people who suffered from hunger up to 2015
1997	Foundation of Fairtrade International (originally Fairtrade Labelling Organizations International)
2000	Millennium Declaration establishes MDG 1 to 'eradicate extreme poverty and hunger' by 2015
2009	Reform of the CFS (institutional participation of non-governmental actors) and establishment of High Level Panel of Experts on Food Security and Nutrition as science-policy interface of the CFS
2012	UN Summit on Sustainable Development (Rio+20) connects food and nutrition security to sustainable agriculture
2014	Second International Conference on Nutrition
2015–2030	Agenda 2030 and SDG 2 (Zero hunger)

Article 11 of the ICESCR recognizes 'the fundamental right of everyone to be free from hunger' and outlines individual and collective policy measures, including specific programs:

(a) To improve methods of production, conservation and distribution of food by making full use of technical and scientific knowledge, by disseminating knowledge of the principles of nutrition and by developing or reforming agrarian systems in such a way as to achieve the most efficient development and utilization of natural resources; (b) Taking into account the problems of both food-importing and food-exporting countries, to ensure an equitable distribution of world food supplies in relation to need. (UN, 1966: 4)

The ICESCR directed international agri-food governance towards ending hunger by means of increasing global food production, trade and aid. But the wording of Article 11 of the ICESCR captured early on the tension between the fundamental human right of everyone to be free from hunger and the policy focus on controlling food surpluses and deficits, partly driven by the big powers' self-interested aid and trade objectives (Khoo, 2010: 37).

McKeon (2015) identifies three milestones of post-Second World War agri-food governance: the creation of the FAO in 1945, the World Food Conference in 1974 and the reform of the CFS in 2009. The FAO is the UN agency that has led international efforts in global agri-food governance for several decades now. The first World Food Conference was held in Rome in 1974 under the auspices of the FAO, in the wake of the devastating famine in Bangladesh over the preceding two years. The conference sought to establish methods to help poor countries finance food purchases, to induce rich countries to provide capital and technical aid to help the developing countries improve domestic production, and to create an international grain reserve system to prevent local famines (Biwas and Biwas, 1975: 20). In the same year, the CFS was established as an intergovernmental body to serve as an additional forum in the UN system to review and follow up on policies concerning world food security including production and physical and economic access to food (Duncan, 2015).

By that time, in the 1970s, food security was understood as a national concept, with the nation-state being food secure when there was sufficient food available for its citizens (Roberts, 2021: 64). The 1996 World Food Summit fundamentally changed this concept and shifted the focus from nation-states to individuals (Roberts, 2021: 65). With the 'right to adequate food', the summit delegates launched a rights-based approach to food security (Roberts, 2021: 65). They defined food security as existing 'when all people, at all times, have *physical and economic access* to sufficient, safe and nutritious food to meet their dietary needs and food preferences for an active and healthy life' (FAO, 1996; emphasis added).

With the Rome Declaration on World Food Security and a related Plan of Action, delegates formulated the goal of reducing world hunger by half no later than the year 2015. According to the Rome Declaration, food security rests on four pillars: food availability, access, utilization and stability (FAO, 1996). With this concept, it was increasingly acknowledged that food insecurity is a problem of structural poverty, markets and market structure, and the relative affordability of different types of food rather than only an issue of food availability (Battersby and Crush, 2016). However, the Plan of Action was unable to reconcile the market-oriented approach to food security prioritizing global agri-food trade, which 'deepened an agrarian crisis in the Global South among small-scale farmers, who had lost price supports and food subsidies via Structural Adjustment loan conditions' (Canfield et al, 2021: 4).

With the advancing globalization of agri-food trade since the 1990s, North/South asymmetries intensified in the agri-food sector, with implications on human development. The majority of undernourished people live in the Global South. The volatile global prices of agricultural commodities, especially cash crops such as coffee and cocoa, often force farmers to sell their products at below production costs. This, compounded by smallholders' lack of access to social infrastructure and services and frequently insecurity of land tenure, leads to further disadvantages (Matthews, 2015). Asymmetries also persist owing to subsidies for domestic agriculture, especially in the EU and the US, and to trade distortions through colonial heritage at the nation-state and international levels (Battersby and Crush, 2016). Global South groups have already raised issues of North/ South asymmetries in international agri-food markets since the 1970s. So-called Worldshops (or Third World or Fair Trade shops) started to sell 'fair trade' commodities. Diverse national labelling organizations joined forces and established the Fairtrade Labelling Organizations International, now called Fairtrade International, to harmonize standards in 1997 (Barratt Brown, 2007).

Once again, reaffirming the significance of food security for global agri-food governance, the 2000 UN Millennium Declaration referred to food security in Millennium Development Goal 1, which was aimed at eradicating extreme poverty and hunger. Like the 1996 Rome Declaration, target 1.3 envisaged halving the proportion of people who suffered from hunger until 2015, compared to the baseline year of 1990. In 2009, in response to global food price spikes in 2007–8, the Declaration of the World Summit on Food Security officially reaffirmed an extension of the 1996 Rome definition of food security by adding the word 'social' to the phrase 'physical, social and economic access' (CFS, 2012: 5). In the same year, the food security governance space institutionally broadened with the reform of the CFS. The committee was opened up to the participation of non-state actors, such as NGOs, research institutions and representatives of philanthropic foundations. The CFS was meant to turn into '*the* international platform for the discussion and coordination of food security policy', striving for a world free from hunger (Duncan, 2015: 86). Moreover, the 2012 UN Conference on Sustainable Development (Rio+20) reaffirmed 'the right to an adequate standard of living, including *the right to food*' (UN, 2012: 2; emphasis added). However, in 2015, MDG 1.3 was narrowly missed as the proportion of undernourished people in countries of the Global South fell by just under half, from 23.3 per cent in 1990–92 to 12.9 per cent in 2014–16 (UN, 2015b).

7.2.2 The spread of production- and technology-related approaches

The food security norm has been and still is the primary norm underlying global agri-food governance (Breitmeier et al, 2021a). It is closely connected to the historical emergence of a production-oriented approach that aims at enhanced agricultural productivity through intensified large-scale agricultural production relying on monoculture, high-yield varieties of grains, intensive use of synthetic fertilizers (Lu and Tian, 2017: 181) and on-field application of biotechnology, for example the cultivation of crops with resistance to herbicides, pest attacks and glyphosate (Perry, 2016).

A variety of norm entrepreneurs, such as agri-food corporations, international organizations, foundations, bilateral donors and university researchers, have 'helped forge a global food system that is increasingly specialised, dependent on trade, and premised on the need to produce more food with industrial methods – all in the name of improving efficiency' (Clapp and Moseley, 2020: 1408). The so-called Green Revolution, which implemented the production-oriented approach, was a turning point for agri-food systems across many countries in the 1950s (Leach et al, 2020: 7). Agricultural subsidies in industrialized countries further fuelled the intensification of agricultural production (Breitmeier et al, 2021a: 631). Modern agri-food production following the production-oriented approach has been credited for increasing the amount of agricultural output over time (Holt-Giménez, 2011: 316), and global food crises strengthened the approach, most recently, the 2007–8 food crisis (Fouilleux et al, 2017), the COVID-19-pandemic (McKeon, 2021: 6) and the war against Ukraine (Ben Hassen and El Bilali, 2022).

The production-oriented approach is closely linked to approaches promoting technology innovation in the agri-food sector. While, back in the 1950s, the industrialization of agriculture was supposed to increase food production volumes, new technologies today are meant to pave the way for visions of circular and high-tech agri-food systems (Schwindenhammer, 2019; 2023). In recent years, central policy documents such as FAO reports (eg FAO et al, 2022) and the SDG framework (to be discussed later) have stressed the potential of new technologies to transform agri-food systems. In consequence, the spread of research, development and application of synthetic biotechnology, field robotics, drone, sensor and nutrient recovery technologies can be observed (Nuijten et al, 2019; Schwindenhammer, 2020).

However, the production- and technology-oriented approaches were and continue to be basically implemented regardless of environmental and social concerns. The production-oriented approach has actually caused a number of negative impacts, such as air pollution, soil acidification and degradation, water eutrophication, biodiversity loss, the monopolization of seed and chemical inputs by companies from the Global North resulting in raising costs for farmers, and the displacement of millions of peasants to fragile hillsides, shrinking forests and urban slums (Holt-Giménez, 2011: 316; Lu and Tian, 2017: 181–2). In response to these impacts and market distortions, alternative normative frames and trajectories evolved, especially in the late 1960s and 1970s. Social movements, such as La Via Campesina (Spanish for 'The Peasants' Way') since the 1990s, have promoted normative counter-frames (Holt-Giménez, 2011). In particular, the norm of food sovereignty emphasizes people's agency by claiming the right to define their own food systems to ensure their own livelihoods and access to culturally appropriate foods (Clapp et al, 2022: 4). Rather than the result of insufficient volumes, hunger is considered to be caused by privileging access to food, and the movement demands the redistribution of control over systems of production and consumption (Holt-Giménez et al, 2021).

7.2.3 The norm of nutrition in addition to food security

Irrespective of movements' counter-frames, the norm of *nutrition security* emerged in the mid-1990s to add to the dominant norm of food security, and established a new phase of global agri-food governance (El Bilali et al, 2019). The new norm extends and contextualizes the human rights approach to food security by specifying human nutritional and social needs (El Bilali et al, 2019: 3). Among diverse organizations, the Bill and Melinda Gates Foundation stands out as a norm entrepreneur regarding this new norm development phase. The world's largest philanthropic foundation promoted the norm by initiating the Global Alliance for Improved Nutrition in 2002, which has successfully endorsed fortified foods and multivitamins to address malnutrition and undernutrition for two decades now (Moravaridi, 2012).

In 2012 the CFS recommended defining food and nutrition security as something that:

> exists when all people at all times have physical, social and economic access to food, which is safe and consumed in sufficient quantity and quality to meet their dietary needs and food preferences, and is supported by an environment of adequate sanitation, health services and care, allowing for a healthy and active life. (CFS, 2012: 8)

The 2012 *State of Food Insecurity in the World* report stresses that economic and agricultural growth should be 'nutrition-sensitive' (FAO et al, 2012: 20). In 2014, after the Second International Conference on Nutrition in Rome, the HLPE identified five key issues of agri-food policy transformation: (1) healthy nutrition in changing food systems; (2) challenges and opportunities of livestock systems, food security and nutrition; (3) inequalities and food security and nutrition referring to the imperative of addressing the needs of disadvantaged and vulnerable populations; (4) the increasing role of financial markets in food security and nutrition; and (5) pathways to sustainable food systems in the pursuit of human and environmental health for all (HLPE, 2014: 6).

7.2.4 The 'new' norm of sustainable agriculture

In addition to the norms of food and nutrition security, the norm of *sustainable agriculture* has emerged in global agri-food governance and established a third central phase of global agri-food governance since the Second World War (Lang and Barling, 2012). The agri-food norm cluster now also covers issues of environmentally sound practices with regard to the use of land, water, fertilizers and other resources; economic efficiency in input costs; crop productivity; and farm income (Mockshell and Kamanda, 2017). It also refers to socially adequate conditions and impacts of agricultural practices with regard to knowledge preservation, livelihood improvement, distribution of land ownership, and intra- as well as intergenerational equity (Mockshell and Kamanda, 2017).

Sustainable agriculture has a long history. The International Federation of Organic Agriculture Movements was an essential norm entrepreneur. It was founded in 1972 with the aim of stopping the expansion of industrialized agriculture (Schwindenhammer, 2017). Industrialized agriculture and its detrimental impacts on the environment originally provoked a normative counter-frame of 'agroecology'. This frame includes replacing synthetic fertilizers, pesticides and hybrid seeds, which are common to industrialized agriculture, with natural manures, cover crops and animal-based fertilization. Moreover, organic movements aim for more direct producer–consumer relations. They want to reduce farmers' financial risks such as vulnerability to price volatility (Holt-Giménez et al, 2021). Some alternative local networks even try to overcome consumer/producer dichotomies, which are inherent in the global capitalist order. For example, in community-supported agriculture (CSA), people contribute membership fees instead of paying a price per food item (Peuker, 2015).

Since the late 1980s, the norm of sustainable agriculture has gained momentum in the context of the global discourse on sustainable development. The Brundtland Report provided a first conceptual definition of sustainability, which stresses the interconnection between the three dimensions of environmental, economic and social sustainability, and postulates intra- and intergenerational justice (WCED, 1987). In the 2000s growing global concern about the negative environmental impacts of agri-food production based on the production-oriented approach led to intensified attempts of agri-food policy institutions to capture sustainability (Canfield et al, 2021: 6). The 2012 UN Summit on Sustainable Development (Rio+20) officially connected food and nutrition security to sustainable agriculture, and emphasized the need to promote more sustainable agriculture 'that improves food security, eradicates hunger and is economically viable, while conserving land, water, plant and animal genetic resources, biodiversity and ecosystems and enhancing resilience to climate change and natural disasters' (UN, 2012: 22).

FAO included sustainable agriculture in the Strategic Framework 2000–2015 and added agro-ecological intensification to the reviewed Strategic Framework in 2017 (FAO, 2017). After FAO (2014) had already pursued the water–energy–food (WEF) nexus approach, which conceptualizes water, energy and food as inextricably linked and dynamically interacting, since 2014, the reviewed 2017 framework also points to crop diversification, ecosystem service and decent living conditions for rural people and small-scale farmers (FAO, 2014; 2017; Breitmeier et al, 2021a). Finally, in 2018 HLPE (2020: 8) suggested extending the four-pillar concept of food security to a six-pillar concept by incorporating the two dimensions of *sustainability* and *agency* to 'codify what is already incorporated in international legal guidance on the right to food' (Clapp et al, 2022: 8). By doing so, HLPE took up the demands of the food sovereignty movement for more self-determined agri-food systems in addition to environmental concerns.

7.3 SDG 2: squaring the circle of the agri-food norm cluster

Agenda 2030 connects the global norms of food security, improved nutrition and sustainable agriculture in SDG 2 (UN, 2015a). This section outlines how the goal reflects the agri-food norm cluster in global governance. The international community has formally addressed different agri-food norms, but food security still dominates and impacts the realization of new norms such as sustainable agriculture. This robustness of the food security norm perpetuates production- and technology-oriented approaches that set aside environmental issues such as water pollution and climate change.

SDG 2 consists of five targets and three subtargets (see Table 7.2). The first five targets (SDG 2.1–2.5) are directly related to food security and agricultural sustainability, while the last three subtargets (SDG 2a–2c) are market-related measures aimed at increasing agricultural investments and reducing market distortions and price volatility (Gil et al, 2019: 686). Behind each target are indicators by which to track progress. Indicators are stipulated as a way of monitoring and uniformly testing progression of the goals across differing urban landscapes (Merino-Saum et al, 2020).

The ranking of targets corresponds to the historical emergence of the norms and the rise of production- and technology-oriented approaches, which were described earlier. Target 2.1 aims to 'end hunger and ensure access by all people ... to safe, nutritious and sufficient food all year round (by 2030)'. Hence Agenda 2030 first and foremost reinforces norms of food security and improved nutrition by strengthening the goals defined by the

Table 7.2: Sustainable Development Goal 2 (Zero hunger)

2.1 By 2030, end hunger and ensure access by all people, in particular the poor and people in vulnerable situations, including infants, to safe, nutritious and sufficient food all year round.

2.2 By 2030, end all forms of malnutrition, including achieving, by 2025, the internationally agreed targets on stunting and wasting in children under 5 years of age, and address the nutritional needs of adolescent girls, pregnant and lactating women and older persons.

2.3 By 2030, double the agricultural productivity and incomes of small-scale food producers, in particular women, indigenous peoples, family farmers, pastoralists and fishers, including through secure and equal access to land, other productive resources and inputs, knowledge, financial services, markets and opportunities for value addition and non-farm employment.

2.4 By 2030, ensure sustainable food production systems and implement resilient agricultural practices that increase productivity and production, that help maintain ecosystems, that strengthen capacity for adaptation to climate change, extreme weather, drought, flooding and other disasters and that progressively improve land and soil quality.

2.5 By 2020, maintain the genetic diversity of seeds, cultivated plants and farmed and domesticated animals and their related wild species, including through soundly managed and diversified seed and plant banks at the national, regional and international levels, and promote access to and fair and equitable sharing of benefits arising from the utilization of genetic resources and associated traditional knowledge, as internationally agreed.

2.A Increase investment, including through enhanced international cooperation, in rural infrastructure, agricultural research and extension services, technology development and plant and livestock gene banks in order to enhance agricultural productive capacity in developing countries, in particular least developed countries.

2.B Correct and prevent trade restrictions and distortions in world agricultural markets, including through the parallel elimination of all forms of agricultural export subsidies and all export measures with equivalent effect, in accordance with the mandate of the Doha Development Round.

2.C Adopt measures to ensure the proper functioning of food commodity markets and their derivatives and facilitate timely access to market information, including on food reserves, in order to help limit extreme food price volatility.

Source: UN (2015a).

1996 Rome Declaration and the 2000 Millennium Declaration. Target 2.2 emphasizes the need to end *malnutrition* in particular. However, unlike the 2012 UN Summit on Sustainable Development (Rio+20), there is no official mention of the human right to food (Vivero Pol and Schuftan, 2016; see also Mehta et al, Chapter 6 in this volume).

Instead of linking 'zero hunger' to problems of structural poverty, markets and market structure, and the relative affordability of different types of food (Battersby and Crush, 2016), SDG 2 falls back on a narrow problem definition of food availability in conjunction with production- and technology-oriented approaches. Target 2.3 aims to 'double the agricultural productivity'. Incomes of small-scale food producers are mentioned but not linked to distributional issues and structural asymmetries (including colonial heritage). Following the production-oriented approach, it is suggested that 'productive resources and inputs, knowledge, financial services, markets and opportunities for value addition and non-farm employment' could compensate for persistent disadvantages.

In line with the norm of sustainability agriculture, target 2.4 mentions the need to ensure systems and practices that 'increase (agricultural) productivity and production, that help maintain ecosystems, that strengthen capacity for adaptation to climate change, extreme weather, drought, flooding and other disasters and that progressively improve land and soil quality'. Because of this target, Gupta and Vegelin (2016: 441–2) consider SDG 2 to be environmentally relevant (see also Partzsch, Chapter 1 in this volume). Demonstrating links between organic and sustainability agriculture, Germany indicates the share of organic farming land in the country as the (only) indicator to track progress on target 2.4 (9.7 per cent in 2021) (DeStatis, 2023). However, other countries such as France (Cling et al, 2019), the UK (UK Government, 2022) and the US (US Government, 2022), are still 'exploring data sources' and do not report on this target at all. This lack of a commonly agreed indicator demonstrates the deferral of the environmental subtarget in global sustainability governance.

Finally, the fifth target as well as three additional subtargets (2.A, 2.B and 2.C) reveal a deep belief in the powers of the market to manage to achieve zero hunger. By contrast, again, there is no mention of historically grown asymmetries and disadvantages of small-scale producers in the Global South. Instead, in line with the production- and technology-oriented approaches, there is the aim to 'increase investment … in … technology development and plant and livestock gene banks in order to enhance agricultural productive capacity in developing countries, in particular least developed countries'.

In sum, SDG 2 opens up new perspectives for integrated and cross-sectoral policies (Sachs et al, 2019), but at the same time demonstrates contested norms in global agri-food governance. With the fourth subtarget, theoretically, Agenda 2030 requires transforming traditional policies in such a way that ecosystem protection is acknowledged to be a prerequisite for food security (Breitmeier et al, 2021a). However, unlike especially the first target on food security, there is as yet no commonly used indicator to even monitor the fourth target (Gil et al, 2019: 685). In consequence, SDG 2 is likely to invoke multiple synergies and trade-offs with the green goals of Agenda 2030. In particular, as mentioned earlier, SDG 2 has high negative interactions with SDG 6. Moreover, agri-food production competes with the expansion of energy infrastructure (SDG 7) for land and water (Fader et al, 2018; Dabla and Goldthau, Chapter 8 in this volume).

7.4 Conclusion

Building on research on global norms in international relations and agri-food governance, this chapter reflects on the normative foundations of SDG 2. It points to current sustainability challenges, and contributes to research on global norm stability and change, conflicting perceptions and alternative frames in global agri-food governance. As this chapter has shown, the global agri-food norm cluster consists of aligned, but distinct, historically evolved norms and approaches that collide and are continuously contested. Food security turns out to be the core of the global agri-food norm cluster, in addition to improved nutrition and sustainable agriculture. The normative priority of global agri-food governance is to ensure food security following production and technology-oriented approaches. This means that the international community aims to increase production volumes on the basis of technological innovation but without regard for the environmental consequences and the social origins of hunger. The robustness of the food security norm, which has even been strengthened by the COVID-19-pandemic and reinforced by the Russian invasion of Ukraine, hence has detrimental environmental and social impacts, which continue to exist with SDG 2.

As has been shown, SDG 2 refers to the global agri-food norm cluster and formally connects the norms of food security, improved nutrition and sustainable agriculture. However, the food security norm implies that production- and technology-oriented approaches continue to dominate, and therefore the norm collides with other agri-food norms, especially sustainable agriculture. SDG 2 also invokes multiple trade-offs with the green goals of Agenda 2030, in particular SDG 6 and SDG 13. Hence, to implement SDG 2, there is an urgent need to transform the current global agri-food systems. Future agri-food policies and solutions have to consider the environmental and social impacts of agriculture and food issues against the backdrop of existing norm collisions and norm contestations in the global agri-food norm cluster.

Conventional agri-food systems have been shown to be unable to respond to broader contextual demands including environmental change. In consequence, alternative normative frames such as food sovereignty and agroecology have increasingly gained in popularity. They have entered norm contestation, but this norm contestation is characterized by the unequal representation of different norms. Alternative norm entrepreneurs, such as organic and food sovereignty movements, advocate alternatives that add to but do not replace the dominant norms and practices.

All in all, this chapter's findings substantiate the need for further research on the impact and resolution of the collision of old and new norms in agri-food governance and on how alternative actors, knowledge and developments can enter global norm contestation to promote greater environmental protection as well as social justice. An ideationally enriched research perspective allows for the capture of the conditions that warrant equal representation of different types of knowledge in norm contestation. For the phase after 2030, global agri-food governance must be prevented from leaving behind those who already have a disadvantaged voice in agri-food systems, especially local actors and social movements that have long advocated for environmentally salient agri-food systems.

References

Acharya, A. (2004) 'How ideas spread: whose norms matter? Norm localization and institutional change in Asian regionalism', *International Organization*, 58(2): 239–75.

Barratt Brown, M. (2007) '"Fair Trade" with Africa', *Review of African Political Economy*, 34(112): 267–77.

Battersby, J. and Crush, J. (2016) 'The making of urban food deserts', in J. Crush and J. Battersby (eds) *Rapid Urbanisation, Urban Food Deserts and Food Security in Africa*, New York: Springer, pp 1–18.

Ben Hassen, T. and El Bilali, H. (2022) 'Impacts of the Russia–Ukraine war on global food security: towards more sustainable and resilient food systems?', *Foods*, 11(15): 2301.

Biwas, M. and Biwas, A.K. (1975) 'World Food Conference: a perspective', *Agriculture and Environment*, 2: 15–37.

Breitmeier, H., Schwindenhammer, S, Checa, A., Manderbach, J. and Tanzer, M. (2021a) 'Politicized sustainability and agricultural policy: comparing norm understandings of international organizations', *Journal of Comparative Policy Analysis: Research and Practice*, 23(5–6): 625–43.

Breitmeier, H., Schwindenhammer, S., Checa, A., Manderbach, J. and Tanzer, M. (2021b) 'Aligned sustainability understandings? Global inter-institutional arrangements and the implementation of SDG 2', *Politics and Governance*, 9(1): 141–51.

Canfield, M., Anderson, M. D. and McMichael, P. (2021) 'UN Food Systems Summit 2021: dismantling democracy and resetting corporate control of food systems', *Frontiers in Sustainable Food Systems*, 5: 661552.

CFS (Committee on World Food Security) (2012) *Coming to Terms with Terminology: Food Security, Nutrition Security, Food Security and Nutrition, Food and Nutrition Security*, CFS 2012/39/4, Rome: CFS, Available at https://www.fao.org/3/MD776E/MD776E.pdf [Accessed 21 June 2023].

Clapp, J. and Scott, C. (2018) 'The global environmental politics of food', *Global Environmental Politics*, 18(2): 1–11.

Clapp, J. and Moseley, W.G. (2020) 'This food crisis is different: COVID-19 and the fragility of the neoliberal food security order', *The Journal of Peasant Studies*, 47(7): 1393–417.

Clapp, J., Moseley, W.G., Burlingame, B. and Termine, P. (2022) 'Viewpoint: the case for a six-dimensional food security framework', *Food Policy*, 106: 102164.

Cling, J.-P., Eghbal-Teherani, S., Orzoni, M. and Plateau, C. (2019) 'France and the Sustainable Development Goals', in *The French Economy*, Available from: https://www.insee.fr/en/statistiques/fichier/4190103/ECOFR A19_D01_ODD.pdf [Accessed 20 August 2022].

DeStatis (2023) Homepage, Available from: https://sdg-indikatoren.de/ 2-4-1/ [Accessed 20 August 2022].

Duncan, J. (2015) *Global Food Security Governance: Civil Society Engagement in the Reformed Committee on World Food Security*, New York: Routledge.

El Bilali, H., Callenius, C., Strassner, C. and Probst, L. (2019) 'Food and nutrition security and sustainability transitions in food systems', *Food and Energy Security*, 8: e00154.

Fader, M., Cranmer, C., Lawford, R. and Engel-Cox J. (2018) 'Toward an understanding of synergies and trade-offs between water, energy, and food SDG targets', *Frontiers in Environmental Science*, 6(112).

FAO (Food and Agriculture Organization) (1996) 'Rome Declaration on World Food Security', 'World Food Summit Plan of Action', 13–17 November 1996, Rome: FAO, Available from: https://www.fao.org/3/ w3613e/w3613e00.htm [Accessed 21 June 2023].

FAO (2014) *The Water–Energy–Food Nexus: A New Approach in Support of Food Security and Sustainable Agriculture*, Rome: FAO, Available at https:// www.fao.org/3/bl496e/bl496e.pdf [Accessed 21 June 2023].

FAO (2017) 'Reviewed Strategic Framework', C 2017/7 Rev.1, Rome: FAO, Available from: https://www.fao.org/3/ms431reve/ms431r eve.pdf [Accessed 21 June 2023].

FAO, WFP (World Food Programme) and IFAD (International Fund for Agricultural Development) (2012) *The State of Food Insecurity in the World 2012: Economic Growth is Necessary But Not Sufficient to Accelerate Reduction of Hunger and Malnutrition*, Rome: FAO, Available from: https://digitallibr ary.un.org/record/3927777?ln=en [Accessed 21 June 2023].

FAO, IFAD, UNICEF (United Nations Children's Fund), WFP and WHO (World Health Organization) (2022) *The State of Food Security and Nutrition in the World 2022. Repurposing Food and Agricultural Policies to Make Healthy Diets More Affordable*, Rome: FAO, Available from: https://www.fao.org/ documents/card/en/c/cc0639en [Accessed 21 June 2023].

Fehl, C. and Rosert, E. (2020) 'It's complicated: a conceptual framework for studying relations and interactions between international norms', PRIF Working Paper No. 49, Frankfurt: Peace Research Institute Frankfurt.

Finnemore, M. and Sikkink, K. (1998) 'International norm dynamics and political change', *International Organization*, 52(4): 887–917.

Fouilleux, E., Bricas, N. and Alpha, A. (2017) 'Feeding 9 billion people: global food security debates and the productionist trap', *Journal of European Public Policy*, 24(11): 1658–77.

Gil, J.D.B., Reidsma, P., Giller, K., Todman, L., Whitmore, A., and van Ittersum, M. (2019) 'Sustainable Development Goal 2: improved targets and indicators for agriculture and food security', *Ambio*, 48: 685–98.

Griggs, D. J., Nilsson, M., Stevance, A., and McCollum, D. (eds) (2017) *A Guide to SDG Interactions: From Science to Implementation*, Paris: International Council for Science (ICSU).

Gupta, J. and Vegelin, C. (2016) 'Sustainable development goals and inclusive development', *International Environmental Agreements: Politics, Law and Economics*, 16(3): 433–48.

HLPE (High Level Panel of Experts on Food Security and Nutrition) (2014) 'Note on critical and emerging issues for food security and nutrition', Prepared for the Committee on World Food Security, Rome: HLPE, Available from: https://www.fao.org/fileadmin/user_upload/hlpe/hlpe_documents/Critical_Emerging_Issues/HLPE_Note-to-CFS_Critical-and-Emerging-Issues_6-August-2014.pdf [Accessed 21 June 2023].

HLPE (2020) *Food Security and Nutrition: Building a Global Narrative towards 2030*, Summary and Recommendations of the 15th report by the HLPE, Rome: HLPE, Available from: https://www.fao.org/3/ca9731en/ca9731en.pdf [Accessed 21 June 2023].

Holt-Giménez, E. (2011) 'Food security, food justice, or food sovereignty? Crises, food movements, and regime change', in A. H. Alkon and J. Agyeman (eds) *Cultivating Food Justice: Race, Class, and Sustainability*, Cambridge, MA: MIT Press, pp 309–30.

Holt-Giménez, E., Shattuck, A. and Van Lammeren, I. (2021) 'Thresholds of resistance: agroecology, resilience and the agrarian question', *The Journal of Peasant Studies*, 48: 715–33.

Khoo, S. (2010) 'The right to food: legal, political and human implications for a food security agenda', *Trócaire Development Review*, 33–50.

Lang, T. and Barling, D. (2012) 'Food security and food sustainability: reformulating the debate', *The Geographical Journal*, 178(4): 313–26.

Lantis, J. S. and Wunderlich, C. (2018) 'Resiliency dynamics of norm clusters: norm contestation and international cooperation', *Review of International Studies*, 44(3): 570–93.

Leach, M., Nisbett, N., Cabral, L., Harris, J., Hossain, N. and Thompson, J. (2020) 'Food politics and development', *World Development*, 134(105024): 1–19.

Lu, C. and Tian, H. (2017) 'Global nitrogen and phosphorus fertilizer use for agriculture production', *Earth System Science Data*, 9: 181–92.

Matthews, A. (2015) 'The Common Agricultural Policy and development', in J. A. McMahon and M. N. Cardwell (eds) *Research Handbook on EU Agriculture Law*, Cheltenham: Edward Elgar, pp 484–504.

McKeon, N. (2015) *Food Security Governance: Empowering Communities, Regulating Corporations*, New York: Routledge.

McKeon, N. (2021) 'Global food governance', *Development*, 64: 48–55.

Merino-Saum, A., Halla, P., Superti, V., Boesch, A. and Binder, C. R. (2020) 'Indicators for urban sustainability: key lessons from a systematic analysis of 67 measurement initiatives', *Ecological Indicators*, 119: 106879.

Mockshell, J. and Kamanda, J. (2017) 'Beyond the agroecological and sustainable intensification debate: is blended sustainability the way forward?', IDOS Discussion Papers 16/2017, Bonn: German Institute of Development and Sustainability.

Moravaridi, B. (2012) 'Capitalist philanthropy and the new green revolution for food security', *International Journal of Sociology of Agriculture & Food*, 19(2): 243–56.

Nuijten, R. J. G., Kooistra, L. and De Deyn, G. B. (2019) 'Using unmanned aerial systems (UAS) and object-based image analysis (OBIA) for measuring plant-soil feedback effects on crop productivity', *Drones*, 3(3): 54.

Perry, M. (2016) 'Sustaining food production in the Anthropocene: Influences by regulation of crop biotechnology', in A. Kennedy and J. Liljeblad (eds) *Food Systems Governance: Challenges for Justice, Equality and Human Rights*, New York: Routledge, pp 127–42.

Peuker, B. (2015) 'Community supported agriculture Macht in und durch die Aushandlung alternativer Landwirtschaft', in L. Partzsch and S. Weiland (eds) *Macht und Wandel in der Umweltpolitik: Sonderband der Zeitschrift für Politikwissenschaft*, Baden-Baden: Nomos, pp 137–60.

Roberts, M. T. (2021) 'Understanding modern history of international food law is key to building a more resilient and improved global food system', *Journal of Food Law & Policy*, 17(1): 56–70.

Rockström, J., Steffen, W., Noone, K., Persson, Å., Chapin, F. S., Lambin, E. F. et al (2009) 'A safe operating space for humanity', *Nature*, 461: 472–5.

Sachs, J.D., Schmidt-Traub, G., Mazzucato, M., Messner, D., Nakicenovic, N. and Rockström, J. (2019) 'Six transformations to achieve the Sustainable Development Goals', *Nature Sustainability*, 2: 805–14.

Sandholtz, W. (2023) 'Is winter coming? Norm challenges and norm resilience', in H. Krieger and A. Liese (eds) *Tracing Value Change in the International Legal Order: Perspectives from Legal and Political Science*, Oxford: Oxford University Press, pp 45–63.

Schwindenhammer, S. (2017) 'Global organic agriculture policy-making through standards as an organizational field: when institutional dynamics meet entrepreneurs', *Journal of European Public Policy*, 24(11): 1678–97.

Schwindenhammer, S. (2019). 'Agricultural governance in the Anthropocene: a research agenda', in T. Hickmann, L. Partzsch, P. Pattberg and S. Weiland (eds) *The Anthropocene Debate and Political Science*, New York: Routledge, pp 146–63.

Schwindenhammer, S. (2020) 'The rise, regulation and risks of genetically modified insect technology in global agriculture', *Science, Technology and Society*, 25(1): 124–41.

Schwindenhammer, S. (2023) 'The future we want? Interlinking global sustainability norm change, technology innovation, and regime complexity', in H. Krieger and A. Liese (eds) *Tracing Value Change in the International Legal Order: Perspectives from Legal and Political Science*, Oxford: Oxford University Press, pp 286–304.

UK Government (2022) Sustainable Development Goals: Goal 2, Available from: https://sdgdata.gov.uk/2 [Accessed 20 August 2022].

UN (United Nations) (2012) 'Resolution adoopted by the General Assembly on 27 July 2012. 66/288 The future we want', A/RES/66/288, New York: United Nations, Available from: https://www.un.org/en/deve lopment/desa/population/migration/generalassembly/docs/globalcomp act/A_RES_66_288.pdf [Accessed 21 June 2023].

UN (2015a) 'Transforming our world: the 2030 Agenda for Sustainable Development', A/RES/70/1, New York: United Nations.

UN (2015b) *The Millennium Development Goals Report 2015*, New York: United Nations Publications.

UN (1966) 'International Covenant on Economic, Social and Cultural Rights', adopted and opened for signature, ratification and accession by General Assembly Resolution 2200A (XXI) of 16 December 1966 entry into force 3 January 1976, in accordance with Article 2, Available from: https://www.ohchr.org/sites/default/files/cescr.pdf [Accessed 21 June 2023].

US Government (2022) Sustainable Development Goals: Zero hunger, Available from: https://sdg.data.gov/zero-hunger/ [Accessed 20 August 2022].

Vivero Pol, J.L. and Schuftan, C. (2016) 'No right to food and nutrition in the SDGs: mistake or success?', *BMJ Global Health*, 1: e000040.

WCED (World Commission on Environment and Development) (1987) *Our Common Future*, Oxford: Oxford University Press.

Wiener, A. (2014) *A Theory of Contestation*, Heidelberg: Springer.

Zimmermann, L. (2016) 'Same same or different? Norm diffusion between resistance, compliance, and localization in post-conflict states', *International Studies Perspectives*, 17(1): 98–115.

Interview with Sandra Schwindenhammer: Partnering with the Enemy to Achieve SDG 2?

Laura Kräh and Ruth Krötz

Agro-ecology is seen as an alternative approach to achieving SDG 2 but is seldom practised. Why is the transition so hard and which dependencies need to be contested for its realization?

Schwindenhammer: The SDGs reflect a general normative reference framework. While deciding on sustainability goals in general is easy for state representatives, the implementation and measurement of concrete targets is much more challenging and contested. SDG 2 aims at combating world hunger by increasing food availability and, while this is important, conventional approaches still widely neglect the social and environmental dimensions of the global hunger challenge. We face various complex problems that cannot be solved by more production and increased technology alone. Rather, the voices of farmers from the Global South need to be heard more. Agroecology and localized approaches to agri-food production can help to ensure self-sufficiency, but smallholder farmers do want not only to produce for themselves but to be equally treated as production partners. The global agri-food market creates unequal outcomes as there are subsidies and large corporations that influence the broad lines of how the food system is organized. For instance, smallholder farmers from the Global South who want to enter the European organic market face several access barriers. It must become easier for them to enter the market and we need standards that fit their situation. Farmers from the Global South face specific social and environmental challenges; barriers to agro-ecology in Germany and Rwanda are not the same.

What kind of governance and cooperation is needed to contest the trade-offs with green SDGs?

Schwindenhammer: Governance actors, especially state representatives, need to be aware of these trade-offs. There is a need for more integrated governance and discourse between different government departments to overcome the sectoral logic of policy making. Civil society actors face the impacts of the trade-offs on the ground, but also have an important role in making agri-food policy challenges visible: we need civil society's voice to bring attention to the issue. We also need to end the blame game: while state representatives call for the business sector to transform, the business sector demands that market policies be changed, but who is responsible? The truth is that all groups need to contribute.

So is there a possibility of getting companies on board for sustainable agriculture?

Schwindenhammer: Business actors have emerged as political actors and influence today's agri-food policies and governance frameworks. From a critical perspective, the rise of corporate norm entrepreneurship limits the influence and authority of states to set and implement rules. But there is hope! Getting companies on board allows us to hold them more accountable and to take advantage of their resources to develop and implement private standards. However, what counts is that solutions are sustainable, no matter whether they are implemented by Nestlé, Kraft or environmental organizations so partnering with the enemy is an entry point. Nevertheless, the question of greenwashing needs to be continually addressed.

8

Clean Energy Services: Universal Access as Enabler for Development?

Nopenyo E. Dabla and Andreas C. Goldthau

According to a joint report by International Energy Agency (IEA), International Renewable Energy Agency (IRENA), United Nations Statistics Division (UNSD), the World Bank and WHO (IEA et al, 2021), some 759 million people lack access to electricity and 2.6 billion are without clean cooking solutions. Clearly, in addition to the human security dimension, universal access to modern and clean energy services comes with a development imperative.[1] Yet, while economic development correlates with higher energy consumption, closing the energy access gap must not come at a cost of future emissions. To fight climate change and to adapt to its effects that have already been manifested, energy demand increments must be covered by clean sources.

The Sustainable Development Goal 7 sets out to ensure 'affordable, reliable, sustainable and modern energy for all' by 2030 (UN, 2015). SDG 7 speaks directly to matters of environmental governance, which is concerned with ensuring the effective functioning of environmental systems that are socially equitable, responsive to social conditions and persistent (Bennett and Satterfield, 2018). It also speaks to other SDGs that are pertinent to environmental governance, notably regarding sustainable production and consumption (SDG 12).

What is the state of play on SDG7? Who are the actors driving action in this domain? And where is the latter effective? Departing from a thorough analysis of the key tenets of the SDG 7 challenge, the present chapter, first, recaps how policy making in developing countries has in the past decade shifted from a sectoral view to the perception of energy as an enabler for broader socio-economic goals. In this context, the chapter discusses some of the pertinent linkages with other SDGs and, with it, environmental governance. Second, the chapter reviews what has worked in developing countries, singling Kenya out as a case to illustrate policy success. Third, the chapter zooms in on clean cooking. As it will show, countries struggle to implement adequate policies and market interventions that make an impact despite the multifaceted co-benefits in terms of health, gender or environmental protection. Finally, the chapter presents an outlook for 2030 and beyond, identifying key policy action points from the perspective of the carbon neutrality target and the 1.5 °C scenario, and in light of the latest Intergovernmental Panel on Climate Change *Sixth Assessment Report*.

8.1 The energy for all challenge

8.1.1 SDG 7: what's the state of play?

SDG 7 rests on the three key tenets of universal access to energy (SDG 7.1), a high share of renewable energy in the global energy mix (SDG 7.2) and improving energy efficiency (SDG 7.3). In terms of SDG 7.1, the number of people without access to electricity dropped from 1.2 billion in 2010 to 759 million in 2019, as stated by the joint *2021 Energy Progress Report* (IEA et al, 2021). Within a decade, some 1.1 billion people across the globe were connected to the electricity grid, and it is estimated that the rate of electricity access worldwide increased from 83 per cent in 2010 to 90 per cent in 2019. This represents a remarkable achievement.

That said, access remains uneven. While access rates in Latin America and across most of Asia reach almost 100 per cent, sub-Saharan Africa trails behind, standing at 46 per cent in 2019. The region accounts for the bulk of people lacking access to electricity globally. Worse, access rates in sub-Saharan Africa have reportedly dropped in 2020, for the first time since 2013 (IEA, 2021: 175). This is argued to be a function of the adverse effects of the COVID-19 pandemic but also of a generally suboptimal investment environment for private actors (KfW et al, 2020), which are estimated to have provided for 86 per cent of total renewable energy investment between 2013 and 2018 (IRENA and CPI, 2020). By some estimates, $35 billion in annual investment is needed to reach full electricity access by 2030 (IEA, 2021). As joint research by IRENA and the Climate Policy Initiative (CPI) shows, the bulk of renewable energy investment happened in countries in the OECD and in East Asia. By contrast, renewable energy investment flows to the remaining regions including Central Asia, Latin America and most of Africa amounted to only 15 per cent of the total (IRENA and CPI, 2020). Even international public financial flows, which are considered to be central to achieving SDG 7, remain below the necessary levels and are lopsided.

Moreover, there persists a strong rural/urban divide in energy access (see Figure 8.1). As IEA et al (2021) estimated, some 84 per cent of all people without access to electricity live in rural areas. A key problem lies in the capital

Figure 8.1: Rural electricity access (percentage of population)

© World Bank. Population estimates based on UN population data.

The boundaries, colours, denominations and other information shown on any map in this work do not imply any judgment on the part of the custodian agencies concerning the legal status of or sovereignty over any territory or the endorsement or acceptance of such boundaries.

Indicator: Electricity access rate (%)

<10 10 - 49 50 - 99 100

Unit: % of population

Source: World Bank (2020c). License: CC BY-4.0.

costs of setting up sufficient infrastructure to connect rural communities to the grid. In addition to supporting grid access, a key challenge therefore consists in fostering off-grid solutions. Yet, as the IEA estimates in both their Stated Policies Scenario (STEPS) and their Announced Pledges Scenario (APS), 670 million people will remain without electricity access in 2030. Again, the access rate in sub-Saharan Africa remains far below average, standing at 60 per cent (IEA, 2021).

A crucial – and, as will be argued, also neglected – aspect in this context is clean cooking. The World Health Organization categorizes solar, electric, biogas, natural gas, liquefied petroleum gas (LPG) and alcohol fuels (including ethanol) as clean cooking fuels and technologies owing to them being clean for health at their point of use and low on fine particle matter and carbon monoxide (WHO, 2022). While LPG is considered a clean cooking solution, it does not represent a renewable energy technology and is associated with greenhouse gas emissions (EPA, 2014).

Globally, unequal access results in 2.6 billion people, predominantly in rural areas, not having clean cooking solutions. Of these, 910 million people who reside in sub-Saharan Africa account for 35 per cent of the total (IEA et al, 2021). Developing Asia, notably China and India, accounts for most of the rest. Once again, this absolute figure has been increasing in sub-Saharan Africa, from less than 600 million in 2000 (WHO, 2021). In 2020, 50 million people went back to traditional cooking techniques based on biomass or kerosene (IEA, 2021), which are believed to cause 2.5 million premature deaths annually, as well as environmental degradation and persisting gender inequality (ESCAP, 2021).

In terms of renewable energy and energy efficiency, the two other SDG 7 elements, the picture is mixed as well. Clearly, a substantial increase in renewables in global energy has not been achieved over the past decade. Though renewables have come to represent 28 per cent of global electricity supply in 2020, their overall share in the energy mix remains at a mere 12 per cent – up from 9 per cent in 2010 (IEA, 2021). This is mainly due to overall energy consumption going up significantly, leading to an incremental supply of all fuels, including from fossil sources. Energy efficiency improvements remain a far cry from the SDG 7.3 target of doubling the rate seen between 1990 and 2010. In 2018 and 2019, energy efficiency improvements stood at 1.5 per cent and 1.6 per cent respectively, and in 2020 they fell to half that level (IEA, 2020a). As a general trend, industrialized economies have decoupled their economic growth from energy consumption – in part also, of course, thanks to their farming out energy-intensive production to third countries (Jiborn et al, 2018) – whereas emerging economies and developing nations have so far not done this. Western Asia, northern Africa and sub-Saharan Africa stood out, judged by their simultaneous growth in total energy supply and GDP (IEA et al, 2021), resulting in stalling energy intensity levels. Overall, meeting the SDG 7 challenge is a multifaceted endeavour. Clearly, 'at today's rate of progress, the world is not on track to achieve SDG 7' (IEA et al, 2021: 1).

8.1.2 Energy access in international development and the clean transition

Energy access, arguably the most prominent element of SDG 7, has for long been discussed as a standalone policy goal. Universal access to adequate levels of energy services at affordable costs represents the flip side of energy poverty, which is generally defined as the lack thereof (UN, 2018). A country that acted successfully here is China, which lifted hundreds of millions of people out of energy poverty over the past few decades. As the case of China also clearly demonstrates, energy access is deeply intertwined with economic development. As China grew into a upper- to middle-income economy, energy poverty rates declined, though they still persist in some rural parts of the country (Jiang et al, 2020). Similar findings were made for India, where a strong correlation exists between energy access rates and economic development (Acharya and Sadath, 2019). Moreover, clean energy access has been found to interact positively with education and economic growth so as to reduce income inequality (Acheampong et al, 2021). Replacing traditional fuels such as biomass with cleaner alternatives, in turn, brings about clear health benefits, as exposure to indoor combustion declines (Maji and Kandlikar, 2020). Against the backdrop of these experiences, the policy debate on energy access has shifted from a sectoral view to perceiving energy as an enabler for broader socio-economic goals.

As depicted in Figure 8.2, the broader context here is that there are clear interlinkages between SDG 7 and other SDGs. Universal and clean energy access and a higher share of renewables in the energy mix have been found to strongly correlate with SDG 1 (No poverty) and SDG 3 (Good health and well-being). The causal dynamics at work here are straightforward: energy access empowers people economically and allows them to engage in higher added value activity. Clean energy sources, in turn, reduce respiratory diseases, enhance air quality at home and across metro areas, and thus enhance human well-being. As an indirect effect, healthier people tend to be economically more productive (McCollum et al, 2018). Moreover, policies aimed at ending energy poverty tend to bring about technology innovation and may therefore yield green growth (Zhao et al, 2021). Notably, advances in SDG 7 also contribute to SDGs that touch on key aspects of environmental governance by, among other means, making cities more sustainable (SDG 11), sustaining the efficient use of natural resources (SDG 12) and, as an overall co-benefit, contributing to combating dangerous climate change (SDG 13) (McCollum et al, 2018).

Much less clear are the linkages to other SDGs (Figure 8.2). For example, it has often been alleged that clean energy access helps address persisting gender

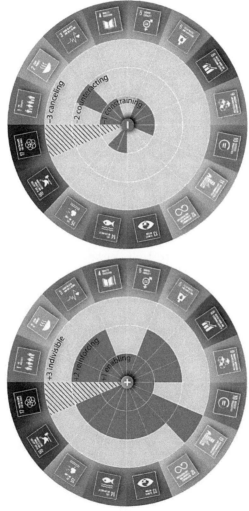

Figure 8.2: Trade-offs in SDGs

Source: Adapted from McCollum et al (2018) under Creative Commons Attribution 3.0 license.

inequality (SDG 5). Yet the evidence here is scarce and inconclusive. In the case of India, females have not been found to be the primary beneficiaries of clean energy access, and gender equity was described as a function of broader intra-household power dynamics (Rosenberg et al, 2020). In sub-Saharan Africa, gender is found to intersect with other determining factors when it comes to the adoption and use of solar home systems, including class, age and geographical location (Ojong, 2021).

Finally, there clearly also exist trade-offs between SDG 7 and other SDGs and central concerns of environmental governance. For example, the expansion of renewables in the global energy mix comes with increased land use, as does bioenergy. This may put solar, wind or biofuels in direct conflict with other land uses such as food production. Moreover, renewable energy may impact water resources and local habitat. For example, large-scale hydropower has been shown to conflict with the traditional ways in which communities use and manage local resources (Erlewein, 2013). This establishes a link between livelihoods and the much discussed water–energy–food nexus which, in addition to entailing significant trade-offs, is often viewed as key to sustainable development (Biggs et al, 2015). Moreover, biomass-based energy may run counter to efforts aimed at preserving land or biodiversity, and may also have negative consequences for climate mitigation. The problem occurs in the context of indirect land-use change (ILUC), for example when biomass is grown on deforested land (Partzsch, 2020). The manufacturing of renewable energy technologies comes with a draw on local ecosystems, for example in the shape of rare earth extraction, and may also impact water security (Sovacool et al, 2021). It may also lead to new divides between urban and rural communities, leaving the latter deprived and raising questions of energy justice (Sovacool et al, 2020). Research suggests that SDG 2 (Zero hunger), SDG 6 (Clean water and sanitation) and SDG 13 (Climate action) and SDG 7 may in part be at odds with each other (Nerini et al, 2018).

8.1.3 Actors and institutions

Actors and institutions engaged in attempts to achieve SDG 7 and its objectives are diverse in nature and can be found at global, continental, subregional and national levels. Pertinent initiatives, programs or projects foster policy and strategy development, financing, technical assistance, financing, capacity building for renewable energy deployment, access to sustainable electricity and/or access to clean cooking technologies.

The United Nations through its agencies and programs have been instrumental here. Indeed, the 1972 United Nations Conference on the Environment in Stockholm was the first global conference that recognized the environment as a major policy issue. The resulting declaration, also referred to as the Stockholm Declaration, contains a set of 'common principles to inspire and guide the peoples of the world in the preservation and enhancement of the human environment' (Sohn, 1973: 435). These principles were not legally binding but were merely regarded as a set of guidelines for maintaining and improving the natural environment while supporting people and their overall needs throughout the process. Yet they set the course of the paradigm shift in global policy making towards environmental sustainability and environmental governance. As an example, the creation of the United Nations Environment Programme (UNEP) is one of the direct outcomes to the Stockholm Conference. Twenty years later, the United Nations Conference on Environment and Development of 1992, also known as the Earth or Rio Summit, reaffirmed the direction established in Stockholm through the Rio Declaration on Environment and Development. The 27 principles included in the declaration continue to serve as a compass for the international action on environment today. Though, at the time of the Rio Summit, renewable energy did not explicitly feature, many of these principles still speak to what is being achieved through renewable energy technologies today.

At the global level, aside from the United Nations and its agencies and programs, organizations such as the World Bank Group (WBG), the International Renewable Energy Agency, the Green Climate Fund, Sustainable Energy for All (or SEforAll) and the Clean Cooking Alliance are active in the energy access domain. WBG, notably through its main arms of the International Bank for Reconstruction and Development, the International Finance Corporation and the Multilateral Investment Guarantee Agency, has undoubtably been at the forefront of public investments towards achieving SDG 7. From 2016 to 2020, it is estimated that the WBG financed 34 gigawatts of renewable energy. In addition to providing finance, the WBG is also involved in the development of targeted energy access programs (eg in Ethiopia or Kenya), capacity-building programs and the provision of technical assistance. Established as a financing arm within the United Nations Framework Convention on Climate Change, the Green Climate Fund provides grants and concessional loans for the implementation of adaptation and mitigation projects.

To be sure, the historical record of multilateral development agencies is at best mixed when it comes to simultaneously catering to energy access and the climate goals and ensuring sustainable environmental governance at the same time. Having ended its support for coal in 2010, the World Bank did not end its support for the oil and gas infrastructure until 2019 (World Bank, 2017). Moreover, Washington Consensus-informed policy packages sought to reform energy subsidies in many countries (Vagliasindi, 2013), in part impacting on energy access for less well-off segments of society. Some internationally supported projects aimed at transitioning resource economies to a low-carbon future were also found as continuing fossil pathologies

(Günel, 2019). That said, it is fair to argue that international development agencies have embraced the goal of a fossil fuel phase-out and are supportive of the SDG 7 agenda.

Initially launched in 2011 as an initiative championed by the UN secretary general, SEforAll became an organization in its own right in 2016 with the stated objective to serve as a platform for private and public actors, financiers and civil society to drive action for the achievement of SDG 7. Its key goals were what was translated into the objectives of SDG 7 when the 2030 Agenda for Sustainable Development was set in 2015. In that capacity, SEforAll support a large number of developing countries to set up country Action Agendas coherent with national targets as well as Investment Prospectuses with the aim of driving investment towards the implementation of these Action Agendas.

Finally, established in 2011, the International Renewable Energy Agency has the mandate to support governments worldwide to adopt adequate policies for scaling up renewable energy. In doing so, IRENA provides policy advice and technical assistance, capacity building and knowledge transfer to its members countries.

The aforementioned institutions are just a few of the dozens of bodies operating at global level to support the achievement of SDG 7. For all the SDGs, custodian agencies have been appointed to monitor and track the progress made towards achieving the identified relevant objectives. At regional and subregional levels, bodies such as the European Union have been setting the pace among the countries with clear renewable energy targets, as has the Association of Southeast Asian Nations or, as will be explained, relevant regional bodies in the African context.

While the global and regional actors and initiatives set the course to follow, the most immediate quantifiable actions are taken at national levels. For this reason, an institutional framework conducive to achieving SDG 7 and climate goals is central. In terms of designing and adopting adequate policies to foster the attainment of sustainable development targets, national parliaments and governments act in tandem, with the latter often operating through dedicated ministries of energy and environment. Typically, specialized renewable energy and rural electrification agencies help with the implementation of these policies. Independent regulatory authorities ensure that the appropriate regulations (subsidies, tariff setting, grid priority, etc) are put in place and followed to create a favourable context for renewable energy. In a number of countries, the rural electrification agencies play a dual role of policy implementation and regulation in off-grid areas or small-scale projects.

Other groups of actors that have been and continue to be instrumental in setting the agenda for the achievement of SDGs are civil society organizations active in the advocacy space; academic institutions, which provide the scientific basis for determined action on climate and energy access; and – very importantly – the private sector, which plays a crucial role in investment.

8.2 Renewable energy and electricity access: the case of Sub-Sharan Africa

8.2.1 Background

Sub-Saharan Africa stands out by being home to three quarters of those without access to electricity worldwide (IEA et al, 2021). In many ways, it epitomizes the energy for all challenge. Bringing electricity to underserved areas has long relied on the extension of national power grids or the installation of diesel-powered distributed generation. The economics of the two solutions have made energy access prohibitively expensive for most African countries. In sub-Saharan Africa, it costs an average of $20,000 to extend the distribution grid with an 11 kV power line by one kilometre (Longe et al, 2017), making it lose economic sense beyond a certain number of kilometres and often limiting grid extension to areas in the vicinity of major cities. At the same time, diesel-powered distributed generation was mostly reserved for critical services, or for those who could afford it but still had to rely on a supply chain that was not always dependable.

However, with the momentum renewable energy solutions have gained since the 2000s, technology innovations and cost motivations have opened new avenues by creating a market for renewables-powered mini- and micro-grids and by making other derivative solutions such as solar home systems more widely available. At a global scale, the period also corresponded to the era of the Millennium Development Goals which, despite not listing energy access as one of the goals, were dependent on available and affordable energy materializing.

In this context and with these new technological opportunities becoming increasingly available, the policy and regulatory landscape of energy access across sub-Saharan African countries has progressively adapted to these new opportunities. A central regional policy initiative here is the New Partnership for Africa's Development (NEPAD)'s all-encompassing drive to increase access to energy supply and its reliability, to reverse the environmental degradation linked to the use of traditional fuels and tointegrate power grids (African Union, 2001). Building on NEPAD, regional economic communities such as the Economic Community of West African States (ECOWAS) put out the White Paper on a Regional Policy for Increasing Access to Energy Services in Peri-Urban and Rural Areas aiming to achieve 66 per cent of electricity access by 2015 and 100 per cent access to improved cooking solutions by the same deadline (ECOWAS, 2006). Similar regional policies spurring in East and Central Africa were developed around the same dates. In the year 2000, the electricity access rate in sub-Saharan Africa stood at 24 per cent (IEA, 2020b), which meant that a little over 500 million people lacked access to electricity. By the year 2019, this rate had increased to 48 per cent, which meant that around 576 million people still lacked access to electricity. It would therefore appear at first glance that all the efforts that have been put in place by governments and all relevant stakeholders failed to even offset the demographic growth.

However, one needs to look closer. Very often, while regional policies create momentum and set countries within a specific region or continent on a certain trajectory, their success relies to a large extent on what is done at the national level, and this is also where the progress tracking is done. In this regard, the latest projections suggest that key countries such as Côte d'Ivoire, Ghana, Kenya and Rwanda, are on course to meeting the electricity access component of SDG 7. This is the result of these countries having taken determined policy action and been able to establish clear strategies to address the energy access challenges. While in Côte d'Ivoire and Ghana, priority was given to an approach favouring grid extension (KfW et al, 2020), a more phased approach with distinct strategies for off-grid areas was followed in Kenya and Rwanda. In those two countries, publicly available electrification plans with clear frameworks for mini-grids and standalone systems have been set out. Key to the success of these countries has been the incentivization of private participation (notably through capital subsidies) and mechanisms to address the affordability of the electricity provided. It therefore is highly instructive to assess the experience of select countries and which key policy and regulatory lessons they can share with other countries. The next section delves into the example of Kenya.

8.2.2 Example of success in achieving SDG 7 in Africa: Kenya

At the start of the millennium, the national electricity access rate in Kenya stood at 15 per cent but it grew dramatically to reach 70 per cent in 2019 (World Bank, 2020a). In rural areas, this rate was about 6 per cent in 2000 and close to 62 per cent today (World Bank, 2020b). With close to 21 million people living in rural areas in 2021 (about 80 per cent of the overall population, as in most Africa countries), Kenya's fight for universal electricity access will be won or lost there. Against the backdrop of renewables becoming widely available at affordable costs, Kenya has been able to develop a market for decentralized renewable energy solutions that is the fastest growing on the continent and that is powering electricity access. The Kenyan government was able to achieve such a feat by combining policy and fiscal levers to send out the market signals to attract the private sector, both local and overseas, and multilateral financing institutions to support attaining its objectives.

One of the first concrete steps the government took was the enactment of the Energy Act in 2006, which saw the establishment of the Energy Regulatory Commission, to create a strong enabling environment for the country's energy sector. Until it was transformed into the Energy and Petroleum Regulatory Authority under the 2019 Energy Act, this independent body was responsible for the economic and technical regulation of electric power, tariff setting and review, licensing, enforcement of compliance, dispute settlement and approval of power purchase and network service contracts (Government of Kenya, 2006).

The same 2006 Energy Act also saw the establishment of the Rural Electrification Authority (REA), which had the mandate to accelerate the pace of rural electrification, with its array of competences encompassing the development and update of the rural electrification masterplan, fund sourcing and the promotion of renewable energy sources, as well as tendering, licensing and permits for rural electrification. REA was also put in charge of the management of the Rural Electrification Fund, which was established through the same Act with the objective of supporting the electrification of rural and other areas considered economically unviable for electrification by licences. After the 2019 Energy Act came into force, the REA was changed to the Rural Electrification and Renewable Energy Corporation, with a broader mission to promote the national renewable energy agenda in addition to its rural electrification remit (Government of Kenya, 2019).

However, the establishing of institutions, while important, cannot lead to meaningful tangible outcomes. Only together with the adequate policy (overarching direction and target) and regulatory (appropriate instruments to steer the market toward achieving the policy objectives) structure can they allow the creation of an overarching enabling environment in which relevant players can play their role. In Kenya, this enabling environment was facilitated by the government putting in place sound long-term energy planning practices for both on-grid electricity and rural electrification. Long-term planning in general, when done properly, helps create consensus among stakeholders, avoid costly investment mistakes, reduce uncertainties in policy directions (Miketa, 2019) and, most importantly, send clear signals to investors on the type and quantity of investments needed.

With all that in place, Kenya has also been able to enact a stick and carrot approach in the shape of fiscal measures to move away from fossil fuels and to boost private participation in the provision of clean energy technologies, which are primarily aimed at supporting rural electrification. The country has exempted quality-assured solar and other renewable energy products from taxes and tariffs. It has no kerosene or diesel subsidies (which can negatively impact the deployment of clean alternatives by making them less competitive) and has introduced a tax on kerosene (Micheni, 2016). Kenya's good standing on the regulatory front to drive electricity access is reflected by the country's high score on the World Bank's Regulatory Indicators for Sustainable Energy (RISE)[2] index. Kenya ranks fourth overall in Africa for electricity access frameworks, but first for the scope of officially approved electrification plans, frameworks for grid electrification and frameworks for standalone systems. The vibrant local market responded positively to the country's efforts, which are behind the current picture.

Overall, the lesson to be drawn from Kenya is that aspirations to tackle the SDG 7 challenge need to be backed by actions and clear signals to the market, incentivizing the private sector to play a bigger role. For Kenya to attain universal access, it still needs to bring some 16 million people to the grid, a goal that the 2018 Kenya National Electrification Strategy hoped to accomplish by the end of 2022. In 2023, this target remains unmet, but the country has made such significant progress over the past decade that the target will be met in the near future.

The positive strides that Kenya has made towards electricity access are not unique across the continent. Eight other sub-Saharan African nations figure among the 20 fastest electrifying countries. In addition, countries such as Côte d'Ivoire, Ethiopia, Ghana, Senegal and South Africa are likely to reach universal access by 2030. For the electricity access numbers for sub-Saharan Africa to start moving in the right direction, it is imperative that bigger countries such as Nigeria and the Democratic Republic of the Congo, which are home to 27 per cent of the sub-Saharan African population without access to electricity, upscale their efforts. For most African economies this means juggling priorities across the sustainable development sphere, and including some other objectives under SDG 7, such as access to clean cooking, for which the picture is somewhat bleaker.

8.3 An eye on clean cooking

Despite its multidimensional nature, which touches on the sectors of energy, health, gender and climate change, clean cooking is the most often overlooked by policy makers of the four target areas for SDG 7 (IEA et al, 2021). Clean cooking is also where, arguably, most of the challenge lies when it comes to clean energy access.

8.3.1 State of play

Evidence suggests that households in developing countries consume approximately 1.5 billion tonnes of wood fuel annually, leading to emissions of roughly 0.8 Gt of CO_2 per year, equivalent to 2 per cent of global greenhouse gas emissions (Parker et al, 2015). This in itself represents about the same as sub-Saharan Africa's total reported GHG emissions in 2018. For many sub-Saharan Africa countries, wood fuel emissions represent about half of their GHG emissions and an even higher share of their total primary energy supply (TPES) and household energy consumption. In Mali, for example, a country located in the Sahel region,[3] wood fuel represents 69 per cent of the TPES and charcoal another 7 per cent. Together, they represent 97 per cent of the residential energy consumption (IPCC, 2019). In the Democratic Republic of the Congo, it is estimated that some 90 per cent of the total volume of wood harvested is for fuel (Shapiro et al, 2021). Consequently, forest degradation trends in the country have reach severe levels. From 2002 to 2020, the Democratic Republic of the Congo lost 5.32 million hectares of humid primary forest, making up 34 per cent of its total tree cover loss in the same time period.

This pattern of wood fuel representing a substantial share of the TPES and the energy consumption by households is very common across sub-Saharan Africa despite the region being home to some of the areas most affected by climate change, such as the Sahel. While not all fuelwood is collected in an unsustainable manner and causes deforestation, evidence shows that the use of biomass for energy was associated with deforestation in some dryland areas (Mirzabaev et al, 2019) such as the Sahel owing to human activities. At the same time, populations in the Sahel region need to find new ways to manage the land cover, which has been decreasing because of anthropogenic climate change. From this perspective, clean cooking is therefore both a climate change mitigation and an adaptation issue.

8.3.2 Clean cooking as a multifaceted policy challenge

At the core of the problem impeding effective action by governments and relevant actors is the severe lack of data on the amount of firewood being collected and how sustainably this is being done. Unlike electricity access, the adoption of clean cooking technologies depends to a great degree on the preferences of the final users (in addition to incentivizing measures). To put it simply, without electricity people will miss out on all the opportunities it brings, but without clean cooking technologies people will still be able to cook. Evidence has shown that the adoption of clean cooking technologies does not always accord with the energy ladder theory, which implies that the more households are economically well off, the more their cooking technology choices become cleaner by moving up from plant/animal waste to wood fuel, then from charcoal to LPG/kerosene and ultimately to electricity. In practice, what is most often observed is a fuel-stacking scenario, when several types of cooking technologies are used by the same household.

To compound the issue, clean cooking remains within the remit of different ministries in different countries because of its multidimensional nature. In countries such as Botswana, Burkina Faso, Côte d'Ivoire and Nigeria the Ministry of Environment and Climate Change oversees policy making, implementation and evaluation, while in others it is the Ministry of Energy (Ghana, Kenya and Senegal). In countries such as Kenya and Rwanda, there is also a strong level of involvement of the Ministry of Health, which also runs parallel programmes. What is lacking is the coordination needed to achieve effective policy deployment, which will require multifaceted strategies to build awareness, pull out the right market levers and drive adoption.

Finally, the global policy environment for advancing clean cooking remains demanding. The WHO definition of clean cooking poses a problem in terms of tracking progress: while the technologies it endorses are included in the tracking framework for SDG 7.1.2, most advances in regulations focus on improved cookstoves, which are not tracked as a clean cooking technology (WHO, 2022). Another aspect here is the extent to which improved cookstoves are considered clean. Under standard testing conditions, very few biomass-based stoves are able to achieve the benchmarks set by WHO for clean cooking technologies. As mentioned, clean cooking is a multidimensional policy challenge that touches on energy, health, gender and climate change. This chapter will not go into the implications of the gap between the WHO definition and the efforts around improved cookstoves, nor elaborate on the important gender dimension of clean cooking, which merits a discussion of its own, beyond its focus on the environmental dimension.

As the world moves steadily closer to 2030, it is likely that the target of SDG 7.1.2 will not be achieved. While a few sub–Saharan African countries (Congo, Côte d'Ivoire, Eswatini, Kenya, and Nigeria) are among the 20 countries with the fastest growing populations who have access to clean cooking fuels and technologies, the region is also home to 19 out of the 20 countries with the lowest percentage of the population with access to clean cooking fuels and technologies. Concerted efforts by all the relevant sectors and actors will need to be made to promote clean cooking as an effective means of building climate resilience.

8.4 Outlook

As part of the implementation of the Paris Agreement, parties to the UNFCCC are required to present nationally determined contributions (NDCs), which are their national ambitions to fight climate change and reduce their GHG emissions. From 2015, NDCs are to be updated every five years, and the first major set of updates was done in 2020–21. Worldwide, the energy sector accounts for three quarters of GHG emissions (IEA, 2021). From that perspective, substantially replacing fossil fuels by renewable energy as prescribed by SDG 7.2 corresponds to the mitigation aspect of climate action.

In their NDC updates and in tune with the rest of the world, sub-Saharan African countries count on the deployment of renewable energy to mitigate the anthropogenic footprint on the environment and to help national environmental governance. For example, Kenya aims to reduce GHG emissions by 32 per cent compared to BAU, by 2030, by increasing renewables in the electricity generation mix of the national grid and enhancing energy efficiency. Nigeria, where energy contributes to 60 per cent of the GHG emissions, plans to reduce its GHG emissions by 20 per cent (unconditional) and 47 per cent (conditional to external support) (Federal Ministry of Environment of Nigeria, 2021). Among the mitigation measures listed by the country in its NDC is energy efficiency and the deployment of renewable energy both on and off the grid.

The global policy imperative here is to provide the necessary international support to these countries in their efforts to decarbonize their energy sectors while at the same time enhancing energy access. The global pandemic has deepened the divide between early decarbonizers and the countries lagging behind, in part as a function of available financial resources (Quitzow et al, 2021). There is also the risk of uneven economic development between climate leaders and laggards (Eicke and Goldthau, 2021), possibly impacting the latter's ability to rise to the SDG 7 challenge. This extends to the imperative to mainstream justice aspects into energy transition and renewable energy policies, so as to reconcile some of the trade-offs entailed in energy decisions (Müller et al, 2021; Global Forest Watch, 2022). Determined global action, including financial and technology support, will ensure a just and equitable transition that accounts for the variety of socio-economic attributes of each country or region, and lives up to the goals underpinning effective, just and persistent environmental governance.

Notes

[1] As studies have shown, the challenge of energy access is, in many parts of the world, correlated with colonialism old and new (Enns and Bersaglio 2020; Allan et al, 2022), as is development more broadly (Acemoglu et al, 2001). This chapter abstains from discussing the colonial origins of energy access in more detail and instead focuses on present policies tackling the latter.

[2] RISE is a set of indicators to help compare national policy and regulatory frameworks for sustainable energy. It assesses each country's policy and regulatory support for each of the four pillars of sustainable energy – access to electricity, access to clean cooking (for 55 access-deficit countries), energy efficiency and renewable energy. The access to electricity pillar covers eight indicators, namely electrification planning, scope of officially approved electrification plan, framework for grid electrification, framework for mini-grids, framework for standalone systems, affordability of electricity for consumers, utility transparency and monitoring, and utility creditworthiness.

[3] The Sahel is a semi-arid part of Africa spanning parts of northern Senegal, Eritrea and the extreme north of Ethiopia, passing by southern Mauritania, central Mali, northern Burkina Faso, the extreme south of Algeria, Niger, the extreme north of Nigeria, the extreme north of Cameroon and Central African Republic, central Chad, central and southern Sudan, the extreme north of South Sudan.

References

Acemoglu, D., Johnson, S. and Robinson, J.A. (2001) 'The colonial origins of comparative development: an empirical investigation', *American Economic Review*, 91(5): 1369–401.

Acharya, R.H. and Sadath, A.C. (2019). 'Energy poverty and economic development: household-level evidence from India', *Energy and Buildings*, 183: 785–91.

Acheampong, A. O., Dzator, J. and Shahbaz, M. (2021) 'Empowering the powerless: does access to energy improve income inequality?', *Energy Economics*, 99: 105288.

African Union (2001) *The New Parternship for Africa's Development*, Pretoria: African Union.

Allan, J., Lemaadel, M. and Lakhal, H. (2022) 'Oppressive energopolitics in Africa's last colony: energy, subjectivities, and resistance', *Antipode*, 54(1): 44–63.

Bennett, N.J. and Satterfield, T. 2018. 'Environmental governance: a practical framework to guide design, evaluation, and analysis', *Conservation Letters*, 11 (6): e12600.

Biggs, E.M., Bruce, E., Boruff, B., Duncan, J.M.A., Horsley, J., Pauli, N. et al (2015) 'Sustainable development and the water–energy–food nexus: a perspective on livelihoods', *Environmental Science & Policy*, 54: 389–97.

ECOWAS (Economic Community of West African States) (2006) 'White paper for a regional policy: geared towards increasing access to energy services for rural and periurban populations in order to achieve the Millennium Development Goals', Abuja: ECOWAS.

Eicke, Laima and Goldthau, Andreas (2021) 'Are we at risk of an uneven low-carbon transition? Assessing evidence from a mixed-method elite study,' *Environmental Science & Policy*, 124: 370–9.

Enns, C. and Bersaglio. B. (2020) 'On the coloniality of 'new' mega-infrastructure projects in East Africa', *Antipode*, 52 (1): 101–23.

EPA (United States Environmental Protection Agency) (2014) *Emission Factors for Greenhouse Gas Inventories*, Washington, DC: EPA.

Erlewein, A. (2013) 'Disappearing rivers: the limits of environmental assessment for hydropower in India', *Environmental Impact Assessment Review*, 43: 135–43.

ESCAP (Economic and Social Commission for Asia and the Pacific) (2021) 'Universal access to all: maximizing the impact of clean cooking', Policy brief, Bangkok: United Nations Economic and Social Commission for Asia and the Pacific.

Federal Ministry of Environment of Nigeria (2021) Nigeria's First Nationally Determined Contribution – 2021 Update; https://climatechange.gov.ng/wp-content/uploads/2021/08/NDC_File-Amended-11222.pdf

Global Forest Watch (2022) Homepage, Available from: https://tinyurl.com/mrx8knff [Accessed 20 August 2022].

Government of Kenya (2006) 'Energy Act, No. 12 of 2006', Nairobi. Available from: www.kenyalaw.org/kl/fileadmin/pdfdownloads/LegalNotices/2012/LN44_2012.pdf [Accessed 20 August 2022].

Government of Kenya (2019) 'Energy act', Nairobi. Available from: https://www.epra.go.ke/download/the-energy-act-2019/# [Accessed 20 August 2022].

Günel, G. (2019) *Spaceship in the Desert. Energy, Climate Change, and Urban Design in Abu Dhabi* , Durham, NC: Duke University Press.

IEA (International Energy Agency) (2020a) *Energy Efficiency 2020*, Paris: IEA.

IEA (2020b) *SDG7: Data and Projections. Access to Affordable, Reliable, Sustainable and Modern Energy for All*, Paris: IEA.

IEA (2021) *World Energy Outlook*, Paris: IEA.

IEA, IRENA (International Renewable Energy Agency), UNSD (United Nations Statistics Division), World Bank and WHO (World Health Organization) (2021), *Tracking SDG7: The Energy Progress Report, 2021*, Washington, DC: World Bank.

IPCC (2019) *Sixth Assessment Report: Renewable Readiness Assessment: Mali*, Abu Dhabi: IRENA.

IRENA and CPI (Climate Policy Initiative) (2020) *Global Landscape of Renewable Energy Finance*, Abu Dhabi: IRENA.

Jiang, L., Yu, L., Xue, B., Chen, X. and Mi, Z. (2020) 'Who is energy poor? Evidence from the least developed regions in China', *Energy Policy*, 137: 111122.

Jiborn, M., Kander, A., Kulionis, V., Nielsen, H. and Moran, D.D. (2018) 'Decoupling or delusion? Measuring emissions displacement in foreign trade', *Global Environmental Change*, 49: 27–34.

KfW, GIZ, IRENA and BMZ (2020) *The Renewable Energy Transition in Africa: Powering Access, Resilience and Prosperity*, Frankfurt: KfW, Available from: www.irena.org/-/media/Files/IRENA/Agency/Publication/2021/March/Renewable_Energy_Transition_Africa_2021.pdf?rev=6a9b2f3239be4031b3b074643ec58ca5 [Accessed 21 June 2023].

Longe, O. M., Rao, N. D., Omowole, F., Oluwalami, A.S. and Oni, O.T. (2017) 'A case study on off-grid microgrid for universal electricity access in the Eastern Cape of South Africa', *International Journal of Energy Engineering*, 7(2): 55–63.

Maji, P. and Kandlikar, M. (2020) 'Quantifying the air quality, climate and equity implications of India's household energy transition', *Energy for Sustainable Development*, 55: 37–47.

McCollum, D.L., Echeverri, L.G., Busch, S., Pachauri, S., Parkinson, S., Rogelj, J. et al (2018) 'Connecting the sustainable development goals by their energy inter-linkages', *Environmental Research Letters*, 13(3): 033006.

Micheni, M. (2016) *Five lessons from Kenya's energy access boom*, Thomson Reuters Foundation, Available from: https://news.trust.org/item/2016091 4071101-xf0ln [Accessed 20 August 2022].

Miketa, A. (2019) Central Asia workshop on long-term capacity expansion planning with a high share of renewables, Au Dhabi: IRENA, www.irena. org/events/2019/Mar/Workshop-on-long-term-capacity-expansion-planning-with-a-high-share-of-renewables [Accessed on 20 August 2022].

Müller, F., Claar, S., Neumann, M. and Elsner, C. (2021) 'Assessing African energy transitions: renewable energy policies, energy justice, and SDG 7', *Politics and Governance*, 9(1): 119–30.

Mirzabaev, A., Wu, J., Evans, J., García-Oliva, F., Hussein, I.A.G., Iqbal, M.H., Kimutai, M., Knowles, T., Meza, F., Nedjraoui, D., Tena, F., Türkeş, M., Vázquez, R.J. and Weltz, M. (2019): Desertification. In: *Climate Change and Land: an IPCC special report on climate change, desertification, land degradation, sustainable land management, food security, and greenhouse gas fluxes in terrestrial ecosystems.* Geneva: Intergovernmental Panel on Climate Change.

Nerini, F.F., Tomei, J., To, L.S., Bisaga, I., Parikh, P., Black, M. et al (2018) 'Mapping synergies and trade-offs between energy and the Sustainable Development Goals', *Nature Energy*, 3: 10–15.

Ojong, N. (2021) 'The rise of solar home systems in sub-Saharan Africa: examining gender, class, and sustainability', *Energy Research and Social Science*, 75: 102011.

Parker, C., Keenlyside, P., Galt, H., Haupt, F. and Varns, T. 2015. 'Linkages between cookstoves and REDD+: a report for the Global Alliance for Clean Cookstoves' Washington, DC: Climate Focus.

Partzsch, L. (2020) *Alternatives to Multilateralism: New Forms of Social and Environmental Governance*, Cambridge, MA: MIT Press.

Quitzow, Rainer, Bersalli, German, Eicke, Laima, Jahn, Joschka, Lilliestam, Johan, Lira, Flavio, Marian, Adela, Süsser, Diana, Thapar, Sapan, Weko, Silvia, Williams, Stephen and Xue. Bing (2021) 'The COVID-19 crisis deepens the gulf between leaders and laggards in the global energy transition', *Energy Research & Social Science*, 74: 101981.

Rosenberg, M., Armanios, D.E., Aklin, M. and Jaramillo, P. (2020) 'Evidence of gender inequality in energy use from a mixed-methods study in India', *Nature Sustainability*, 3(2): 110–18.

Shapiro, A.C., Bernhard, K., Zenobi, S., Mueller, D., Aguilar-Amuchastegui, N. and Dannnunzio, R. (2021) 'Proximate causes of forest degradation in the Democratic Republic of the Congo vary in space and time', *Frontiers in Conservation Science*, 2, doi: 10.3389/fcosc.2021.690562.

Sohn, L.B. (1973) 'The Stockholm Declaration on the Human Environment.' *Harvard International Law Journal*, 14(3): 423.

Sovacool, B.K., Hook, A., Martiskainen, M., Brock, A., and Turnheim, B. (2020) 'The decarbonisation divide: contextualizing landscapes of low-carbon exploitation and toxicity in Africa', *Global Environmental Change*, 60: 102028.

Sovacool, B.K. and Turnheim, B., Hook, A., Brock, A. and Martiskainen, M. (2021) 'Dispossessed by decarbonisation: reducing vulnerability, injustice, and inequality in the lived experience of low-carbon pathways', *World Development*, 137: 105116.

UN (United Nations) (2015) 'Resolution adopted by the General Assembly on 25 September 2015: Transforming our world: the 2030 Agenda for Sustainable Development', A/RES/70/1, New York.

UN (United Nations) (2018) *Accelerating SDG7 Achievement – Policy Brief 08 - Interlinkages among Energy, Poverty and Inequalities,* New York: United Nations Department of Economic and Social Affairs. Available from https://sustainabledevelopment.un.org/content/documents/17480PB8.pdf [Accessed 20 August 2022].

Vagliasindi, M. (2013) *Implementing Energy Subsidy Reforms: Evidence from Developing Countries*, Washington. DC: The World Bank.

WHO (2021) 'Global Health Observatory', Geneva: WHO.

WHO (2022) 'Defining clean fuels and technologies', Available from: www.who.int/tools/clean-household-energy-solutions-toolkit/module-7-defining-clean [Accessed 20 August 2022].

World Bank (2017) 'World Bank Group announcements at One Planet Summit', 12 December, Available from: https://www.worldbank.org/en/news/press-release/2017/12/12/world-bank-group-announcements-at-one-planet-summit [Accessed 21 June 2023].

World Bank (2020a) 'Access to electricity (% of population) – Kenya', Available from: https://data.worldbank.org/indicator/EG.ELC.ACCS.ZS?locations=KE [Accessed 21 June 2023].

World Bank (2020b) 'Access to electricity, rural (% of rural population) – Kenya', Available from: https://data.worldbank.org/indicator/EG.ELC.ACCS.RU.ZS?locations=KE [Accessed 21 June 2023].

World Bank (2020c) 'Access to electricity (% of population)', Available from: https://data.worldbank.org/indicator/EG.ELC.ACCS.ZS [Accessed 27 June 2023].

Zhao, J., Shahbaz, M. and Dong, K. (2021) 'How does energy poverty eradication promote green growth in China? The role of technological innovation', *Technological Forecasting and Social Change*, 175: 121384.

Interview with Nopenyo E. Dabla and Andreas C. Goldthau: The Disconnect Between Sustainability and Development

Ettore Benetti and Marco Aurélio Mayer Duarte Neto

What do you see as the main trade-off or issue for energy governance in sub-Saharan Africa?

Dabla: The main problem is a disconnect between sustainability and development, but this is starting to fade away little by little. We're starting to discover many new opportunities that can be created through sustainable energy. We're talking about sustainable development in general, but also about socioeconomic benefits from an energy perspective. Among such benefits, for example, political commitment to an energy transition will create jobs that help people to feel more empowered. In Africa, where the great majority of the population is young, jobs need to be created for the future, and entrepreneurship prospects stemming from the energy transition provide an opportunity for them.

Governments play an important role in providing funding for transforming the energy market. How can they bring private investors into play?

Dabla: Despite the big role governments play, private sector participation is paramount for the energy transformation to materialize. However, private investors need to know whether they will be able to recoup what they invest. In the energy access context in remote areas, for example, consumer levels aren't always enough to attract private operators due to low levels of general income.

However, in several African countries, subsidies on capital expenditure for rural electrification via mini-grids or even subsidies on operating expenditure in some cases, have been ways for governments attract investors into this space. By removing taxes on renewable energy products and limiting risks through mitigation mechanisms, the renewables market gains attractiveness. Long-term planning is also crucial. A government must lay the foundation to receive investments through an energy sector plan that accounts for different risks, showing readiness and willingness to act. This reality can be extended to the on-grid context, where even more capital is required and where risk mitigation instruments and guarantees play a big part in securing the required private sector confidence in the market.

How did perceptions of energy policy change after the war in Ukraine began? What can we expect for the future?

Goldthau: In Europe there are two things going on at the same time. On the one hand, fossil fuels have at least a temporary comeback. Europeans are eager to get coal and gas supplies from countries other than Russia, against the backdrop of increasing concerns over energy security. However, at the same time, we see a rush towards renewables, clean transitions and demand-side measures, putting the European energy transition on steroids. There will be a couple of years where the European energy system remains vulnerable. But the current measures are going to pay off. It is important to understand that the significant investments that are made at present come with a securitization of renewables. Securitization means politicians have taken renewables out of the climate domain and put it into the security domain. This is how extraordinary measures become possible: if there are trade-offs between, say, environmental concerns of enlarging offshore wind and national security, decisions are taken in favour of the latter.

Will Europe's challenges of achieving its 2030 goals and the current energy crisis delay the transition to cleaner energy sources in the rest of the world?

Goldthau: There is a risk. The EU is replacing Russian gas mainly with liquefied natural gas (LNG). For LNG the market is rather inflexible in supply as additional production capacity has long lead times. This puts Europe in direct price competition with China, Japan and South Korea, the other large importers of LNG. Yet, smaller consumers and countries with lower purchasing power risk being priced out. This may impact their energy security. Worse, if they were banking on natural gas as a transition fuel, they may now settle for coal instead. This does not spell good news for the Paris climate targets.

9

From Economic Growth to Socio-ecological Transformation: Rethinking Visions of Economy and Work under SDG 8

Ekaterina Chertkovskaya

In contemporary societies across the world, as well as in global governance frameworks, economic growth is seen as a precondition to solving societal problems, including inequality, ecological degradation and other grand challenges. Within the Sustainable Development Goals adopted by the United Nations too, economic growth is seen as one of the universal goals to be aimed for. In comparison to the Millennium Development Goals preceding them, which 'focused on improving well-being in the developing world, the 17 SDGs address all countries and aim at reconciling economic and social with ecological goals' (Eisenmenger et al, 2020: 1101). They have become a key reference point for governance at all levels, from supranational and national to regional, municipal, organizational and local. However, they contain a problematic assumption, as manifested in SDG 8 (Decent work and economic growth): that perpetual economic growth is socially desirable and ecologically sustainable. This chapter contributes to the discussion on environmental governance by questioning the centrality of economic growth in SDGs and arguing for a reorientation of economies towards socio-ecological transformation.

SDG 8 refers simultaneously to two spheres of life – the economy and (paid) work, implying a strong link between the two. It is formulated the following way: 'Promote sustained, inclusive and sustainable economic growth, full and productive employment and decent work for all' (UN, 2020a). Growth, however, has not always been a societal goal. It is a particular ideology that has taken shape mostly since the end of the Second World War (Barry, 2020). This has been the case both in capitalist countries, with the OECD as a key institution fostering the growth agenda, and in the countries of the Eastern Bloc, such as the Soviet Union, albeit with notably different approaches (Schmelzer, 2016; Chertkovskaya, 2019). It is the pursuit of growth within capitalist economies, however, that is part and parcel of the SDGs, with economic growth being key to accumulation under the capitalist mode of production (Kallis, 2019; Chertkovskaya and Paulsson, 2021). Within these contexts, economic growth has been associated with making possible the welfare states that emerged in the second half of the twentieth century, and is often credited with addressing poverty. However, the pursuit of growth goes in hand with ecological degradation, while inequalities have been rising since the 1980s, with the richest groups benefiting most from economic growth and also contributing more to ecological degradation (Alvaredo et al, 2018; Hickel and Kallis, 2020; Oswald et al, 2020; Wiedmann et al, 2020). While the recognition of the ecological limits to growth was present in international governance frameworks in the 1970s, the adoption of the sustainable development agenda and the 1992 Earth Summit in Rio have marked attempts to reconcile economic growth with sustainability, which persists in the SDGs today (Gómez-Baggethun and Naredo, 2015).

Addressing grand social challenges via what is at the root of the problem will only exacerbate it. Instead, it is necessary to move away from the ideology of growth and to focus on socio-ecological transformation. Such a shift would allow governance models to be devised that can truly reconcile social and ecological objectives, and that are informed by a pluriverse of alternatives to capitalism, development and growth (Kothari et al, 2019). This chapter speaks from this stance, drawing on the research literature on degrowth. Degrowth critiques the centrality of economic growth in today's economies and societies, and seeks their organization with ecological sustainability, social justice and human flourishing at the core (Chertkovskaya and Paulsson, 2016). This implies societies with a lower biophysical throughput while ensuring well-being for all (Kallis, 2018). While degrowth as a concept stems from the European context, being a response to the multiple crisis of capitalism and modernity, it aligns with a pluriverse of alternatives to development across the world, such as *buen vivir*, ecological *swaraj* and many others (Kothari et al, 2014; 2019).

The rest of this chapter will proceed as follows. First, the problems with the pursuit of economic growth, which are left unnoticed under SDG 8, are elaborated. Second, the unsustainability and injustices of how work is organized in growth-centric economies is discussed. To go beyond these critiques and towards devising alternative forms of governance, it is articulated how both economy and work could be reoriented from the focus on economic growth to socio-ecological transformation. The chapter concludes by suggesting how an alternative to SDG 8 could be formulated.

9.1 Problems with the pursuit of economic growth

SDG 8 consists of ten targets, which are focused on economic growth, work or a combination of the two (see Kreinin and Aigner, 2021: table 2). Targets 8.1 and 8.4 give an overall vision for the economy and its relation to environmental degradation, respectively:

> 8.1 Sustain per capita economic growth in accordance with national circumstances and, in particular, at least 7% gross domestic product growth per annum in the least developed countries.

> 8.4 Improve progressively, through 2030, global resource efficiency in consumption and production and endeavour to decouple economic growth from environmental degradation, in accordance with the 10-year framework of programmes on sustainable consumption and production, with developed countries taking the lead. (UN, 2020a)

Target 8.1 positions economic growth, calculated as gross domestic product, as the key to defining and evaluating the economy within SDG 8. With reference to decoupling, target 8.4 positions continuous economic growth as possible without furthering ecological degradation. The problem with these targets, as this section will show, is that they are not achievable without exacerbating ecological devastation and injustices. In addition, they are expected to be achieved via the expansion of sectors that are environmentally and socially destructive.

Before unpacking this argument, it is worth noting two points. First, GDP growth is a problematic measure. It emphasizes economic growth in monetary terms, and thus does not discern harms associated with the pursuit of growth; it also does not include the many non-monetized activities that are essential for maintaining the economy and sustaining life (Dengler and Lang, 2021; Chertkovskaya, 2022). Second, this growth is expected to be perpetual and compound, that is, growing forever and at exponential rates (Kallis et al, 2020). With an annual global growth rate of 3 per cent expected under SDG 8 and in various economic institutions global economies are projected to double in size only in 24 years from now, and to grow sixteen-fold in a century (Hickel, 2019; Kallis et al, 2020). From a degrowth stance, such growth is neither feasible nor desirable. What is needed, instead, is a society that does not depend on growth for living well, where a good life and well-being are available for all within the planetary boundaries, and nobody is living at someone else's expense (Brand et al, 2021).

9.1.1 Growth is unsustainable and unjust

The period since the 1950s, known as the 'great acceleration', has been associated with an exponential rate of economic growth. However, growth has been tightly coupled with use of energy, resources and materials and, as a result, with various environmental problems, from the rise of CO_2 emissions to ocean acidification (see Steffen et al, 2015). The question of the limits to growth had already been raised in the 1970s, with the *Limits to Growth* report being a key publication at the time (Meadows et al, 1972). It is also when the term *décroissance* was coined by André Gorz, which after over 30 years was translated to 'degrowth' (Kallis et al, 2015).

In response to the tight connection of economic growth with environmental degradation, the hypothesis of *green growth* emerged (see Hickel and Kallis, 2020). Its key premise is that it is possible to decouple this connection, and to have continuous economic growth that is ecologically sustainable, via resource efficiency and the development of new technologies and sectors of the economy. The stance on growth can be seen as part of a shift in international governance policies from discussing growth versus the environment since the 1970s to growth *for* the environment since the Brundtland Report in 1987 (Gómez-Baggethun and Naredo, 2015). The green growth hypothesis has been adopted and promoted by many international organizations, such as the OECD, the World Bank and the UNEP, as well as by nation-states and corporate actors.

Green growth is also key to SDG 8, with target 8.4 pointing to 'global resource efficiency' that is expected to make decoupling possible. However, to achieve decoupling at the rates of growth expected under target 8.1, efficiency gains three to six times higher than have ever been known in history are needed (Hickel, 2019). This is therefore extremely unrealistic, contradicting sustainable resource use and environmental goals, such as SDGs 6, 12, 13, 14 and 15 (Hickel, 2019; Eisenmenger et al, 2020). Overall, there is no empirical evidence of decoupling of material throughput and environmental pressures from economic growth on a global scale or over the long term (Parrique et al, 2019; Hickel and Kallis, 2020). While decoupling of only CO_2 emissions from economic growth could theoretically be achieved, it has been estimated not to be fast enough to live up to the goal of the Paris Agreement limiting global warming to 1.5 ° to 2 °C (Hickel and Kallis, 2020). The Nordic countries have been positioned as one example of 'genuine green growth' (Stoknes and Rockström, 2018), but their 'success' has been questioned, as decoupling within these national contexts is dependent on increased environmental pressures in other parts of the world (Tilsted et al, 2021).

Indeed, a systemic inequality has been observed between high- and low-income nations, where the former are net importers of embodied materials, energy, land and labour, gaining a monetary trade surplus, while the latter provide resources but experience monetary trade deficits (Dorninger et al, 2021). Similarly, there is a large inequality in international and intranational energy footprints between different income groups: 39 per cent of energy is consumed by the top 10 per cent, while the share of the bottom 10 per cent is only 2 per cent (Oswald et al, 2020). Affluence, in other words, has been an important driver of environmental pressures (Wiedmann et al, 2020). International development policies since the 1990s have been sidelining this issue, positioning poverty rather than opulence as a problem for ecological sustainability, thus ignoring critiques of growth and framing the poor as responsible for ecological degradation (Gómez-Baggethun and Naredo, 2015; see also Kothari et al, 2014). This neglects how Indigenous, peasant and working class communities have been at the forefront of environmental struggles, caring for and sustaining their environments, and fighting injustices associated with the pursuit of growth and capital accumulation (Akbulut et al, 2019; Barca, 2019; 2020; Martínez-Alier, 2021). The Environmental Justice Atlas has documented thousands of cases around the world where these communities contest socio-ecological degradation that comes with extractivism, industrial activities and waste disposal related to the fossil economy but also to many sectors that are positioned as 'green' (EJAtlas, 2022).

SDG 8, and the Agenda 2030 more generally, can be described as eco-modernist and technologically optimistic, neglecting the more feasible and desirable low-energy demand scenario that is also compatible with increased well-being (Hickel et al, 2021; Vogel et al, 2021). This connects to a broader shift from political to technocratic ways to address international sustainability policy that came with the green growth agenda (Gómez-Baggethun and Naredo, 2015). This holds for target 8.2, which emphasizes productivity via technology development:

> 8.2 Achieve higher levels of economic productivity through diversification, technological upgrading and innovation, including through a focus on high-value added and labour-intensive sectors. (UN, 2020a)

This target underestimates the materiality of technology and the amounts of energy, resources and materials needed in technologically intensive production processes. For example, digitalization is emphasized under SDG 9 and can be seen as one area where technology is rapidly developing. It has been estimated that information and communications technology already takes up about 3.7 per cent of CO_2 emissions, which are expected to rise to 8 per cent by 2025 (Shift Project, 2019) and is supported by very material infrastructures and processes, from undersea water cables that make internet connection possible to rare minerals in technology used in our everyday lives.

All major climate mitigation scenarios rely on negative emissions technologies such as carbon capture and storage (CCS) or bioenergy with CCS for decoupling to be achieved. These technologies, which do not exist on a large scale today, require hierarchical systems of control and cannot be managed by small villages or communities, while their side effects, such as concentrated dioxide pollution due to leakage, will directly affect communities (Muraca and Neuber, 2018).

Finally, there are well-recorded rebound effects associated with technological development, showing that efficiency gains per unit of output tend to result in increases in overall production and consumption of a product, thus partially or totally counterbalancing these gains (York and McGee, 2016; Ruzzenenti et al, 2019). For example, if a car becomes more efficient, producers oriented towards profit and production growth will want to produce more such cars for the market, which will result in higher use of energy and resources, despite the efficiency gains. Consumers, in turn, might drive their more efficient cars more and at higher speeds. In other words, the material and energy throughput associated with technology development further shows how SDG 8 contradicts environmental SDGs for example, SDG 13 (Climate action) and SDG 15 (Life on land) and prioritizes economic growth over sustainable resource use (Eisenmenger et al, 2020).

9.1.2 Growth is expected to be achieved via expansion of already problematic sectors

Within SDG 8, some sectors are specifically emphasized as areas for growth. One of them is tourism, as articulated in target 8.9:

> 8.9 By 2030, devise and implement policies to promote sustainable tourism that creates jobs and promotes local culture and products. (UN, 2020a)

However, tourism as a sector of the economy is highly unsustainable. It relies heavily on the aviation industry, which accounted for about 2.9 per cent of CO_2 emissions in 2018, but also had a threefold contribution to global warming via non-CO_2 emissions (Lee et al, 2021). These emissions cannot be reduced drastically, and there is simply no fossil-free aviation (Stay Grounded and PCS, 2021). Passenger aviation is also an activity that only a small proportion of people participate in, with only 5–20 per cent of the world's population having ever set foot on a plane (IEEP, 2019). While the connection of tourism and aviation is implied in SDG 8, aviation is explicitly promoted in a related SDG 9 (Industries, innovation and infrastructure). Within this goal, in view of the COVID-19 pandemic, it is highlighted how the aviation has been hit and that its recovery is needed as it is associated with economic growth and development (UN, 2020b). What is needed for a sustainable society, however, is a reduction in aviation and the promotion of sustainable modes of transport that are organized in the public interest. Here, the problems created by the pandemic open up spaces to rethink patterns of travelling and modes of travel to reduce the overall level of air traffic while ensuring a just transition for workers and mobility for people (Stay Grounded and PCS, 2021).

Problems with the emphasis on the expansion of tourism under SDG 8 go beyond aviation. In the context of orientation towards growth and profit, tourism comes with violence and dispossessions, as well as inequalities and waste (Büscher and Fletcher, 2017; Devine and Ojeda, 2017). There is a huge difference in who has access to tourism, and a bigger flow of tourists from richer to poorer countries. The gains from tourism in the so-called host countries often stay with international companies that are not even based in these countries, while at best providing precarious jobs to the local working class (Bianchi and De Man, 2021). Thus, a rethink of what sustainable tourism means is needed, rather than defining the success of tourism in terms of growth in tourism numbers (Gössling et al, 2021).

Another sector that is explicitly flagged in SDG 8 is finance, as formulated by target 8.10:

8.10 Strengthen the capacity of domestic financial institutions to encourage and expand access to banking, insurance and financial services for all. (UN, 2020a)

This is a very open formulation of what is expected from finance (Kreinin and Aigner, 2021), and the indicators under this goal refer mainly to access to commercial bank branches and automated teller machines (ATMs), as well as the proportion of adults with a bank account. However, in emphasizing the importance of finance, this sector is assumed to be good in itself. This ignores the experience of the financial crisis, when millions of indebted people and the public sector suffered deprivation, austerity and new waves of neoliberalization as a result of finance boosting economic growth through speculative practices and creation of private debt (eg Marazzi, 2010). Furthermore, positioning financial institutions simply as providers of 'financial services for all' silences how the financial sector keeps investing into the fossil economy (eg RAN, 2019; Bauer and Fontenit, 2021).

In the related SDG 9, industrialization of economies is emphasized:

9.2. Promote inclusive and sustainable industrialization and, by 2030, significantly raise industry's share of employment and gross domestic product, in line with national circumstances, and double its share in least developed countries. (UN, 2020c)

Industry worldwide is already a big source of environmental degradation, as well as environmental justice conflicts, which come with violence to people and their livelihoods (Roy and Martínez-Alier, 2019; Chertkovskaya and Paulsson, 2021; Martínez-Alier, 2021). At the same time, the industry is immensely powerful and actively engaging in lobbying, with the policies targeting the industry often designed to protect rather than transform it (Nilsson et al, 2021).

In sum, SDG 8 and other SDGs related to the economy highlight the importance of sectors such as finance and tourism, and the processes of industrialization and digitalization, but without taking into account their contribution to unsustainability and injustices or the quality of their transformation. Their contribution to GDP, and their growth in the proportion of GDP, are seen as desirable in themselves, without reflection on why this expansion is needed in the first place or how much of it needed, or on its environmental and social consequences.

9.2 Problems with work in a growth-centric economy

Work is a key sphere of life that co-constitutes the economy, with any economic activity relying on human labour as well as nature. The key characteristic of work under SDG 8 is being 'decent': 'Decent work means opportunities for everyone to get work that is productive and delivers a fair income, security in the workplace and social protection for families, better prospects for personal development and social integration' (UN, 2020d). While these are important characteristics of work, there are two problems with this definition. First, it does not refer to ecological sustainability in relation to work. Second, characteristics such as job security and a fair income are eroded in an economy that is oriented to growth at its core. Severe problems associated with work that are highlighted in SDG 8 will not be solved without going away from growth, as they are what enables growth in the first place. In other words, how work is organized contributes to ecological degradation and is rooted in injustices, which will be elaborated in this section.

9.2.1 Unsustainability of work and productivity

Work is not inherently unsustainable, but it is so under the current relations of production, with a lot of work today going into fossil-dependent, productivist and wasteful economic activities. According to Hoffman and Paulsen (2020), four factors make work unsustainable: scale, time, income, work-induced infrastructure, mobility and consumption. First, when it comes to scale, more working hours require more energy and materials, and are generally associated with a higher ecological footprint, even if there are qualitative differences between the kind of work that is undertaken (Hayden and Shandra, 2009). Second, working time poses a constraint on people and frames their consumption patterns outside work. Fast, efficient and easy but also unsustainable consumption is likely to be a key behavioural pattern in the absence of time (Schor, 2005). Third, work is usually associated with getting a monetary income, and higher incomes lead to more consumption and higher ecological footprints (Wiedmann et al, 2020). Finally, work-induced mobility, infrastructure and consumption are associated with different processes surrounding work and their reliance, for example, on driving and flying, energy-consuming office spaces and fast fashion (Hoffman and Paulsen, 2020). For example, employers actively engage in promoting their employer brands and consumption opportunities that come with work (Chertkovskaya et al, 2020), which encourages a consumerist lifestyle and, indeed, may make such a lifestyle a necessary part of one's job.

The emphasis on economic productivity under SDG 8, including through a focus on labour-intensive sectors and in the aforementioned target 8.2, is misplaced. High value added is often not associated with labour-intensive production. Indeed, there is an increased risk of labour-intensive production being replaced by new technologies such as artificial intelligence because it is not as productive as the technology would allow. The widespread use of high value-added technology, as in the gig economy for example, actually enables workers to be replaced more easily, and to be employed under precarious and low-paid conditions.

Furthermore, the emphasis on economic productivity can challenge the quality of work that is done. In much work that is key to well-being in society (for example, of teachers, care workers and nurses) what matters is how work is done rather than how much or how quickly. For example, the emphasis on productivity in neoliberal healthcare often means that healthcare staff have limited time to devote to each person and are pressurized to treat many more people than they can handle, which often happens in understaffed and underfunded public healthcare. At the same time, work like this would usually not be associated with a high value added, which comes with jobs of, say, financiers or consultants. Thus, it is systematically undervalued in the current economic system (Dengler and Lang, 2021; MacGregor and Mäki, Chapter 10 in this volume), and the emphasis on productivity through work will only further this trend.

To sum up, the emphasis on economic productivity under SDG 8 might contribute to economic growth, but it also contributes to ecological degradation and compromises the very decency of work and working conditions that SDG 8 seeks.

9.2.2 Work and injustices

Work is unjustly distributed within and across societies, with the hardest, least valued and most invisibilized work falling on the shoulders of working-class communities, women, people of colour and migrants. Tedious labour-intensive production has increasingly moved to countries in the Global South where labour is cheap, characterized by unequal exchange (Emmanuel, 1972) and often conducted under hazardous, dangerous and unprotected conditions, with wages that are insufficient to live well. Even when work is done in safe working environments, much of it is still alienating, deskilling and lacks meaning, whilst done at accelerated rhythms (Chertkovskaya and Stoborod, 2018).

Modern slavery is a severe form of injustice surrounding work, characterized by forced labour and ownership/control via threat and abuse, dehumanization and commoditization, constraints on freedom and movement, and economic exploitation through underpayment (Crane, 2013: 51). This is still a significant phenomenon of contemporary work, with 16 million people working under these conditions (ILO and Walk Free Foundation, 2017). SDG 8 acknowledges and addresses it:

> 8.7 Take immediate and effective measures to eradicate forced labour, end modern slavery and human trafficking and secure the prohibition and elimination of the worst forms of child labour, including recruitment and use of child soldiers, and by 2025 end child labour in all its forms. (UN, 2020a)

Modern slavery is present in industries such as agriculture, mining and extraction, construction and some forms of manufacturing, as well as in unregulated or poorly regulated service industries (Crane, 2013; ILO and Walk Free Foundation, 2017). For example, in electronics manufacturing, the company Foxconn, the largest industrial employer in China, became notorious after a wave of employee suicides in 2010. It is characterized by flexible contracts, the hiring of migrant workers, highly securitized work settings, extreme intensity of work and punishment for non-compliance (Pun et al, 2016). However, far from being an exception, the model of Foxconn characterizes electronics manufacturing more generally in different parts of the world (Andrijasevic and Sachetto, 2017; Lüthje and Butollo, 2017). The industries where modern slavery has been recorded more often are contributors to productivity and economic growth, and some like mining and electronics manufacturing are the base for technology innovation, which is emphasized under SDG 8.

The issues raised are expected to be addressed under SDG 8 to some extent, via the targets that emphasize safe and secure work environments, fair wages and protection of labour rights:

> 8.5 By 2030, achieve full and productive employment and decent work for all women and men, including for young people and persons with disabilities, and equal pay for work of equal value.

8.8 Protect labour rights and promote safe and secure working environments for all workers, including migrant workers, in particular women migrants, and those in precarious employment. (UN, 2020a)

What is not acknowledged in these targets is that the problems they are addressing are systemic within an economy oriented towards economic growth, productivity and profit rather than sufficiency, quality and well-being. Work under neoliberalism has become more precarious, with employability having replaced job security – the pursuit of economic growth having been an important driver – by allowing employers to adapt to the fluctuations of the market (Standing, 2011; Chertkovskaya et al, 2013). This is also reflected in a shift from state regulation to market instruments in international governance frameworks (Gómez-Baggethun and Naredo, 2015). By focusing on 'decent work', SDG 8 addresses only the symptoms rather than the root of the problem. Both work and economy need, instead, to be reoriented towards socio-ecological transformation.

9.3 Reorienting economy and work towards socio-ecological transformation

Achieving well-being for people and the planet requires a socio-ecological transformation altering the fundamental attributes of a system, and constituting new meanings and practices (Asara et al, 2015: 379). The economy with its practices of production, consumption and waste is a key area to be transformed. What is needed instead of the focus on perpetual and compound economic growth is a transformation to a regenerative economy focused on well-being and equity within planetary boundaries (Raworth, 2017; O'Neill et al, 2018; Kallis et al, 2020). It would focus not on producing more but on producing sufficiently and differently to satisfy human needs (Büchs and Koch, 2017; Kallis, 2018).

To enable socio-ecological transformation, the centrality of economic growth in societies, economies and policy frameworks should be deprioritized and the focus placed on absolute reduction of material and energy throughput. This position, while not featuring in the SDGs, has been gaining wider recognition in sustainability research and also some environmental institutions (eg see EEA, 2019; IPBES, 2019; Ripple et al, 2020), and can thus help devise alternative forms of governance. Reorienting the economy in this way means that the production and consumption of energy-intensive goods will be reduced as much as possible. More fundamentally, the economy needs to be organized to support areas that sustain life both human and non-human, and to focus on social reproduction instead of production (Barca, 2020). This means that some sectors will disappear, others will be substantially transformed and new regenerative sectors will flourish (Hardt et al, 2021). Existing alternative organizing practices, often oriented at just and/or sustainable relations of production, already engage in transforming the economy but face multiple challenges in the shadow of powerful mainstream actors and growth-focused governance frameworks and policies (Chertkovskaya, 2022). Thus, transforming both work and the economy requires visibilizing and supporting such organizations.

The prioritization of absolute reduction of the use of energy and materials, and moving away from economic growth as a central policy goal does not mean that growth will not happen at all. It can still take place in certain sectors or geographical areas. From a degrowth stance, sufficient provision of material needs for everyone is essential, as well as support of spheres of life such as education, health and care, repair and maintenance. These areas may well expand as part of socio-ecological transformation. However, there is no need for perpetual and compound growth to be a goal, and possible growth of these sectors is unlikely to bring the kind of monetary value that comes with speculation, extraction and production for planned obsolescence (see also Kallis, 2019). Notably, provisioning systems such as health and education feature in the Agenda 2030 framework under SDG 3 and SDG 4, but these crucial provisioning systems are framed mainly in terms of access rather than working conditions or available funding. What has been neglected, however, is how these sectors have been hit by years of neoliberalization and austerity.

Work too needs to be reimagined, and organized in both ecologically sustainable and socially just ways. To be aligned with socio-ecological transformation, both the purpose and the organization of work need to change. Instead of pursuing productivity, the purpose of work will be to enable regenerative economies and to focus on care for people and environments (Raworth, 2017; Barca, 2020). This means that work will be focused on shifting from fossil dependency, on using and wasting less, and on repairing and reusing as much as possible. Small-scale and labour-intensive production, where feasible, can be encouraged (Hardt et al, 2021). Labour-intensive production, as mentioned earlier, already features in SDG 8, but it also needs to come together with a larger reorientation of the economy.

This reorientation requires recognition of the diversity of the economy and work – not only wage labour done for and in the market, but invisibilized labour carried out in households, communities and neighbourhoods (Fraser, 2014; Dengler and Lang, 2021). Focusing the economy on the sustainability of life rather than of output and profit also means valuing (in social terms) and visibilizing the care work that sustains life today. This does not mean shifting more unpaid work into the paid sector but, instead, supporting the care commons and challenging the dichotomy between the monetized economy and the invisibilized economy of social provisioning (Dengler and Lang, 2021). At the same time, much work that helps to maintain life and that takes place within the monetized economy needs to move from being underpaid to being generously supported by various institutions of society. In turn, work that is environmentally or socially harmful needs to be discouraged.

Two dimensions can be identified as key to transforming work: liberation from work and liberation of work (Barca, 2019). Liberation from (waged) work refers to changing societies so that people can work less and use the released time for other ideally regenerative activities, where bonds, communities and transformative action can be built. Liberation of work relates to organizing work itself differently, by putting regenerative production and just labour relations at the heart of it. Democratization of work, and worker control and ownership, are often seen as key in relation to the latter. Others have also argued for work as craft, moving from the division and fragmentation of labour to emphasizing the importance of the wholeness of work, as well as a more equal division of labour within and across societies (Cherkovskaya and Stoborod, 2018). Liberation of work is not to become a prerogative of the privileged few but will come with a just transition from both fossil fuels and unequal (labour) relations (Velicu and Barca, 2020).

Reduction of working time is a concrete measure that may help to achieve liberation from work, ideally releasing the time from work that is unsustainable and unwanted to activities that people find meaningful (Hoffman and Paulsen, 2020). Basic income is one measure that can both liberate from work and create possibilities for organizing work on different principles (Chertkovskaya and Stoborod, 2018). It can also be provided in a local currency rather than general purpose money, supporting the local economy and more ecologically sustainable production practices (Hornborg, 2017). Universal basic services (Gough, 2019) is another measure that can liberate from work, by satisfying the basic needs of every person, thus giving them more space to decide the kind of work they would like to do, while not being attached to a monetary income. However, none of these measures necessarily leads to supporting regenerative activities, so being regenerative by design (Raworth, 2017) will be key to such measures. Reorienting economy and work towards socio-ecological transformation will not be just in itself, thus it is crucial to reflect on power relations that may still be reproduced in this process, and to challenge them, striving for class, gender, racial and environmental justice as well as equal footing across societies and spaces.

9.4 Conclusion

From acknowledgement of environmental limits to growth in the 1970s, the hypothesis of green growth that is, reconciling growth with ecological sustainability, came to the forefront of international governance frameworks and sustainable development policies (Gómez-Baggethun and Naredo, 2015), as manifested in the UN's Agenda 2030. While the OECD has been key to positioning economic growth at the centre of international governance in the aftermath of the Second World War (Schmelzer, 2016), it has been taken on board by multiple international organizations, such as the World Bank, the UN and nation-states, in addition to corporate actors. A unique feature of the UN's SDGs is that they are applicable to the whole world (Eisenmenger et al, 2020), different levels of governance, and organizations. In other words, the 17 SDGs of the Agenda 2030 can be found not only in the programme of international organizations, states and corporations, but also in municipalities, community organizations and educational spaces. Through SDG 8, the Agenda 2030 also reinforces the omnipresence of economic growth and the problematic assumptions about the possibility of green growth and about work.

This chapter has unpacked the critiques of economic growth and work under SDG 8, and sketched an alternative that can truly reconcile social and ecological objectives, including for international and other levels of governance. Simply put, it has argued that the focus on growth is both unsustainable and unjust, while emphasizing 'decent work' is not enough without problematizing the very way in which the economy and work are organized. Socio-ecological transformation should be at the forefront of economies and societies, and the economy should be organized for regeneration, instead of focusing on the pursuit of economic growth. Work too can be decent and sustainable only if it is reorganized in line with this goal, through focusing on regenerative activities, working less and being democratized.

From this chapter's perspective, the very framing of 'sustainable development goals' is problematic', as 'growth' is inherent in the pursuit of development, bringing in its wake environmental degradation, injustices within societies and between the Global North and the Global South (Kallis et al, 2015; Sachs, 2019/1992). The degrowth perspective taken here aligns instead with alternatives to development (Kothari et al, 2014). In view of this, the SDGs should be reframed to avoid referring to the problematic concept of development, for example, into Sustainable Well-being Goals (SWGs). Such a framework would need to be informed by different knowledges that help to understand and build well-being beyond growth, development and capitalism, with degrowth being one of them.

In line with degrowth thinking, SWG 8 (an alternative to SDG 8) will need to capture the proposed changes to the economy and work, and could be titled 'Regenerative economy, democratized and sustainable work'. Regenerative economy flags a direction for the economy oriented towards socio-ecological transformation. Choosing 'democratized' and 'sustainable' as the key terms for work indicates how sustainability and justice would be central to how work is organized. This goal will focus on labour-intensive activities within the regenerative economy, which will come together with working less and with work being organized on democratic grounds.

While it is outside the scope of this chapter to provide targets or indicators for the new SWG 8 (for ideas of concrete alternative indicators, see Kreinin and Aigner, 2021), the chapter concludes with some general directions for these. Absolute reduction of biophysical throughput and human needs satisfaction across the globe should inform the core targets of SWG 8. Other key entry points into devising alternative targets would be ensuring equitable exchange across countries and the extent of regenerative economic activities, and monitoring of the phase-out of harmful activities. Some of the key principles of work would be democratization, a shift towards work in the regenerative economy and a phasing out of harmful work, as well as independence from paid labour for subsistence. On the basis of these, new targets could be devised.

References

Akbulut, B., Demaria, F., Gerber, J.-F. and Martínez-Alier, J. (2019) 'Who promotes sustainability? Five theses on the relationships between the degrowth and the environmental justice movements', *Ecological Economics*, 163: 106418.

Alvaredo, F., Chancel, L., Piketty, T., Saez, E. and Zucman, G. (2018) *World Inequality Report, 2018*, Available from: https://wir2018.wid.world [Accessed 20 August 2022].

Andrijasevic, R. and Sacchetto, D. (2017) '"Disappearing workers": Foxconn in Europe and the changing role of temporary work agencies', *Work, Employment & Society*, 31(1): 54–70.

Asara, V., Otero, I., Demaria, F. and Corbera, E. (2015) 'Socially sustainable degrowth as a social-ecological transformation: Repoliticizing sustainability', *Sustainability Science*, 10: 375–84.

Barca, S. (2019) 'An alternative worth fighting for: Degrowth and the liberation of work', in E. Chertkovskaya, A. Paulsson and S. Barca (eds) *Towards a Political Economy of Degrowth*, London: Rowman & Littlefield, pp 175–191.

Barca, S. (2020) *Forces of Reproduction: Notes for a Counter-hegemonic Anthropocene*, Cambridge: Cambridge University Press.

Barry, J. (2020) 'A genealogy of economic growth as ideology and Cold War core state imperative', *New Political Economy*, 25(1): 18–29.

Bauer, F. and Fontenit, G. (2021) 'Plastic dinosaurs: Digging deep into the accelerating carbon lock-in of plastics', *Energy Policy*, 156: 112418.

Bianchi, R.V. and De Man, F. (2021) 'Tourism, inclusive growth and decent work: A political economy critique', *Journal of Sustainable Tourism*, 29(2–3): 353–71.

Brand, U., Muraca, B., Pineault, É., Sahakian, M., Schaffartzik, A., Novy, A. et al (2021) 'From planetary to societal boundaries: An argument for collectively defined self-limitation', *Sustainability: Science, Practice and Policy*, 17(1): 264–91.

Büchs, M. and Koch, M. (2017) *Post-growth and Well-being: Challenges to Sustainable Welfare*, Basingstoke: Palgrave Macmillan.

Büscher, B. and Fletcher, R. (2017) 'Destructive creation: Capital accumulation and the structural violence of tourism', *Journal of Sustainable Tourism*, 25(5): 651–67.

Chertkovskaya, E. (2019) 'Degrowth in theory, pursuit of growth in action: Exploring the Russian and Soviet contexts', in E. Chertkovskaya, A. Paulsson and S Barca (eds) *Towards a Political Economy of Degrowth*, London: Rowman & Littlefield.

Chertkovskaya, E. (2022) 'Degrowth', in L. Pellizzoni, E. Leonardi and V. Asara (eds) *Handbook of Critical Environmental Politics*, Cheltenham: Edward Elgar, pp 116–128.

Chertkovskaya, E. and Paulsson, A. (2016) 'The growthocene: Thinking through what degrowth is criticising', ENTITLE (European Political Ecology Network) blog, 19 February, Available from: http://entitleblog.org/2016/02/19/the-growthocene-thinking-through-what-degrowth-is-criticising [Accessed 20 August 2022].

Chertkovskaya, E. and Stoborod, K. (2018) 'Work', in B. Franks, N. Jun and L. Williams (eds) *Anarchism: A Conceptual Approach*, London: Routledge.

Chertkovskaya, E. and Paulsson, A. (2021) 'Countering corporate violence: Degrowth, ecosocialism and organising beyond the destructive forces of capitalism', *Organization*, 28(3): 405–25.

Chertkovskaya, E., Watt, P., Tramer, S. and Spoelstra, S. (2013) 'Giving notice to employability', *ephemera*, 13(4): 701–16.

Chertkovskaya, E., Korczynski, M. and Taylor, S. (2020) 'The consumption of work: Representations and interpretations of the meaning of work at a UK university', *Organization*, 27(4): 517–36.

Crane, A. (2013) 'Modern slavery as a management practice: Exploring the conditions and capabilities for human exploitation', *Academy of Management Review*, 38: 49–69.

Dengler, C. and M. Lang (2021) 'Commoning care: Feminist degrowth visions for a socio-ecological transformation', *Feminist Economics*, 28(1): 1–28.

Devine, J. and Ojeda, D. (2017) 'Violence and dispossession in tourism development: A critical geographical approach', *Journal of Sustainable Tourism*, 25(5): 605–17.

Dorninger, C., Hornborg, A., Abson, D.J., von Wehrden, H., Schaffartzik, A., Giljum, S. et al (2021) 'Global patterns of ecologically unequal exchange: Implications for sustainability in the 21st century', *Ecological Economics*, 179: 106824.

EEA (2019) *The European Environment State and Outlook 2020*, Copenhagen: European Environmental Agency.

Eisenmenger, N., Pichler, M., Krenmayr, N., Noll, D., Plank, B., Schalmann, E. et al (2020) 'The Sustainable Development Goals prioritize economic growth over sustainable resource use: A critical reflection on the SDGs from a socio-ecological perspective', *Sustainability Science*, 15(4): 1101–10.

EJAtlas (Environmental Justice Atlas) (2022) The global atlas of environmental justice, Available from: https://ejatlas.org [Accessed 20 August 2022].

Emmanuel, A. (1972) *Unequal Exchange: A Study of the Imperialism of Trade*, London: Monthly Review Press.

Escobar, A. (1995) *Encountering Development: The Making and Unmaking of the Third World*, Princeton, NJ: Princeton University Press.

Fraser, N. (2014) 'Behind Marx's hidden abode', *New Left Review*, 86: 55–72.

Gómez-Baggethun, E. and Naredo, J.M. (2015) 'In search of lost time: The rise and fall of limits to growth in international sustainability policy', *Sustainability Science*, 10: 385–95.

Gössling, S., Scott, D. and Hall, C.M. (2021) 'Pandemics, tourism and global change: A rapid assessment of COVID-19', *Journal of Sustainable Tourism*, 29(1): 1–20.

Gough, I. (2019) 'Universal basic services: A theoretical and moral framework', *Political Quarterly*, 90(3): 534–42.

Hardt, L., Barrett, J., Taylor, P.G. and Foxon, T.J. (2021) 'What structural change is needed for a post-growth economy: A framework of analysis and empirical evidence', *Ecological Economics*, 179: 106845.

Hayden, A., and Shandra, J.M. (2009) 'Hours of work and the ecological footprint of nations: An exploratory analysis', *Local Environment*, 14(6): 575–600.

Hickel, J. (2019) 'The contradiction of the Sustainable Development Goals: Growth versus ecology on a finite planet', *Sustainable Development*, 27(5): 873–84.

Hickel, J. and Kallis, G. (2020) 'Is green growth possible?', *New Political Economy*, 25(4): 469–86.

Hickel, J., Brockway, P., Kallis, G., Keyßer, L., Lenzen, M., Slameršak, A. et al (2021) 'Urgent need for post-growth climate mitigation scenarios', *Nature Energy*, 6: 766–8.

Hoffman, M. and Paulsen, R. (2020) 'Resolving the "jobs–environment–dilemma"? The case for critiques of work in sustainability research', *Environmental Sociology*, 6(4): 343–54.

Hornborg, A. (2017) 'How to turn an ocean liner: A proposal for voluntary degrowth by redesigning money for sustainability, justice, and resilience', *Journal of Political Ecology*, 24(1): 623–32.

IEEP (Institute for European Environmental Policy) (2019) 'Linking aviation emissions to climate justice', 9 October, Available from: https://ieep.eu/news/linking-aviation-emissions-to-climate-justice [Accessed 23 June 2023].

ILO (International Labour Organization) and Walk Free foundation (2017) *Global Estimates of Modern Slavery: Forced Labour and Forced Marriage*, Geneva: ILO and Walk Free Foundation.

IPBES (Intergovernmental Platform on Biodiversity and Ecosystem Services) (2019) *The Global Assessment Report on Biodiversity and Ecosystem Services: Summary for Policy-Makers*, Bonn: IPBES.

Kallis, G. (2018) *Degrowth*, Newcastle upon Tyne: Agenda.

Kallis. G. (2019) 'Socialism without growth', *Capitalism Nature Socialism*, 30(2): 189–206.

Kallis, G., Demaria, F. and D'Alisa, G. (2015) 'Introduction: Degrowth', in G. D'Alisa, F. Demaria and G. Kallis (eds) *Degrowth: A Vocabulary for a New Era*, London: Routledge, pp 1–17.

Kallis, G., Paulson, S., D'Alisa, G. and Demaria, F. (2020) *The Case for Degrowth*, Cambridge: Polity Press.

Kothari, A., Demaria, F. and Acosta, A. (2014) 'Buen vivir, degrowth and ecological swaraj: Alternatives to sustainable development and the green economy', *Development*, 57(3–4): 362–75.

Kothari, A., Salleh, A., Escobar, A., Demaria, F. and Acosta, A. (eds) (2019) *Pluriverse: A Post-development Dictionary*, Delhi: Authors Up Front.

Kreinin, H. and Aigner, E. (2021) 'From "decent work and economic growth" to "sustainable work and economic degrowth": A new framework for SDG 8', *Empirica*, 49: 281–311.

Lee, D.S., Fahey, D.W., Skowron, A., Allen, M.R., Burkhardt, U., Chen, Q. et al (2021) 'The contribution of global aviation to anthropogenic climate forcing for 2000 to 2018', *Atmospheric Environment*, 244: 117834.

Lüthje, B. and Butollo, F. (2017) 'Why the Foxconn model does not die: Production networks and labour relations in the IT industry in South China', *Globalizations*, 14(2): 216–31.

Marazzi, C. (2010) *The Violence of Financial Capitalism*, Los Angeles: Semiotext(e).

Martínez-Alier (2021) 'Mapping ecological distribution conflicts: The EJAtlas', *Extractive Industries and Society*, 8(4): 100883.

Meadows, D. H, Meadows, D. L., Randers, J. and Behrens III, W.W. (1972) *The Limits to Growth: A Report for the Club of Rome's Project on the Predicament of Mankind*, New York: Universe Books.

Muraca, B. and Neuber, F. (2018) 'Viable and convivial technologies: Considerations on climate engineering from a degrowth perspective', *Journal of Cleaner Production*, 197(2): 1810–22.

Nightingale, A. J. (2018) 'Geography's contribution to the Sustainable Development Goals: Ambivalence and performance', *Dialogues in Human Geography*, 8(2): 196–200.

Nilsson, L.J., Bauer, F., Åhman, M., Andersson, F.N.G., Bataille, C., de la Rue du Can, S. et al (2021) 'An industrial policy framework for transforming energy and emissions intensive industries towards zero emissions', *Climate Policy*, 21(8): 1053–65.

O'Neill, D.W., Fanning, A.L., Lamb, W.F. and Steinberger, J.K. (2018) 'A good life for all within planetary boundaries', *Nature Sustainability*, 1: 88–95.

Oswald, Y., Owen, A. and Steinberger, J.K. (2020) 'Large inequality in international and intranational energy footprints between income groups and across consumption categories', *Nature Energy*, 5: 231–9.

Parrique, T., Barth, J., Briens, F., Kerschner, C., Kraus-Polk, A., Kuokkanenand, A. and Spangenberg, J.H. (2019) *Decoupling Debunked: Evidence and Arguments against Green Growth as a Sole Strategy for Sustainability*, Brussels: European Environmental Bureau.

Pun, N., Shen, Y., Guo, Y. and Lu, H. (2016) 'Apple, Foxconn, and Chinese workers' struggles from a global labor perspective', *Inter-Asia Cultural Studies*, 17(2): 166–85.

RAN (Rainforest Action Network) (2019) *Banking on Climate Change: Fossil Fuel Report Card 2019*, Available from: https://www.ran.org/wp-content/uploads/2019/03/Banking_on_Climate_Change_2019_vFINAL1.pdf [Accessed 20 August 2022].

Raworth, K. (2017) *Doughnut Economics: Seven Ways to Think Like a 21st-Century Economist*, London: Penguin.

Ripple, W.J., Wolf, C., Newsome, T.M., Barnard, P., Moomaw, W.R. et al (2020) 'Corrigendum: World scientists' warning of a climate emergency', *BioScience*, 70(1): 8–12.

Roy, B. and Martínez-Alier, J. (2019) 'Environmental justice movements in India: An analysis of the multiple manifestations of violence', *Ecology, Economy and Society*, 2(1): 77–92.

Ruzzenenti, F., Font Vivanco, D., Galvin, R., Sorrell, S., Wagner, A. and Walnum, H.J. (2019) 'The rebound effect and the Jevons' paradox: Beyond the conventional wisdom', *Frontiers in Energy Research*, 7: 90.

Sachs, W. (2019) *The Development Dictionary: A Guide to Knowledge as Power*, 3rd edn, London: Zed Books.

Schmelzer, M. (2016) *The Hegemony of Growth: The OECD and the Making of the Economic Growth Paradigm*, Cambridge: Cambridge University Press.

Schor, J.B. (2005) 'Sustainable consumption and worktime reduction', *Journal of Industrial Ecology*, 9(1–2): 37–50.

Shift Project (2019) 'Lean ICT: Towards digital sobriety', Available from: https://theshiftproject.org/wp-content/uploads/2019/03/Lean-ICT-Report_The-Shift-Project_2019.pdf [Accessed 23 June 2023].

Standing, G. (2011) *The Precariat: The New Dangerous Class*, London: Bloomsbury.

Stay Grounded and PCS (2021) 'A rapid and just transition of aviation: Shifting towards climate-just mobility', Discussion paper, February, Available from: https://stay-grounded.org/wp-content/uploads/2021/02/SG_J ust-Transition-Paper_2021.pdf [Accessed 23 June 2023].

Steffen, W., Richardson, K., Rockström, J., Cornell, S.E., Fetzer, I., Bennett, E.M. et al (2015) 'Planetary boundaries: Guiding human development on a changing planet', *Science*, 347: 1259855.

Stoknes, P.E. and Rockström, J. (2018) 'Redefining green growth within planetary boundaries', *Energy Research and Social Science*, 44: 41–9.

Tilsted, J.P., Bjørn, A., Majeau-Bettez G. and Lund, J.F. (2021) 'Accounting matters: Revisiting claims of decoupling and genuine green growth in Nordic countries', *Ecological Economics*, 187: 107101.

UN (United Nations) (2020a) '8. Decent work and economic growth', Available from: https://sdgs.un.org/goals/goal8 [Accessed 23 June 2023].

UN (2020b) '9. Industry, innovation and infrastructure', Available from: https://unstats.un.org/sdgs/report/2020/goal-09 [Accessed 23 June 2023].

UN (2020c) '9. Industry, innovation and infrastructure', Available from: https://sdgs.un.org/goals/goal9 [Accessed 8 July 2023]

UN (2020d) 'Decent work and economic growth: Why it matters', Available from: https://www.un.org/sustainabledevelopment/wp-content/uploads/ 2016/08/8_Why-It-Matters-2020.pdf [Accessed 23 June 2023].

Velicu, I. and Barca, S. (2020) 'The just transition and its work of inequality', *Sustainability: Science, Practice and Policy*, 16(1): 263–73.

Vogel, J., Steinberger, J.K., O'Neill, D.W., Lamb, W.F. and Krishnakumar, J. (2021) 'Socio-economic conditions for satisfying human needs at low energy use: An international analysis of social provisioning', *Global Environmental Change*, 69: 102287.

Wiedmann, T., Lenzen, M., Keyßer, L.T. and Steinberger, J. (2020) 'Scientists' warning on affluence', *Nature Communications*, 11: 3107.

York R. and McGee, J.A. (2016) 'Understanding the Jevons paradox', *Environmental Sociology*, 2(1): 77–87.

Interview with Ekaterina Chertkovskaya: Economic Growth Cannot Be the Answer

Elanur Alsac, Wonyoung Cho and Emmanuel Dahan

Growth for the sake of growth has been criticized at least since the publication of the Limits to Growth *report in the 1970s. What are the reasons for the continued sacralization of economic growth and the disregarding of degrowth arguments in the SDGs?*

Chertkovskaya: Economic growth has been promoted for decades as a solution to all sorts of societal problems, and this way of thinking will take time to desacralize. The logic of growth is at the core of some of the most powerful institutions in society: corporations, states and international organizations. When states and international organizations frame policies, they assume that economic growth is always good and has to continue. The idea behind this is that monetary gains will bring prosperity and 'trickle down' to different social groups. However, it doesn't really happen this way, with immense wealth concentrated in the hands of the few, and inequalities only increasing across the world. Moreover, the pursuit of economic growth has been going hand in hand with environmental degradation. The tight entanglement between powerful actors, for example, corporate lobbying of national and supranational policies contributes to growth-critical perspectives being disregarded while sustainability-oriented concepts that do not question growth, such as the circular economy, are promoted. The discipline of economics contributes to this too, by not questioning the motive of making profit and producing more. After the financial crisis in 2008–9, however, a vast movement started to rethink economics. Alternative economics, for example, ecological economics, or Kate Raworth's *Doughnut Economics* (2017) sees the economy as material, and tries to reorient it towards the well-being of people and the planet rather than economic growth and profit making.

You mentioned that efficiency gains are oftentimes outweighed by increased production and consumption. How can this be countered?

Chertkovskaya: Processes to improve efficiency are not problematic in themselves, but we must ask what these processes are for. If growth or profit is the answer, then it's a problem. For example, replacing all petrol cars with electric cars will produce a huge rebound effect because of the energy and materials needed for making and running electric vehicles. Rather than blaming car users only, we should think how policies can be designed to reduce car use and to provide public transport infrastructure for everyone. Another example of questionable choice guided by capital accumulation is the introduction of electric scooters in cities, which are positioned as 'green' but need a lot of energy and materials to produce and to keep running, and they have a short lifetime too. It is important, following bottom-up principles, to pressurize decision-makers at different institutional levels to think about what is really needed within societies.

Who can create the real political power to achieve the socio-ecological transformation? Is the leadership of international institutions, governments and parties able to reorient the economy towards degrowth, sustainability and justice?

Chertkovskaya: I would say, no, those are not the powers I believe in. It is the power of the people organizing in social movements, in alternative forms of economic activities and communities. We talk about bottom-up transformation from a degrowth perspective, meaning the crisis is already being addressed locally, and various alternatives exist. However, if this driving force does not change institutions by putting pressure on them, then institutions will act as barriers. In addition, if we think from the bottom up about organizing inclusively, with direct democracy mechanisms, this is a process that takes time. A challenge we are facing is thus to find out how to build strong networks and overcome this time issue, while still living up to the democratic bottom-up principles where different voices are heard.

10

'We Do Not Want to Be Mainstreamed into a Polluted Stream': An Ecofeminist Critique of SDG 5

Sherilyn MacGregor and Aino Ursula Mäki

We caution against developing another set of reductive goals, targets and indicators that ignore the transformational changes required to address the failure of the current development model rooted in unsustainable production and consumption patterns exacerbating gender, race and class inequities. We do not want to be mainstreamed into a polluted stream. We call for deep and structural changes to existing global systems of power, decision-making and resource sharing. This includes enacting policies that recognize and redistribute the unequal and unfair burdens of women and girls in sustaining societal wellbeing and economies, intensified in times of economic and ecological crises. (Statement by WGC, DAWN and WECF, 2013)[1]

Feminist organizations from around the world have long tried to lobby the United Nations to take gender issues seriously by moving beyond the simplistic add-women-and-stir approach and to attack unjust power asymmetries at their roots (see Box 10.1). As the extract makes clear, the UN has a poor track record on gender-sensitive policy making in the sustainable development (SD) field. In 2013, as the end of the 2015 Agenda and the Millennium Development Goals was approaching, and in the run-up to the drafting of the Agenda 2030 Sustainable Development Goals, feminist activists called on UN officials to change their tune. Nearly a decade later, there is an SDG specifically calling for gender equality. But questions remain. Have feminists succeeded in swimming against the current to achieve deep policy change or have token drops of 'gender equality' been poured into the same polluted stream?

This chapter evaluates SDG 5, 'Achieve gender equality and empower all women and girls', from an ecofeminist perspective. It outlines the risks and possibilities associated with linking developmental and environmental goals with the pursuit of gender equality through a focus on women. Its interrogation of SDG 5, and Agenda 2030 more broadly, is organized around two broad clusters of criticisms, concerned with their reductionist view of gender and the production/reproduction dualism that underpins their green economic vision, that are also relevant for the field of global environmental governance at large. In response to these criticisms, it is suggested that the road to achieving SD must be paved, not with liberal concepts of equality and empowerment, but with ambitious political goals and concrete policies for a fundamental transformation in the relations and responsibilities of socio-ecological reproduction to redress the structural dimensions of gendered and racialized injustice.

Box 10.1: Feminist organizations involved in the UN SDGs process

The Women's Major Group (WMG) has been an official participant in the UN sustainable development process since the 1992 Rio Earth Summit. It is a large network made up of feminist civil society organizations from around the world that promote women's human rights, political empowerment and gender equality. The WMG played an active role in the development of the SDGs from 2013 to 2015, and since then has continued each year to monitor progress towards key targets. While feminist organizations have generally welcomed the inclusion of gender equality as a standalone goal (SDG 5), as well as its mention in other goals (eg SDGs 4 and 8), they have also voiced criticisms of its failure to tackle root causes and so continue to lobby for more systemic transformative action at all levels (see Gabizon, 2016; CGSHR, 2017).

The chapter is organized as follows. It provides a brief introduction to the authors' critical ecofeminist perspective, then situates Agenda 2030 within decades of UN sustainable development discourse, which helps to contextualize criticisms of SDG 5. The final section outlines, explains and illustrates an alternative ecofeminist approach to environmental and gender justice, which entails a caring economy and sustainable livelihoods.

10.1 What is a critical ecofeminist perspective?

An ecofeminist perspective is broadly concerned with connections between the exploitation of the living environment and the oppression of humans under the interconnected systems of capitalism, colonialism and patriarchy. Ecofeminism draws on more than three decades of scholarship on the gender–environment nexus, which challenges hierarchical binaries in western philosophical thought that legitimate those oppressive and destructive systems (for an overview see MacGregor, 2017a). These binaries, including reason–emotion, culture–nature, individual–collective and man–woman have shaped and sustained not only big social structures but also the micro-politics of everyday life. A core aim of ecofeminist thinking is to find strategies for transcending binaries to create integrated and egalitarian alternatives (Plumwood, 1992; Barca, 2021).

The word 'critical' signals the incorporation of tools of critical social theory into the ecofeminist project. This means it draws on counter-hegemonic and heterodox theoretical traditions, including Marxism and post-structuralism, to develop 'ruthless criticism of the existing order' (Marx, 1843) and to pursue a project of political and economic transformation. It is not interested in problem solving per se, if that entails taking the world as it is (to paraphrase Cox, 1981:129). Instead, critical ecofeminist theory adopts a questioning stance that foregrounds the violence of existing power relations to disrupt them, insists on an intersectional (as opposed to a single-axis) understanding of the world[2] and makes normative commitments to just human–nature relations (see eg Cudworth, 2014).

A common project of ecofeminist thinkers is to integrate feminist goals for gender justice with the goals of environmental movements for sustainable, liveable communities on a finite planet (Gaard, 2017). Whereas many involved in environmentalism tend to homogenize humans into a global 'we', contemporary ecofeminist scholars use an intersectional lens to understand diverse identities and experiences of people of all genders as well as to avoid essentialism (ie the idea that biological sex determines gender identity or sexuality). Although some ecofeminist activists in the past have used rhetorical strategies that appear to support the notion that women are inherently 'closer to nature', charges that ecofeminist theory is essentialist are largely unfounded (Thompson and MacGregor, 2017). While being concerned to understand and improve the status of women as a group, 21st-century ecofeminist scholarship overwhelmingly seeks to avoid conflating sex and gender, and does not suggest that women have a unique or shared perspective due to their biology (MacGregor, 2017a). On the contrary, challenging and politicizing such claims has historically been one of the core areas of interest for critical ecofeminist analysis and activism, and as an approach it is thus uniquely well equipped to dismantle essentialist thinking.

Critical analysis of the SDGs here draws on the tools of ecofeminist political economy (EPE), which is a subfield within ecofeminism that examines the role of gender inequality in economic development and theorizes a material link between how nature, women and all things feminine have historically been externalized and exploited in capitalist economic systems (Mellor, 2006:140; see also Mellor, 2017). Critical of the logical structure of western industrialist modernity and Eurocentric values of the Enlightenment, this perspective regards the climate crisis as stemming from the same root causes that drive the plethora of gendered, racialized and class inequalities in the world today (Merchant, 1989). Through this lens, EPE problematizes the contradictions evident in mainstream SD that arise from attempting to protect the environment and further equality while also pursuing economic growth. To the extent that it reproduces neoclassical economic models and neoliberal ideology, the day-to-day functioning of SD depends on the devaluation of caring labour performed in families and communities as well as of the natural resources generated in the environment (Folbre, 2021). And yet, because these free subsidies are invisibilized and externalized by hegemonic economic paradigms and political discourse, their exploitation and depletion is unlikely to be challenged by mainstream approaches to environmental governance, regardless of their claims to the language of sustainability (Bauhardt, 2014).

10.2 Centring gender equality in sustainable development debates

Connecting the pursuits of gender equality and environmental protection has a long history in the field of SD as well as of global environmental governance more broadly. Within the UN, the link between gender equality and environmental sustainability formally dates back to the 1992 UN Conference on Environment and Development in 1992, where feminist advocacy resulted in a highly visible role for women in SD (Braidotti et al, 1994; Harcourt, 1994). Agenda 21, the outcome document of the conference, had women's empowerment at its centre and foregrounded the role of women in pursuing environmental and developmental goals because of their unique skills and know-how in environmental conservation and biodiversity protection.

In subsequent decades, conceptualizations of the links between gender, environment and development have taken different forms in SD discourse. A common narrative is that gender is relevant because, by their secondary social status in most countries, women are disproportionately affected by ecological destruction and climate change (Buckingham, 2020). Evidence has been collected to support the claim that women's roles in unpaid domestic and care work, such as gathering firewood, carrying water and tending to crops, make them most reliant on land for their livelihoods and incomes, and therefore more vulnerable than men to the deterioration and destruction of natural resources and environmental quality (Resurrección, 2021). Another claim that emerges from this association is that women's care taking and stewardship roles make them highly aware of, concerned about and inclined to take action in the face of catastrophic ecological change. For these reasons, elevating women from their lower status by educating and including them in decision-making – in short, empowering them – has been adopted by UN agencies among others for over three decades (MacGregor, 2019).

The various ways in which women have been inserted into global environmental governance have been hotly debated by feminists. Some who subscribe to more liberal values and single-axis analysis of gender argue that mainstreaming gender into SD is an important strategy that will yield benefits for marginalized women, especially those in the Global South, and that empowering women will both reduce inequality and protect the environment as well as furthering other development aims (see eg Mary Robinson Foundation, 2015). Others who are critical of liberal mainstream feminism are sceptical. For example, critical ecofeminists argue that this strategy relies on a number of problematic assumptions about the interconnections of gender, environmental destruction and climate change (MacGregor, 2017b). They most want to challenge the construction of a 'materialist-essentialist linkage' between women and the environment, which manifests itself in a selection of harmful stereotypes and norms pertaining to women's special relationship to nature (Resurrección, 2013; Arora-Jonsson, 2017). According to this perspective, all versions of this narrative implicitly maintain that women are somehow naturally closer to the environment than men, whether as vulnerable victims of climate change, as altruistic sustainability saviours and 'Earth mothers' or as culprits who exhaust natural resources by having too many children (MacGregor, 2010a; 2010b; 2019; Arora-Jonsson, 2011; Resurrección, 2013; Lau et al, 2021). This narrative is harmful, because it universalizes and reinforces rather than challenges socially constructed and historically specific gender norms that responsibilize women for social and ecological reproduction (Hirshman, 2003; Nightingale, 2006; see also Mohanty, 1991). In practice, such an approach has led to harnessing women's roles and responsibilities for externally determined programs and policies for the benefit of states and private actors, thereby intensifying their burdens in both paid and unpaid work (Jackson, 1994).

The valorization of women's supposedly natural skills and responsibilities in environmental management has thus deepened gender hierarchies, but ignoring women's pre-existing workloads is not a solution either. Policies that leave out care and unpaid domestic work invisibilize many of women's activities and contributions, and also make it difficult to problematize gendered divisions of labour (MacGregor et al, 2022). Approaches such as women's economic empowerment (WEE) that target women primarily to incorporate them into the market to achieve poverty reduction and other state goals demonstrate another way for the vocabulary of feminist advocacy to be co-opted to support the status quo (MacGregor et al, 2022). In the context of this longstanding struggle to challenge rather than affirm business as usual, ecofeminists are cautious about linking global sustainability goals to the pursuit of gender equality, and believe that it is always necessary to be aware of the positive and negative synergies between them. With these insights in mind, the chapter interrogates the dangers and possibilities in the varied conceptualizations of SD made possible by visions offered in Agenda 2030.

10.3 What's wrong with SDG 5?

SDG 5 has been met with mixed reviews by feminist activists and academics. On the positive side, many agree that the SDGs as a whole are a vast improvement over the MDGs, which were very weak on women's rights and gender inequality (Carant, 2017). A standalone goal for gender equality with nine specific targets (see Box 10.2) is a notable success in itself, and contains some hard-won improvements by feminist lobby groups. Mainstreaming gender and dedicating one out of 17 goals to gender issues is a significant achievement, as is the inclusion of unpaid care and domestic work. SDG 5 aims to 'achieve gender equality and empower all women and girls', and in all includes many relevant targets such as ending discrimination, violence and other 'harmful practices' against all women and girls. Targets 5.5 and 5.6 continue in the same vein as previous development discourse with regards to women, stressing the importance of education, participation and opportunities for leadership in decision making.

While it is important to acknowledge the positive aspects of SDG 5, it is nevertheless unavoidable that most feminist organizations are highly critical of the negative aspects, such as the uncritical focus on individual empowerment and acceptance of orthodox economic models of development at the expense of attending to structural power relations and the underlying causes of poverty and inequality (Consortium on Gender, Security and Human Rights (CGSHR), 2017). From an ecofeminist perspective, there are several noteworthy points of contention in SDG 5 and the other goals and targets of Agenda 2030 that require attention. This section discusses two specific problems that are visible when SDG 5 is seen through an ecofeminist lens and that prevail within environmental governance in general.

10.3.1 Problem 1: adopting a binary approach to gender

First, SDG 5 upholds a simplistic view of gender as being synonymous with women. Why is this problematic? It is true that women are denied

---

Box 10.2: SDG 5: Gender equality

5.1 End all forms of discrimination against all women and girls everywhere

5.2 Eliminate all forms of violence against all women and girls in the public and private spheres, including trafficking and sexual and other types of exploitation

5.3 Eliminate all harmful practices, such as child, early and forced marriage and female genital mutilation

5.4 Recognize and value unpaid care and domestic work through the provision of public services, infrastructure and social protection policies and the promotion of shared responsibility within the household and the family as nationally appropriate

5.5 Ensure women's full and effective participation and equal opportunities for leadership at all levels of decision-making in political, economic and public life

5.6 Ensure universal access to sexual and reproductive health and reproductive rights as agreed in accordance with the Programme of Action of the International Conference on Population and Development and the Beijing Platform for Action and the outcome documents of their review conferences

5.A Undertake reforms to give women equal rights to economic resources, as well as access to ownership and control over land and other forms of property, financial services, inheritance and natural resources, in accordance with national laws

5.B Enhance the use of enabling technology, in particular information and communications technology, to promote the empowerment of women

5.C Adopt and strengthen sound policies and enforceable legislation for the promotion of gender equality and the empowerment of all women and girls at all levels. (UN, 2022)

---

the same rights and opportunities that men enjoy in most countries around the world. Women as a group are under-represented as both citizens and leaders within political, corporate, academic and cultural sectors; they earn less, own less and carry out more hours of unpaid subsistence and care work than men. If these disparities are to be addressed successfully, the complex processes that create and maintain them need to be understood holistically and relationally. But, rather than take this approach, SDG 5 focuses on the empirical category 'women and girls', without regard for the complex nexus of power relations between women/girls and men/boys that produce social inequality. It conflates gender and sex and maintains a strictly binary view of gender that ignores people who do not identify as either. Moreover, the goal does not sufficiently address relations between men, women and others, but instead treats women and girls as isolated objects of violence and disempowerment instead of subjects with variable forms of agency and subjectivity in relation to others. This individualistic focus obscures women's relative lack of power in society and the gender-specific harms and structural constraints that result from the cultural overvaluation of men and masculinity and the mistreatment of women's bodies.

Seen in this light, advocating for the empowerment of women and girls through increased opportunities and participation in the market and/ or mainstream politics without addressing these fundamental barriers is insufficient to achieve gender equality (Esquivel, 2016). There is marked inconsistency in mainstreaming gender and pursuing women's empowerment while setting contradictory targets in other areas, such as commitment to a macroeconomic policy that contributes to increasing financialization, trade liberalization and a growing role for transnational corporations and public–private partnerships (Razavi, 2016). Well-documented sources of gendered injustice and contributors to the economic model that structurally devalues women and their work: these policies have 'pushed women behind' by exacerbating instead of resolving intersecting forms of discrimination (GADN, 2019). SDG 5 therefore veers towards an apolitical and tokenistic view of gender equality as 'smart economics' wherein women are regarded in a solutionist manner, as untapped potential that can act as 'a source of growth' and raise productivity (Roberts and Soederberg, 2012).

10.3.2 Problem 2: accepting the production/reproduction binary

A second point of contention is that SDG 5 (and other SDGs) perpetuates a dichotomy between production and reproduction, which has been extensively criticized by those concerned with the devalorization and invisibilization of women's work and the feminization of reproductive labour (Ferber and Nelson, 1993; Salleh, 1997; Mies, 1998). Why is unpaid and domestic work not included under SDG 8 on sustainable economic growth, SDG 16 on peaceful and inclusive societies for sustainable development or SDG 12 on sustainable consumption and production patterns, but instead relegated to gender equality? This location of unpaid and domestic work under gender equality rather than under other goals associated with conventionally more masculine domains reflects the subordinate status of women's work and reinforces the marginalization of social reproduction as a policy issue. It is assumed to be solely associated with women in the private sphere of the home and not relevant to broader questions of power, politics and economics. Even though target 5.5 appears to 'recognize and value' unpaid care and domestic work, it involves no commitment to their reduction or redistribution beyond the 'promotion of shared responsibility within the household and the family', and thus frames the unequal division of unpaid care and domestic labour as a private household matter instead of as a structural relationship of exploitation connected to inequalities in paid work (O'Manique and Fourie, 2016; see also Elson, 2016).

Agenda 2030 thereby implicitly presents gender equality as a policy issue isolated from the wider framework that is mainly relevant to the feminized area of reproduction. The devaluation of women and areas, activities and identities coded as feminine is also evident in the area of sexuality. Even though the commitment under targets 3.7 and 5.6 to universal access to sexual and reproductive health and reproductive rights is a commendable result of decades of campaigning and activist work, the omission of sexual rights deepens the erasure and lack of power of already marginalized communities, such as LGBTQI+ persons and sex workers (Logie, 2021). The UN's SD discourse has also been more widely criticized for maintaining a heteronormative bias and for consolidating western norms with regards to forms of gender, sex and family by colluding with the Malthusian rhetoric that has historically legitimized racist and violent policies targeting women of colour (Foster, 2011; 2014; see also Corrêa and Reichman, 1994; Hartmann, 2016).

The agenda architecture is thus conceptually divided into differently valued and gendered sectors. The form of SD discourse currently encompassed in the SDGs is actually more masculinist than ever. Overall, there are very few mentions of women or gender outside of SDG 5, even in areas where it would be particularly necessary, such as under SDG 13 on climate change or SDG 7 on energy. This omission has been observed by the Women's Environment & Development Organization (WEDO) (2014), who have criticized the document for its lack of ambition, accountability and funding regarding both climate action and women's human rights. Instead, there is a growing emphasis on 'science, technology and innovation', as illustrated in the launch of a 'Technology Facilitation Mechanism' and the agenda's means of implementation being divided into areas titled 'finance', 'technology', 'capacity-building', 'trade' and 'systemic issues', which are by no means gender neutral in themselves (see UNEP, 2015: 10-15; WMG, 2015).

Arguably, then, compared to previous UN SD documents, the majority of the SDGs are oriented around stereotypically masculine discourses of environmental security and ecological modernization, which frame sustainability as a techno-scientific problem to be solved through elite intervention and expertise (MacGregor, 2010a; 2010b). In contrast to the SD discourse of the 1992 Earth Summit, the onus in the SDGs is no longer on empowering women to save 'Mother Earth' – an essentialist trope in itself, as pointed out earlier – but on 'technological innovation, green economy and technocratic management', constituting 'a shift in emphasis from feminized nature-knowers to masculine technology-shapers' that invisibilizes women altogether (Foster, 2017: 224–5).

Box 10.3: What is intersectionality?

Intersectionality is a tool for analysing how power relations and different systems of inequality that sustain power, such as class, race/ethnicity, gender, sexual orientation and so on intersect to create specific effects. One important reason for applying this tool is that, because these systems are intersecting and mutually reinforcing, they need to be analysed and addressed simultaneously; tackling just one will likely leave the others intact. The concept of intersectionality, most often attributed to the scholarship of the Black feminist legal scholar Kimberlé Crenshaw (1989), has, over the past 20 years, become highly influential within the social sciences. Critical ecofeminst theory has always been intersectional because of its analysis of the logics of domination, in addition to its concern with interconnections between ecological, economic and social crises (Plumwood 1992; MacGregor 2017b; Barca 2021). It is now common to see calls for an intersectional understanding of gender to be embedded in gender equality policy, and many mainstream institutions of governance including the UN and EU have adopted this language in recognition that single-axis approaches are less effective or may even exacerbate the problem. The WMG and affiliated feminist organizations such as WEDO and WGC all promote intersectionality as a necessary component of gender-just environmental policy, including SD (WEDO, 2020).

10.4 Gender–transformative sustainability? Ecofeminist pathways

In response to the criticisms of SDG 5 in particular, and of Agenda 2030 in general, an ecofeminist approach to SD is proposed that transcends the binaries criticized earlier to be transformative rather than affirming of the status quo. These are pathways rather than solutions in that they suggest directions for change and invite inclusive debate within environmental governance. This is an approach that is increasingly being taken by activist organizations calling for a 'Feminist Green New Deal' and other 'eminist climate justice alternatives to the policy aims and targets of SDG 5 (WEDO, 2020; Cohen and MacGregor, 2021; Sultana, 2021).

10.4.1 Transformation instead of empowerment, justice instead of equality

First, an ecofeminist outlook requires a more nuanced view of gender, power and social relations. Ecofeminist thinkers have long embraced intersectionality, a concept that highlights interactions between gender, race, class and other categories of difference as an approach to analysis of power and knowledge production that avoids essentialization of social categorizations and structures (see Box 10.3 and Kaisjer and Kronsell, 2014, for a useful discussion). Intersectional analysis is a valuable tool in transcending the binary models of gender because it makes visible the complex interconnections between the production of different social categories, structures of oppression and inequality constitutive of the climate crisis (Di Chiro, 2008; Perkins, 2019). Beyond this important alternative to a single-axis understanding of gender that homogenizes women and treats gender as synonymous with women, an ecofeminist response to the shortcomings of SDG 5 also proposes a rethinking of the very aims of policy: what if liberal understandings of empowerment and equality do not yield the most desirable outcomes?

Instead of a solutionist approach that potentially instrumentalizes women for exogenous goals, the empowerment of women has to be seen as an objective with intrinsic value. Making the business case for feminism and treating gender equality as smart economics in pursuit of economic growth is not a sustainable path towards achieving more equitable relations between men and women or environmental goals (Chant, 2016). Ending discrimination and violence against women and girls and improving their socio-economic and political status should be a moral imperative in itself, not a lever that does the heavy lifting to achieve more important goals. Talking of empowerment without a focus on power relations (Cornwall and Rivas, 2015) will serve to further entrench pre-existing inequalities, and may be co-opted for policies that are counter-productive to realizing women's rights. Instead, empowerment would entail a more profound transformation in relations of social and ecological reproduction, not just 'mainstreaming into a polluted stream' by adding women into unjust and exploitative systems (WMG, 2013).

From our critical ecofeminist point of view, gender equality cannot be achieved without challenging structural domination and the end of all oppressions, including imperialism, colonialism and racism, which rely on hierarchical dualisms that devalue women's work, the environment, and social and ecological reproduction. Women's empowerment and gender inequality thus cannot be discussed as separate from the varied set of social and spatial relations, institutional constraints and biophysical environments within which they take place. This is because equality is not something to be imposed from the top down but instead involves processes of change driven by women themselves (Chant, 2016). According to Naila Kabeer, empowerment involves 'the expansion in people's ability to make strategic life choices in a context where this ability was previously denied to them' (1999: 437). Having a share of wealth and power that is equal to men's will enable women eventually to transform the norms and structures that drive their subordination, and having the same educational and health outcomes as boys will enable girls to grow up with expectations of equal treatment as well the resources to fight injustice. But empowerment is a complex process that requires attention to multiple actors, sites and timescales. Adult women who have lived their lives as second-class citizens may feel empowered by earning better wages for insecure, low-status jobs, whereas girls being raised by economically independent mothers may have greater freedom than their mothers to choose their careers and determine their life patterns (MacGregor, 2019).

At the same time, it is important to acknowledge the high risk of backlash in societies where men do not want to create a level playing field by yielding some of their power to women. This risk is becoming a reality in an ever growing number of states passing laws to curb reproductive freedom, ban same-sex relationships and hinder stopping violence against women. It is for these reasons that feminist organizations (such as those involved in the WMG) should be the ones leading both the empowerment process and the implementation of gender equality goals. To do so they need secure funding to operate independently of state and corporate agendas. New pathways are needed in the 21st century because the development paradigm that has dominated thus far has been a cause of both gender inequality and environmental damage (see Leach, 2015). Critical ecofeminist perspectives can push back against unrealistic visions and identify the tensions and trade-offs as well as the opportunities presented by SD. Gender equality is achieved when women and men (and people who don't identify as either) have the same rights and opportunities in society, across all spheres and sectors, and the same level of participation in decision making. Its realization requires that different gendered needs, practices and aspirations are valued equally. Feminist scholars argue that equality is insufficient if it is simply about balanced numbers or a measure of sameness, or if it involves women becoming like men (Fraser, 1997). Justice is a broader term than equality, which incorporates both sameness and difference and allows unequal treatment (ie giving disadvantaged people a bigger share) to correct an imbalance. Therefore, *gender justice* may have been preferable to gender equality as a SDG (MacGregor, 2019).

10.4.2 Beyond the three pillars: redefining sustainability

Second, in response to the enduring dichotomy between production and reproduction and the patriarchal gender order furthered by the SDGs, an ecofeminist alternative involves fundamental transformation in the gendered values, ideas and social relations that underpin mainstream economics. This transformation demands the dismantling of the binary logic that separates production from reproduction, paid work from unpaid care and humanity from nature. Insofar as the SDG architecture divides environmental, social and economic goals into three silos, it perpetuates a fragmented worldview that fails to account for the interdependence and connectedness of activities taking place across the society–environment nexus. It also obscures the path dependencies, trade-offs and synergies involved in pursuing potentially contradictory goals.

The construction of social, environmental and economic categories as separate pillars is founded on masculinist and productivist biases that prioritize some discourses and concepts over others and further marginalize policy areas that have been historically neglected. To illustrate, the Rio+20 flagship concept of the 'green economy' embedded in the UN framework has been widely criticized for neglecting the social pillar of SD, routinely leaving out considerations of care, social reproduction or gender, and deepening the commodification and financialization of natural resources by attributing the environmental crisis to a misallocation of capital instead of structural domination (Unmüßig et al, 2012; Schalatek, 2014; Harcourt and Nelson, 2015; Herman, 2015). The green economy concept accepts global capitalism and reduces sustainability to a calculus of ecology and natural resources. It is incapable of grasping the shared origins of the current climate and care crisis resulting from the pursuit of infinite economic growth and accumulation at the expense of life-making processes in households, communities and the broader environment (Floro, 2012; Bauhardt, 2014). This 'original contradiction' between capital and nature mediated by women's work (Salleh 1995; 2012) is fundamentally unsustainable, as it leads to the intensifying depletion of social and ecological reproduction – in other words, of humans' differentiated capacities to labour and of non-human nature's capacity to act as a source of inputs and as a sink for the outputs of capitalist production (Rai et al, 2014; Fraser, 2017). It is precisely the ongoing dichotomization taking place in relation to gendered and racialized social categories, of the shared (re)productivity of labour and the environment in mainstream modes of valuation, that has created the contemporary socio-ecological crises of food, fuel and finance that threatens planetary life (Biesecker and Hofmeister, 2010).

As an alternative to neoclassical thinking and neoliberal hegemony, critical ecofeminists have combined concerns for nature with intra- and intergenerational equality to produce a significant body of work that redefines the meaning of sustainability. This meaning includes the sustaining of not only the productive economy, but also of its social and biophysical context embodied and embedded in the ecosystem and in systems of care (O'Hara, 1995; Mellor, 1997; Pietilä, 1997; Bauhardt and Harcourt, 2018). These accounts offer a more holistic view of social and ecological (re)production, not as an inferior gendered responsibility but as a unified process taking place within nature through the forces of reproduction the agencies of the 'racialized, feminized, waged and unwaged, human and non-human labours ... that keep the world alive' (Barca, 2021: 18). In these terms, centring women's unpaid care work and questioning the feminization and devaluation of social reproduction represents a political and analytical entry point for environmental governance towards developing alternatives to unjust and unsustainable relations.

10.5 Conclusion: out of the polluted stream, paving the way towards the world we need

In the decade since the writing and adoption of the SDGs, there has been a renaissance in critical ecofeminist thinking about mainstream and alternative programmes for a sustainable future. Once dismissed as utopian at best, irrational fluff at worst, ecofeminist ideas about political economy, democracy and justice now inform exciting new visions for the kind of 'future we want' (Wichtericht, 2012; see also Akbulut, 2017; Dengler and Lang, 2022). With a record number of young women and gender non-binary activists participating actively in the global movement for climate justice, the connections between gender politics and sustainable development have never been so radical or so visible (WEDO, 2020; Feminist Economic Justice for People & Planet Action Nexus, 2021a).

By way of conclusion, then, the chapter point to campaigns and policy visions that articulate ecofeminist alternatives to the vision embodied in SGD 5. Perhaps the best examples are the growing number of *feminist green new deals* (FGND) that have been developed in Global North countries in recent years (WEDO, 2020). Originating in the US in 2019, a coalition of organizations from all over the world signed up to a set of principles that together represent a radical vision of a 'a transformative feminist agenda that centers the leadership of women, and acknowledges and addresses the generational impacts of colonization and anti-Black racism' (FGND, 2021), creating a 'new paradigm that forges active links between climate change, racialized and gendered labor exploitation, trade rules and economic structures that reproduce inequalities both within and among nations' and 'recognizes that the ecological collapse we are experiencing in climate change is the direct result of an unequal social contract in which these hierarchies shape our social and economic relations' (Feminist Economic Justice for People & Planet Action Nexus, 2021b: 1). In 2020 UK feminists followed suit, articulating a policy road map that follows an 'intersectional approach [that] enables the recognition and inclusion of the needs and concerns of a diversity of constituencies (such as women, BAME people, im/migrants, LGBTQI people, youth, elders, disabled people, etc.)' (Cohen and MacGregor, 2021: 1). Feminists in the EU are organizing to produce their own context-specific versions (Heffernan et al, 2021). What these FGND visions have in common is a rejection of mainstream SD, with its desire for gender equality within a capitalist system that retains all of its hetero-patriarchal, colonial and exclusionary features. Instead, they have climbed out of the polluted stream to chart a new course down a different river to a gender- and climate-just future where gender norms are radically transformed, where care work is seen as a collective public good and where the domination of a minuscule minority over the majority of the planet's humans and other species is no longer tolerated.

At COP 26 in Glasgow in November 2021, the Women's and Gender Constituency of the UNFCCC, a major group consisting of over 40 civil society organizations, made it clear that, for all its commitments and goals on gender equality, the UN continues to fail women. There was no hiding the facts that the vast majority of delegates and decision makers at the COP were men, that the UN has done almost nothing to ensure adequate financing for SDG 5 (eg only 3 per cent of the climate overseas development aid actually targets women's rights and gender equality) and that the progress made on prioritizing human rights and environmental safeguards was shameful (WGC, 2021). In a closing press release, the WGC, the world's loudest collective ecofeminist voice, pledged to continue to 'unapologetically and boldly call for the change we demand, call out false solutions, and pave the way toward the world we need'. They said: 'Civil society and feminist movements know that there is no choice but to continue pushing for the action and justice that our communities and our world needs. And we will continue to do so, together and with fierce care for people and the planet' (WGC, 2021).

In the aftermath of the COVID-19 pandemic, activists, academics and decision makers around the world have been devising strategies for a gender- and climate-just post-pandemic recovery (see Friends of the Earth International et al, 2020; GADN, 2022). The overlapping crises of COVID, care and climate have engendered new vulnerabilities while exacerbating old ones, and further foreground the need to challenge dominant development models and their constitutive injustices in favour of radical intersectional politics that values socio-ecological reproduction instead of prioritizing profit and growth at the expense of care and the environment (Oxfam International, 2020; Leach et al, 2021; Sultana, 2021). This present historical moment presents not only an opportunity for the UN to 'build back better' (UN Women, 2021), but also reinforces the urgency of the need for the kind of deep-rooted transformation within global environmental governance that critical ecofeminists have championed for decades. This transformation is necessary, not only to reclaim the failed promises of the SDGs and to recover from the pandemic, but also to dismantle the oppressions caused by centuries of racial and patriarchal capitalism.

Notes

[1] This statement was delivered by women's rights organizations at the international NGO conference 'Advancing the Post-2015 Sustainable Development Agenda', in Bonn, Germany, 20–22 March 2013. Available online at: https://www.awid.org/fr/node/1121.

2   A single-axis understanding is one that sees the world only through one social category. Marxism can be considered single axis in that class is the primary axis and some types of feminism – notably White liberal feminism – has been labelled single axis for their focus on gender.

References

Akbulut, B. (2017) 'Carework as commons: towards a feminist degrowth agenda', Available from: www.resilience.org/stories/2017-02-02/carew ork-as-commons-towards-a-feminist-degrowth-agenda [Accessed 20 August 2022].

Arora-Jonsson, S. (2011) 'Virtue and vulnerability: Discourses on women, gender and climate change', *Global Environmental Change*, 21: 744–51.

Arora-Jonsson, S. (2017) 'Gender and environmental policy', in S. MacGregor (ed) *The Routledge Handbook of Gender and Environment*, London: Routledge, pp 289–303.

AWID (Association for Women's Rights in Development) 2016 Evaluation Report: 2016 AWID International Forum

Barca, S. (2021) *Forces of Reproduction: Notes for a Counter-hegemonic Anthropocene*, Elements in Environmental Humanities, Cambridge: Cambridge University Press.

Bauhardt, C. (2014) 'Solutions to the crisis? The green new deal, degrowth, and the solidarity economy: Alternatives to the capitalist growth economy from an ecofeminist economics perspective', *Ecological Economics*, 102(2014): 60–68.

Bauhardt, C. and Harcourt, W. (2018) *Feminist Political Ecology and the Economics of Care: In Search of Economic Alternatives*, London: Routledge.

Biesecker, A. and Hofmeister, S. (2010) 'Focus: (Re)productivity', *Ecological Economics*, 69(8): 1703–11.

Braidotti, R., Charkiewicz, E., Häusler, S. and Wieringa, S. (eds) (1994) *Women, the Environment and Sustainable Development: Towards a Theoretical Synthesis*, London: Zed Books.

Buckingham, S. (2020) *Gender and Environment*, 2nd edn, London: Routledge.

Carant, J.B. (2017) 'Unheard voices: A critical discourse analysis of the Millennium Development Goals' evolution into the Sustainable Development Goals', *Third World Quarterly*, 38(1): 16–41.

CGSHR (Consortium on Gender, Security and Human Rights) (2017) *Feminist Critiques of the Sustainable Development Goals: Analysis and Bibliography*, Available from: https://genderandsecurity.org/projects-resour ces/annotated-bibliographies/feminist-critiques-sustainable-development-goals [Accessed 20 August 2022].

Chant, S. (2016) 'Women, girls, and world poverty: Empowerment, equality or essentialism?', *International Development Planning Review*, 38(1): 1–24.

Cohen, M. and MacGregor, S. (2021) *A Draft Roadmap for a Feminist Green New Deal*, Women's Budget Group (WBG) and the Women's Environmental Network (WEN), Available from https://wbg.org.uk/wp-content/uploads/2021/03/FINAL-Roadmap-Feminist-Green-New-Deal.pdf [Accessed 23 June 2023].

Cornwall, A. and Rivas, A. (2015) 'From "gender equality" and "women's empowerment" to global justice: Reclaiming a transformative agenda for gender and development', *Third World Quarterly*, 36(2): 396–415.

Corrêa, S. and Reichman, R. (1994) *Reproductive Rights and Population: Feminist Voices from the South*, London: Zed Books.

Cox, R. W. (1981) 'Social forces, states and world orders: Beyond international relations theory', *Millennium*, 10(2): 126–55.

Crenshaw, K. (1989) 'Demarginalizing the intersection of race and sex: A Black feminist critique of antidiscrimination doctrine, feminist theory and antiracist politics', *University of Chicago Legal Forum*, 1(8): 139–167.

Cudworth, E. (2014) 'Feminism', in C. Death (ed) *Critical Environmental Politics*, London: Routledge, pp 91–9.

Dengler, C. and Lang, M. (2022) 'Commoning care: feminist degrowth visions for a socio-ecological transformation', *Feminist Economics*, 28(1): 1–28.

Di Chiro, G. (2008) 'living environmentalisms: coalition politics, social reproduction, and environmental justice', *Environmental Politics*, 17(2): 276–98.

Elson, D. (2016) 'Plan F: Feminist plan for a caring and sustainable economy', *Globalizations*, 13(6): 919–21.

Esquivel, V. (2016) 'Power and the Sustainable Development Goals: A feminist analysis', *Gender & Development*, 24(1): 9–23.

Feminist Economic Justice for People & Planet Action Nexus (2021a) *A Feminist Agenda for People and Planet: Principles and Recommendations for a Global Feminist Economic Justice Agenda*, Feminist Blueprint for Action June 2021 Generation Equality Forum, Available from: https://wedo.org/wp-content/uploads/2021/06/Blueprint_A-Feminist-Agenda-for-People-and-Planet.pdf [Accessed 20 August 2022].

Feminist Economic Justice for People & Planet Action Nexus (2021b) *A Feminist and Decolonial Global Green New Deal: Principles, Paradigms and Systemic Transformations.* Issue Brief, Action Nexus for Generation Equality, Available from: https://wedo.org/wp-content/uploads/2021/06/FemEcon Climate-ActionNexus_Brief_FemGND-1.pdf [Accessed 20 August 2022].

Ferber, M. and Nelson, J. (1993) *Beyond Economic Man: Feminist Theory and Economics*, Chicago: University of Chicago Press.

FGND (Feminist Green New Deal) (2021) 'Principles', Available from: https://feministgreennewdeal.com/principles [Accessed 20 August 2022].

Floro, M.S. (2012) 'The crises of environment and social reproduction: Understanding their linkages', *Journal of Gender Studies*, 15: 13–31.

Folbre, N. (2021) *The Rise and Decline of Patriarchal Systems: An Intersectional Political Economy*, London: Verso.

Foster, E. (2014) 'International sustainable development policy: (Re) producing sexual norms through eco-discipline', *Gender, Place & Culture*, 21(8): 1029–44.

Foster, E. (2017) 'Gender, environmental governmentality, and the discourses of sustainable development', in S. MacGregor (ed) *Routledge Handbook of Gender and Environment*, London: Routledge, pp 216–28.

Foster, E. A. (2011) 'Sustainable development: Problematising normative constructions of gender within global environmental governmentality', *Globalizations*, 8(2): 135–49.

Fraser, N. (1997) *Justice Interruptus: Critical Reflections on the 'Postsocialist' Condition*, New York: Routledge.

Fraser, N. (2017) 'Crisis of care? On the social re-productive contradictions of contemporary capitalism', in T. Bhattacharya and L. Vogel (eds) *Social Reproduction Theory: Remapping Class, Recentering Oppression*, London: Pluto Press, pp 22–36.

Friends of the Earth International, World March of Women and REMTE (Red Latinoamericana de Mujeres Transformando la Economía) (2020) *Feminist Economics and Environmentalism for a Just Recovery: Outlooks from the South*, Available from: https://marchemondiale.org/index.php/2020/11/12/feminist-economics-and-environmentalism-for-a-just-recovery-outlo oks-from-the-south [Accessed 20 August 2022].

Gaard, G. (2017) *Critical Ecofeminism*, Lanham, MD: Lexington Books.

Gabizon, S. (2016) 'Women's movements' engagement in the SDGs: Lessons learned from the Women's Major Group', *Gender & Development*, 24(1): 99–110.

GADN (Gender and Development Network) (2019) 'Push no one behind: How current economic policy exacerbates gender inequality', Available from: https://gadnetwork.org/gadn-resources/push-no-one-beh ind-how-current-economic-policy-exacerbates-gender-equality [Accessed 20 August 2022].

GADN (2022) 'Feminist macroeconomic proposals: rebuilding more equitable, just and sustainable economies post-COVID-19', Available from: https://gadnetwork.org/gadn-resources/feminist-macroecono mic-proposals-rebuilding-more-equitable-just-and-sustainable-econom ies-post-covid-19 [Accessed 20 August 2022].

Harcourt, W. (1994) *Feminist Perspectives on Sustainable Development*, London: Zed Books.

Harcourt, W. and Nelson, I. (2015) *Practising Feminist Political Ecologies: Moving beyond the 'Green Economy'*, London: Zed Books.

Hartmann, B. (2016) *Reproductive Rights and Wrongs: The Global Politics of Population Control*, Chicago: Haymarket Books.

Heffernan, R., Heidegger, P., Köhler, G., Stock, A. and Wiese, K. (2021) *A Feminist European Green Deal: Towards an Ecological and Gender Just Transition*, Bonn: Friedrich-Ebert-Stiftung.

Herman, C. (2015) 'Green new deal and the question of environmental and social justice', Global Labour University Working Paper No. 31, Geneva: International Labour Organization.

Hirshman, M. (2003) 'Women and development: A critique', in M.H. Marchand and J.L. Parpart (eds) *Feminism/ Postmodernism/ Development*, London: Routledge, pp 42–55.

Jackson, C. (1994) 'Gender analysis and environmentalisms', in M. Redclift and T. Benton (eds) *Social Theory and the Global Environment*, London: Routledge, pp 113–49.

Kabeer, N. (1999). 'Resources, agency, achievements: Reflections on the measurement of women's empowerment', *Development and Change*, 30(3): 435–64.

Kaijser, A. and Kronsell, A. (2014) 'Climate change through the lens of intersectionality', *Environmental Politics*, 23(3): 417–33.

Lau, J.D., Kleiber, D., Lawless, S. and Cohen, P.J. (2021) 'Gender equality in climate policy and practice hindered by assumptions', *Nature Climate Change*, 11(3): 186–92.

Leach, M. (ed) (2015) *Gender Equality and Sustainable Development*, New York: Routledge.

Leach, M., MacGregor, H., Scoones, I. and Wilkinson, A. (2021) 'Post-pandemic transformations: How and why COVID-19 requires us to rethink development', *World Development*, 138: 105233.

Logie, C.H. (2021) 'Sexual rights and sexual pleasure: Sustainable Development Goals and the omitted dimensions of the leave-no-one-behind sexual health agenda', *Global Public Health*, 18: 1–12.

MacGregor, S. (2010a) 'A stranger silence still: The need for feminist social research on climate change', in B. Carter and N. Charles (eds) *Nature, Society and Environmental Crisis*, Oxford: Wiley Blackwell, pp 124–40.

MacGregor, S. (2010b) '"Gender and climate change": from impacts to discourses', *Journal of the Indian Ocean Region*, 6(2): 223–38.

MacGregor, S. (2017a) 'Gender and environment: An introduction', in S. MacGregor (ed) *The Routledge Handbook of Gender and Environment*, London: Routledge, pp 1–24.

MacGregor, S. (2017b) 'Moving beyond impacts: More answers to the "gender and climate change" question', in S. Buckingham and V. Le Masson (eds) *Understanding Climate Change through Gender Relations*, London: Routledge, pp 16–30.

MacGregor, S. (2019) 'Goal #5: Gender equality', SDG Online, Available from: www.taylorfrancis.com/sdgo/goal/GenderEquality [Accessed 20 August 2022].

MacGregor, S., Arora-Jonsson, S. and Cohen, M. (2022) 'Caring in a changing climate: Centering care work in climate action', Oxfam, Available from: https://policy-practice.oxfam.org/resources/caring-in-a-chang ing-climate-centering-care-work-in-climate-action-621353 [Accessed 20 August 2022].

Marx, K. (1843) 'Letter from Marx to Arnold Ruge', Letters: Letter from Marx to Arnold Ruge, Available from: www.marxists.org/archive/marx/ works/1843/letters/43_09-alt.htm [Accessed 20 August 2022].

Mary Robinson Foundation (2015) *Women's Participation: An Enabler of Climate Justice*, Available from: www.mrfcj.org/resources/womens-partic ipation-an-enabler-of-climate-justice [Accessed 20 August 2022].

Mellor, M. (1997) 'Women, nature and the social construction of "economic man"', *Ecological Economics*, 20: 129–40.

Mellor, M. (2006) 'Ecofeminist political economy', *International Journal of Green Economics*, 1 (1–2): 139–50.

Mellor, M. (2017) 'Ecofeminist political economy', in S. MacGregor (ed) *Routledge Handbook of Gender and Environment*, London: Routledge, pp 86–100.

Merchant, C. (1989) *The Death of Nature: Women, Ecology, and the Scientific Revolution*, New York: Harper & Row.

Mies, M. (1998) *Patriarchy and Accumulation on a World Scale: Women in the International Division of Labour*, London: Zed Books.

Mohanty, C. T (1991) 'Under western eyes: Feminist scholarship and colonial discourses', in C. T. Mohanty, A. Russo and L. Torres (eds) *Third World Women and the Politics of Feminism*, Bloomington: Indiana University Press, pp 51–80.

Nightingale, A.J. (2006) 'The nature of gender: Work, gender, and environment', *Environment and Planning D: Society and Space*, 24: 165–85.

O'Hara, S. (1995) 'Sustainability: Social and ecological dimensions', *Review of Social Economy*, 53(4): 529–51.

O'Manique, C., and Fourie, P. (2016) 'Affirming our world: Gender justice, social reproduction, and the Sustainable Development Goals', *Development*, 59(1–2): 121–6.

Oxfam International (2020) *Climate, Covid and Care: Feminist Journeys*, Available from: www.oxfam.org/en/take-action/campaigns/climate/clim ate-zine-feminist-journeys [Accessed 20 August 2022].

Perkins, P. (2019) 'Climate justice, gender, and intersectionality', in T. Jafry (ed) *Routledge Handbook of Climate Justice*, New York: Routledge, pp 349–58.

Pietilä, H. (1997) 'The triangle of the human economy: Household cultivation industrial production. An attempt at making visible the human economy in toto', *Ecological Economics*, 20(2): 113–27.

Plumwood, V. (1992) *Feminism and the Mastery of Nature*, London: Routledge.

Rai, S., Hoskins, C. and Thomas, D. (2014) 'Depletion: The cost of social reproduction', *International Feminist Journal of Politics*, 16(1): 86–105.

Razavi, S. (2016) 'The 2030 Agenda: Challenges of implementation to attain gender equality and women's rights', *Gender & Development*, 24(1): 25–41.

Resurrección, B. (2013) 'Persistent women and environment linkages in climate change and sustainable development agendas', *Women's Studies International Forum*, 40: 33–43.

Resurrección, B. (2021) 'Gender, climate change and disasters: Vulnerabilities, responses, and imagining a more caring and better world', Background paper, UN Women Expert Group Meeting, Available from: https://www.unwomen.org/sites/default/files/Headquarters/Attachments/Sections/CSW/66/EGM/Background%20Papers/Bernadette%20RESURECCION_CSW66%20Background%20Paper.pdf [Accessed 20 August 2022].

Roberts, A. and Soederberg, S. (2012) 'Gender equality as smart economics? A critique of the 2012 World Development Report', *Third World Quarterly*, 33(5): 949–68.

Salleh, A. (1995) 'Nature, woman, labor, capital: Living the deepest contradiction', *Capitalism Nature Socialism*, 6(1): 21–39.

Salleh, A. (1997) *Ecofeminism as Politics: Nature, Marx, and the Postmodern*, London: Zed Books.

Salleh, A. (2012) 'Green economy or green utopia? Rio+20 and the reproductive labor class', *American Sociological Association*, 8(2): 141–5.

Schalatek, L. (2014) *The Post-2015 Framework: Merging Care and Green Economy Approaches to Finance Gender-Equitable Sustainable Development*, Washington, DC: Heinrich-Böll-Stiftung, Available from: https://us.boell.org/sites/default/files/schalatek_gender-equitable_post-2015_sustainable_development.pdf [Accessed 23 June 2023].

Sultana, F. (2021) 'Climate change, COVID-19, and the co-production of injustices: A feminist reading of overlapping crises', *Social & Cultural Geography*, 22(4): 447–60

Thompson, C.M. and MacGregor, S. (2017) 'The death of nature: Foundations of ecological feminist thought', in S. MacGregor (ed) *Routledge Handbook of Gender and Environment*, London: Routledge, pp 43–53.

UN (United Nations) (2022) *Goal 5: Achieve gender equality and empower all women and girls*. Available from: https://www.un.org/sustainabledevelopment/gender-equality/ [Accessed 20 August 2022].

UNEP (United Nations Environment Programme) (2015) 'Sustainable Development Goals and the 2030 Agenda: Why environmental sustainability and gender equality are so important to reducing poverty and inequalities', *UNEP Perspectives 17*. Available from: https://wedocs.unep.org/handle/20.500.11822/7464 [Accessed 20 August 2022].

Unmüßig, B., Sachs, W. and Fatheuer, T. (2012) 'A critique of the green economy: Toward social and environmental equity', Heinrich-Böll-Stiftung, Available from: https://us.boell.org/en/2012/06/07/critique-green-economy-toward-social-and-environmental-equity [Accessed 20 August 2022].

UN Women (2021) *Pathways to Building Back Better: Advancing Feminist Policies in COVID-19 Response and Recovery*, Available from: https://www.unwomen.org/sites/default/files/Headquarters/Attachments/Sections/Library/Publications/2021/Think-piece-Pathways-to-building-back-better-en.pdf [Accessed 20 August 2022].

WEDO (Women's Environment & Development Organization) (2014) *Lost & Found in the Proposed Sustainable Development Goals A view from WEDO*, Available from: https://wedo.org/lost-found-in-the-proposed-sustainable-development-goals-a-view-from-wedo [Accessed 20 August 2022].

WEDO (2020) *Global Feminist Frameworks for Climate Justice Town Hall: Frameworks Reader*, Available from: https://wedo.org/global-feminist-frameworks-for-climate-justice-town-hall-reader [Accessed 20 August 2022].

WGC (Women's and Gender Constituency) (2021) 'Press release: The power is with us: COP26 fails people & planet', 13 November, Available from: https://womengenderclimate.org/press-release-the-power-is-with-us-cop26-fails-people-planet [Accessed 23 June 2023].

WGC, DAWN (Development Alternatives with Women for a New Era) and WECF (Women International for a Common Future) (2013) *Statement for endorsement: We will not be mainstreamed into a polluted stream: Feminist statement on the 2015 development agenda*, Bonn, 22 March 2013. Available from: https://dawnnet.org/publication/statement-for-endorsement-we-will-not-be-mainstreamed-into-a-polluted-stream-feminist-statement-on-the-2015-development-agenda-bonn-22-march-2013/ [20 August 2022].

Wichterich, C. (2012) *The Future we Want: A Feminist Perspective*, Berlin: Heinrich-Böll-Stiftung.

WMG (Women's Major Group) (2013) *Gender Equality, Women's Rights and Women's Priorities: Recommendations for the proposed Sustainable Development Goals (SDGs) and the Post-2015 Development Agenda*, Available from: www.wecf.org/gender-equality-womens-rights-and-womens-priorities-recommendations-for-the-sustainable-development-goals-sdgs-and-post-2015-development-agenda [Accessed 20 August 2022].

WMG (2015) 'Compilation of Women Major Group policy recommendations on monitoring and means of implementation for the Post-2015 Sustainable Development Agenda', Available from: www.wecf.org/womens-major-group-position-paper-compilation-of-policy-recommendations [Accessed 20 August 2022].

Interview with Sherilyn MacGregor and Aino Ursula Mäki: Environmental Policy Needs to Stop Being Indifferent to Difference

Jana Beier, Maike Laengenfelder and Rosa-Lena Lange

You criticize the narrative that women are more affected by climate change than men. Even though it is true, what makes it problematic?

MacGregor: It implies that gender is relevant to climate change only because women are more affected. This reinforces the stereotype that women are vulnerable victims needing help from UN agencies, which often takes on a problematic Global North versus Global South dynamic. Women in the Global South do not want to be considered as victims. We must not ignore statistics indicating that women may be more or differently affected by climate change than men, but these shouldn't be the only reason to inject gender analysis into the sustainable development discourse.

In practical terms, you are in favour of involving more women in policy making?

MacGregor: I'm not against it, but again I resist simplistic arguments. We could have more women at the decision-making table, but what if they were all like Marine Le Pen or Liz Truss? The narrative that women will 'save the planet' because of their inherent ethic of care, or because they give birth to the next generation, is essentialist. Ecofeminism tries to tackle such views. We don't just want an equal number of women involved in policy: we want more people of all genders who subscribe to feminist, anti-racist, anti-fascist views. Environmental policy needs to embed an intersectional approach at a structural level to stop being indifferent to differences and to embrace diversity.

Mäki: In some ways, the women as agents of change discourse adopted by the UN and others, which is aimed at increasing the participation of women, can be seen as a kind of a reversal of the women as victims trope, because it is also based on a single axis, a somewhat simplistic understanding of gender relations. Simply including or empowering women without other forms of political change does not challenge the structural features of capitalist racist patriarchy which reproduce injustice, environmental destruction and inequality. Although involving more women in policy making is of course necessary, on its own it is not enough to guarantee any form of deep-seated socio-ecological transformation.

You also point out how feminist green new deals are an alternative to 'the polluted stream' that facilitate the way towards a climate-just future. Can you illustrate what makes your vision for a FGND in the UK feminist?

MacGregor: We analysed existing green new deal plans through an ecofeminist lens and found three main problems: they aren't inclusive, they are mainly about a transition led by technological change, and the process is very elite driven. In response, we developed a vision of what we want as ecofeminists. This vision starts from an intersectional analysis of society and a commitment to social justice. We developed five categories of recommendations. First, instead of focusing only on high tech and construction jobs dominated by men, we call for a broader definition of green jobs. Second, the care economy and investing in collective social infrastructure are central: it's about rebuilding and greening the welfare state after decades of neoliberalism. Third, we suggest how cities and housing could be redesigned to be both gender and eco friendly. Fourth, we want the democratization of money and finance through innovations like community banking, co-ops and green investment. And, finally, we recommend devolving power and deliberation to the local level, which is what an ecofeminist approach to the policy process itself would look like.

11

Realizing Sustainable Consumption and Production

Sylvia Lorek, Maurie Cohen and Eva Alfredsson

To make our societal systems sustainable, tremendous reductions are required in the throughput of resources as well as in the volume of greenhouse gases and other harmful emissions released into the environment. On the basis of recent assessments, wealthier countries will need to curtail their current GHG emissions by 68–86 per cent by 2050. Even less affluent countries will need to cut their prevailing emissions by 76 per cent to meet the targets of the Paris Agreement (Akenji et al, 2021). However, reductions of a similar magnitude in resource consumption more generally will also be necessary (Bringezu, 2015).

Sustainable Development Goal 12 seeks to 'ensure sustainable consumption and production patterns' but developments to date do not even come close to meeting this objective. As originally recommended by Lebel (2004), SDG 12 conjoins environmental goods and services, individuals, households, organizations and states through linkages in which energy and materials are transformed. These circumstances create a profound need for coordinated global environmental governance, and this requirement is compounded by progress on this goal affecting the achievement of the other 16 SDGs. As recent evaluations from the Organisation for Economic Co-operation and Development (OECD, 2022) show, we are not even moving in the right direction with respect to most indicators. The challenges that lie ahead are truly daunting and will enormously strain our capacity for both innovation and resilience (Bengtsson et al, 2018).

Section 11.2 briefly analyses the ambitions of SDG 12 and its underlying targets. Section 11.3 then discusses why efforts to decouple resource and energy use from GDP growth will not be sufficient, and makes the case for dematerialization that transcends issues of continual economic expansion.

That household income is closely correlated with resource consumption is also highlighted: 'simply' striving to raise the prices of energy and materials (as many policy makers and others contend) is an inadequate strategy. Section 11.4 focuses on various elements of system change to achieve sustainable patterns of consumption and production. The prospects of the currently popular concept of a circular economy and the potential of supplementing profit-driven enterprises with cooperatives and smaller-scale community-based businesses to foster SCP are considered. The section also discusses how reduced working hours, quality public services and demand-side solutions can contribute to achieving sustainable lifestyles. Finally, section 11.5 concludes the chapter with a brief discussion on how to make meaningful action on SCP a more salient issue on policy agendas.

11.1 SDG 12 and its targets

This section assesses from a critical perspective how the targets of SDG 12 address the challenges of SCP, which is a cross-cutting issue in the framework and includes 50 of the 169 SDG targets and 13 different goals. Given the centrality of this objective, it is of concern that progress to date has been extremely limited and several indicators show that relevant developments are on the wrong trajectory (OECD, 2022). One reason for the slow pace is that the outcome of the SDG 12 negotiations reflects a production- and design-centred perspective and is dominated by a faith in solutions through new technologies (Gasper et al 2019). Another explanation is that efforts to achieve SCP are in conflict with the relentless pursuit of economic growth. (Chertkovskaya, Chapter 10 in this volume, reflects on the shortcomings of SDG 8, which calls for maintaining per capita economic growth.)

Furthermore, there are many aspects of a comprehensive plan for pursuing SCP that are not represented in SDG 12. Perhaps most glaring is that the overall goal limits itself to the environmental dimensions of sustainability, which is a salient misrepresentation of what the SDGs are meant to enable, namely equitable human flourishing in a shared biosphere. To create more effective synergies between human welfare and ecological sustainability, SCP researchers have proposed a number of reforms including a radical re-envisioning of the relationship between material accumulation and well-being. The first facet of this conception involves recognizing how care work – in its paid and unpaid forms – provides the basis for reproducing capacity for future productive activities. The second aspect involves a shift from owning physical products to sharing the services provided by products. The final feature entails acknowledging the need to creatively blend reliance on *efficiency* improvements with new commitments to *sufficiency*. Wider acceptance of these corrective measures could help to more effectively connect SCP to other goals in the SDG framework (Bengtsson et al, 2018).

The main problem is that the targets undergirding SDG 12 mainly aim to enhance the productivity of provisioning systems while ignoring the need for reduced consumption, a shift to lifestyles predicated on self-limited utilization of energy and resources, and a renewed commitment to well-being among the world's affluent consumers (O'Neill et al, 2018). The architects of SDG 12 also contend that SCP is based to a large degree on information disclosure by product manufacturers and conscientious decisions by individual consumers to act on this guidance as, for example, formulated in target 12.8: 'By 2030, ensure that people everywhere have the relevant information and awareness for sustainable development and lifestyles in harmony with nature' (UN, 2015).

The targets, furthermore, suffer from a limited understanding of the underlying drivers and weak formulation of SDG 12. Waste reduction remains the primary focus, as evidenced by the longstanding and continuing emphasis on measuring progress by recycling rates and raising public awareness about the importance of this practice. It is notable that only target 12.3 on reducing food waste has a quantified objective. It aims to 'halve per capita global food waste at the retail and consumer levels and reduce food losses along production and supply chains, including post-harvest losses' (UN, 2015). Accordingly, there have been commendable efforts – France is a prominent example – to implement policies to establish binding responsibilities that prevent supermarkets from destroying unsold food products and instead to divert overstocked supplies to donation channels (Zero Waste Europe, 2020).

The most important target of SDG 12 to reduce material footprints on an unprecedented scale – has received little consideration. An emphasis on efficiency improvements and a general voicing of salutary regard for 'less unsustainable' modes of consumption is what remains predominant. Policies that seek to seriously ensure SCP will need to go far beyond the current technology-based formulations in SDG 12 that mostly evade the true scale and scope of the challenge.

11.2 Failure of neoliberal approaches to achieve SCP

11.2.1 The limits of decoupling and the need for dematerialization

This section explores the dynamics between efficiency and scale and the potential for decoupling environmental pressure from the growth of GDP. It shows how strong policies providing incentives for efficiency, on the one hand, have the potential to reduce GHG emissions and resource consumption but, on the other hand, increase energy and material use in absolute terms. The result is that greater overall throughput often outweighs the environmental gains from enhanced efficiency. This unfortunate phenomenon is known as Jevons Paradox, and it makes the rapid reductions needed for holding global warming to 1.5 °C difficult even under ambitious climate policies. Limiting global warming to well below 2 °C, preferably to 1.5 °C, compared to preindustrial levels is the target the international community agreed on in 2015 (see Marquardt and Schreurs, Chapter 2 in this volume).

All production of goods transforms physical resources in a process that requires energy and the use of various materials. The common measure of GDP captures the total value of goods and services produced in a society (generally calculated for a country). As the volume of production grows, so does GDP, and this outcome is generally regarded, especially in political terms, as an unambiguously positive development. Industrial ecologists introduced the notions of social metabolism (Fischer-Kowalski and Haberl, 1997) and industrial metabolism (Ayres and Simonis, 1994) in the 1990s to refer to the fundamental character of the interactions between society and nature that underpin human development. Numerous studies have demonstrated that there are generally and on a global level strong correlations between GDP growth, resource use and GHG emissions (Parrique et al, 2019; Haberl et al, 2020; Vadén et al, 2020). In other words, increases in scale due to economic growth can undermine environmental gains achieved through technological or process improvements (Brockway et al, 2021).

A remarkable number of studies have evaluated the evidence on the decoupling of GDP from resource use and GHG emissions, and have built a solid basis for various summarizing review studies. For instance, a review (Haberl et al, 2020) of over 800 articles analyzing the empirical evidence for absolute decoupling shows different patterns for lower-income and higher-income countries. In lower-income countries (generally defined as having GDP of less than $10,000 per capita), there is notable coupling between GDP, resource use (including energy) and GHG emissions; when the economy grows, so does consumption of resources and carbon emissions. In contrast, for higher-income countries (with GDP greater than $10,000 per capita), GHG emissions from within their own borders on average show a small absolute decoupling (Haberl et al, 2020). This means that carbon emissions are decreasing to some extent even when the economy grows. However, when GHG emissions are accounted for in relation to the production of goods imported by higher-income countries (generally in the form of carbon that is embodied by physical products), only relative decoupling in the vast majority of cases can be observed. More tangibly, carbon emissions continue to grow as GDP increases but at a less rapid rate (Haberl et al, 2020).

Another review by Vadén et al (2020), based on 179 articles covering the two-decade period between 1990 and 2019, reached similar results, and Schandl et al (2016) found absolute decoupling of CO_2 from GDP but concluded that material use more than trebled between 1970 and 2010. Similarly, world energy use also roughly increased threefold during the same period (IEA, 2013). Again, to emphasize the prior point, the general pattern is that, as GDP grows, resource use and GHG emissions increase as well. And, in general, higher-income countries are characterized by both high GHG emissions and high material footprints (Wiedmann et al, 2020).

Analysts now contend that economic growth needs to be decoupled – or dissociated – from escalating resource use and adverse environmental impacts to secure long-term sustainability for humankind and the planet (Brockway et al, 2021). A critical, but still outstanding, question is to what extent this is actually possible.

To transition to a sustainable economy, global GHG emissions will need to be cut to close to zero by 2050 and this achievement must then be followed by an extended period of negative carbon release (IPCC, 2021). In other words, the carbon currently in the atmosphere will need to be removed with various kinds of technological apparatus. Sustainability and climate stabilization will also require reduced environmental impact from resource utilization. The volume of renewable materials appropriated to meet human needs will need to be limited to its long-term sustainable level which, for biomass, equates approximately to its annual production rate. Furthermore, the use of non-renewable resources will need to be substantially curtailed, ultimately to a level that is close to 100 per cent recycling for long-term sustainability. Achieving these goals will require drastic restructuring of major features of contemporary economies (Alfredsson et al, 2018).

In terms of decoupling, for countries with relatively low per capita income and wealth the initial aim is to achieve economic growth while slowing the rate of increase of natural resource exploitation and GHG emissions (*relative decoupling*) and then for environmental impacts to decrease over time in absolute terms (*absolute decoupling*). For the most affluent countries, generally defined as members of the OECD, the challenge is to pursue absolute decoupling and to refocus societal aspirations away from further economic growth towards an emphasis on the quality of growth (and perhaps even degrowth) (Martinez-Alier et al, 2010).

While there is, as noted, a strong correlation between GDP and energy use (see also Steinberger et al, 2013), the connection between GDP and carbon emissions is not as strong. There are substantial variations between countries depending on their specific energy mix (the particular combination of sources they use to generate electricity and operate transport networks) and other factors (Steinberger et al, 2013). However, the share of fossil fuels in the world's total energy mix remains as high as it was a decade ago. As recently as 2019, approximately 84 per cent of global primary energy was still being derived from fossil sources, that is, coal, oil and natural gas (IEA, 2021).

By increasing the share of renewable energy, the carbon intensity of the energy system can be reduced (Tudor and Sova, 2021). However, if the demand for energy increases, GHG emissions are apt to increase despite a lower carbon intensity (Haberl et al, 2020). According to the IEA's latest forecast, the demand for all fossil fuels is set to grow significantly in future years. Coal demand alone, because of its relatively low cost, is projected to increase by 60 per cent more than all renewables combined, underpinning a rise in carbon emissions of almost 5 per cent, or 1,500 metric tonnes (IEA, 2021). This unfortunate situation illustrates the largely disregarded necessity of pursuing strategies that encourage absolute decoupling (as well as associated modes of consumption reduction) (Alfredsson et al, 2018).

Le Quéré et al (2018) show that, in some higher-income countries where GHG emissions have been reduced in absolute terms, this phenomenon has contributed to absolute decoupling when measured from a production perspective. Further analysis demonstrates that the cause of this decoupling is strong climate policies, but the decoupling is not as high as it needs to be to reach prevailing climate goals. Hickel and Kallis (2020) conclude that current empirical evidence of small absolute decoupling rates will not reduce emissions rapidly enough to bring carbon budgets into alignment with the 1.5 °C target. They show that GDP growth of 3 per cent per year requires a decoupling of 10.5 per cent per year for 1.5 °C. A growth rate of zero requires an annual decarbonization rate of 6.8 per cent (Hickel and Kallis, 2020). But are these decoupling rates achievable?

Hickel and Kallis conclude, on the basis of a model by Schandl et al (2016), that decoupling can happen by at most 3 per cent each year given strong policies, including a global carbon price of US$50 per tonne starting in 2015 and reaching US$236 by 2050. This demonstrates that, while technological innovation can contribute to the reduction of resource use and emissions, it is nevertheless outpaced by increases in economic growth and consumption (Dyrstad et al, 2019). Studies by researchers in the field of sustainable consumption increasingly argue for a turn towards 'strong sustainable consumption governance' (Lorek and Fuchs, 2013) that aims to reduce the volume of materials and energy resources consumed substantially while maintaining levels of well-being.

Such a concept of dematerialization argues that current environmental problems (such as climate change and biodiversity loss) are closely related to the volume of material and energy used in the production of goods and services; if the input decreases, the overall environmental impact will decrease as well. The calculations call for absolute reductions of material flows by a factor of four (Weizsäcker et al, 1998) or a factor of ten (Schmidt-Bleek, 2008), depending on which materials are considered (Lettenmeier, 2018).

11.2.2 The unintended consequences of 'getting the price right'

One of the common solutions to reduce unsustainable consumption patterns is 'getting the price right' through internalizing the environmental costs of products and services (Parry et al, 2014). In a market that does not account for the costs of unsustainable production, more sustainable alternatives are more expensive and thus less attractive. Financial policies implemented to date often fail to properly target those parties that are most responsible for high emissions and material footprints (and also have the greatest capacity to carry the burden), and thus consequently undermine achievements for SDG 12 (Gore and Alestig, 2020). The most significant determinant of a person's environmental footprint is their income. Accordingly, the richest 10 per cent of the global population is responsible for approximately half of total consumption-related emissions, while the poorest 50 per cent account for only about 10 per cent (Gore and Alestig, 2020). To ensure fairness when moving towards SCP, policies need to: (1) put the burden on the individuals and organizations that are most responsible and have the greatest capacity while compensating vulnerable groups for the financial burden of climate and resource use policies; (2) target luxury goods (eg sport utility vehicle) while guaranteeing and improving basic needs satisfaction (public transport); and (3) tax high income and wealth as resources for financing a just transition while supporting green living options (eg affordable low-carbon housing) (Lorek et al, 2021b).

There are good examples where unfair outcomes have been deliberately avoided via effective policy design (Lamb et al, 2020). Notable instances include the carbon tax implemented in the Canadian province of British Columbia, which recycles revenues as lump-sum transfers to households to offset negative distributional outcomes, and the UK Warm Front energy efficiency program and Saving In-House initiative in Greece, which involve thermal efficiency subsidy programmes that allocate a higher proportion of funds to lower-income households. Such schemes bring together climate policy with social policy and directly address the associated contradictions (Lamb et al, 2020).

11.3 System changes for sustainable consumption and production

11.3.1 Shifting to a circular economy

Since adoption of the 2030 Agenda in 2015, one of the major challenges to progressing towards SDG 12 has been to develop policies and business models towards a circular economy that highlight the benefits for countries in both the Global South and Global North. As a group of scientific and political experts summarized it during a United Nations high-level political forum on the circular economy the business community, on the one hand, is energized about new circular economy opportunities along value chains, but they need policy support for markets, access to finance (particularly to facilitate innovation and the upskilling of workers) and investor engagement to better understand and encourage relevant initiatives. Consumers, on the other hand, are disposed to enacting more sustainable practices and to using products that derive from circular manufacturing processes, but they often do not know whether these products exist and are available at affordable prices. While policy makers agree that shifting to responsible consumption and production patterns is necessary, they have difficulties prioritizing where and how to design circularity into extant systems and how to monitor, measure and assess their outcomes and impact (UN, 2021).

Substantive policies in this context entail, for example, financial incentives such as lower value-added taxes on the repair of products and surcharges for single-use plastic bags (One Planet Network 2020; 2021). Regulatory measures include mandatory building codes and minimum requirements on energy performance in buildings. These are politically challenging undertakings, and they are often sidestepped in favour of more modest initiatives focused on improving the provision of information to consumers and others about, for instance, building materials and product reusability, recyclability and durability (One Planet Network 2022a; 2022b).

Extended producer responsibility (EPR) is a financial and operational policy instrument that aims to internalize environmental externalities related to end-of-life product management and waste. Such programs assign significant responsibility to producers to develop interventions to take back, recover, treat and dispose of post-consumer products and waste. The aim is to incentivize efforts to minimize resource utilization at the source and to promote more environmentally conscious product design (Schröder and Raes, 2021).

With all this in mind, it nevertheless should not be forgotten that the concept of a circular economy faces various limitations. First, according to the basic rule of thermodynamics, transformation processes always lead to a loss and, under the best of circumstances, only a limited amount of material can be reused in circular systems. Second, the examples of successful circularity of products are mainly limited to local or regional cases that cannot be transferred to the global scale. Related to this situation are challenges with respect to the governance and management of circular economy processes that depend on the common decision making of a variety of independent political and business actors. Finally, the circularity principle shares key features with the notion of efficiency where the reductions achieved can be outweighed by increases in absolute terms if no absolute cap is imposed (Korhonen et al, 2018). For all these reasons, the member countries of the European Union are still some distance away from becoming a circular economy despite the explicit goal (EEA, 2020).

11.3.2 Alternatives to profit-driven enterprises

Broad parts of the business system are organized through profit-driven enterprises where the interests of oftentimes veiled owners, investors and shareholders define the strategic objectives and performance targets of companies. By contrast, cooperatives, worker-owned companies, community ownerships and small-scale businesses rooted in particular localities can directly generate benefits for those involved (Cohen, 2017a; 2017b). These organizations often provide multiple benefits for livelihoods, strength of community bonds and trust, and cohesion for members and consumers. With respect to ecological side effects, they contribute to a reduced need for motorized transport and the people involved often demonstrate an ability to take greater responsibility for the environmental impacts that result from business-related activities (Cohen, 2017a; 2017b).

The most familiar of these alternative business models in the context of responsible consumption and production are coffee-producing cooperatives involved in the fair trade movement. Despite large geographical distances, there is typically close collaboration between producers, traders and consumers that reaches far beyond ensuring payment of fair prices. Fair trade businesses aim to overcome the model of profit primacy to develop holistic practices of fair exchange that become inherent in resultant business models. These firms reinvest the majority of their revenues into furthering a social mission and deploying mission primacy in their governance. Built to make management and investment decisions that favour workers, farmers and artisans, they are effective in some of the most challenging contexts (Doherty et al, 2020). Fair trade businesses also increasingly drive ecological practices that protect the environment (Partzsch et al, 2021). Doherty et al (2020) summarize key insights from a study carried out in cooperation with the World Fair Trade Organization (WFTO): 92 per cent of WFTO member organizations reinvest all profits in their social mission and 52 per cent are led by women. They are also four times less likely to declare bankruptcy and 85 per cent of these organizations report actively sacrificing financial goals to pursue social or environmental objectives while retaining commercial viability (Doherty et al, 2020). To tackle inequality, end poverty and protect the planet, it is essential to foster these models of SCP-oriented businesses.

11.3.3 Reduced working hours restructuring the boundaries between consumption and production

While current volumes of production are pressing against ecological limits, many people in wealthy countries (as well as consumer elites in the Global South) experience overwork and other nations struggle with unemployment and underemployment. This situation creates an opportunity for affluent countries to use their improvements in labour productivity to free up time for other activities. From the perspective of the SDGs, reduced working hours could have a number of benefits including improved health and a revitalized civil society.

Writing in support of this kind of post-growth society, Paech (2018) contends that the only way to mitigate the overconsumption of consumers in the Global North is to reduce industrial output that will in turn impel changes in lifestyles and supply patterns. By reconfiguring consumption and production levels, need satisfaction could be achieved through a new found system of reliance on local provisioning, a regional economy and a greatly reduced industrial sector. To cushion the transformation, working time in the paid economy could be limited to 20 hours per week with proportional adjustments in income. Further needs fulfilment would be achieved by devoting a further (approximately) 20 hours of unpaid work in self-production and unpaid care work in the family, proximate neighbourhood or other social or environmental settings (Paech, 2018).

While Kallis and colleagues (2013) do not go as far in their recommendations, they analyse the potential of the widely discussed option of a four-day work week. In addition to the likely strong benefits that such arrangements would have in terms of the quality of peoples' lives, there is also the potential that reductions in working hours would absorb some unemployment, especially in the short run. Environmental benefits are a probable feature of this proposal but will ultimately depend on complementary policies or social conditions to ensure that newly created free time is not directed to resource-intensive or environmentally harmful consumption (Kallis et al, 2013). A crucial element here will be whether the reduced working time is coupled with reduced income, at least when earnings rise above a pre-specified level. Insights from research on voluntary work-time reduction suggest that the underlying motives behind the employees' decisions to cut back their working hours are crucial (Hanbury et al, 2019). A beneficial climate-saving effect tends to arise only among employees who dedicate their newly gained time to activities that require a certain degree of commitment, such as parenting and further education. In contrast, people who reduce their working hours from a desire for more recreational time tend to increase the adverse environmental impacts of their lifestyles owing to their uptake of carbon-intensive leisure activities (Hanbury et al, 2019).

In sum, reducing working hours in the paid economy needs to go hand in hand with an adequate appreciation how a rebalancing of wage labour and unpaid (care) work could strengthen resilience and promote well-being in the necessary shift away from high material-consumption societies.

11.3.4 Quality public services

Public services that are accessible and affordable to all can support inclusive well-being while moderating the need for private consumption and ownership, resulting in lower environmental pressure. Research on universal basic services (UBS) is important in this context. UBS represents a form of consumption that is public and shared rather than private and individualized. Services in this context are activities that are essential for enabling people to meet their needs and contribute to the public interest. Areas of relevance include healthcare, childcare, adult social care, schooling and social work, along with other services such as housing, transport and access to digital information and communications (Coote, 2021). Fare-free public transport is the most prominent example of a UBS-related policy that has been established for social as well as environmental reasons. Successful cases already exist within member states of the European Union such as Luxembourg, though for others such measures exist mostly at the local level. Community-based products and various kinds of sharing initiatives also belong to this category. At the national level, policies for robust welfare systems with free education and universal health insurance can contribute to reduced inequality and status-driven consumerism (Lorek et al, 2021b).

11.3.5 Demand-side solutions

For adequate progress to be potentially realizable, it will be necessary to develop, implement and enforce ambitious policies that tackle the scale and patterns of consumption though adequate demand-side policies (Martin et al, 2021). Households are not given adequate attention in prevailing SCP policies and the focus is largely on incremental changes in supply-side conditions (Creutzig et al, 2016). Recent research has emphasized the potential of the consumption, or demand, side of the system, recognizing that, through the lens of equity, there are distinct implications for different contexts. To achieve 1.5 °C lifestyles, which would entail reducing household carbon footprints to compatibility with the Paris Agreement while improving quality of life, it will be necessary to cut global per capita GHG emissions by 50 per cent by 2030 (Ivanova and Wood, 2020). The most significant areas for action include reducing individual automobile use and air travel, switching to plant-based diets and adopting more 'space-sufficient' housing alternatives (Ivanova et al, 2020). These changes will not happen on their own and there is a growing body of work on the essential need for behaviour changes by individuals (Khanna et al, 2021). However, the dematerialization of lifestyles cannot be achieved through behaviour change alone. It will require the implementation of mutually reinforcing interventions by the public and business sectors to support these adjustments and to modify individual and societal value systems (Newell et al, 2021).

Aside from common assumptions, research shows that low-carbon and materially sufficient lifestyles do not preclude a 'good life' (Millward-Hopkins et al, 2020), and there are indications that even absolute energy reductions would not impede human well-being (Steinberger et al, 2020). Nevertheless it needs to be recognized that fulfilling basic needs requires minimum levels of consumption while limited resources and sink capacities (eg for carbon emissions) require establishing a maximum threshold for consumption. The spaces between minimally and maximally acceptable consumption have been described as 'consumption corridors', where individuals and households can determine the specific parameters of their lifestyles (Defila and Di Giulio, 2020). Moving the entire global population into this space would greatly improve life for billions while requiring significant changes in the ability of high-consuming elites to accumulate wealth and material possessions. Consumption corridors are intended to serve as a guide for people whose consumption exceeds the acceptable maximum and will need to be established through democratic processes that embrace a commitment to social equity (Fuchs et al, 2021) so that those experiencing the most disruption from climate change because of a lack of resources are not forced to incur additional burdens as a result of demand-side policies.

11.4 Conclusion

From a sectoral and consumption-based perspective the key provisioning areas with the highest environmental relevance are food, mobility and housing (with the latter also including heating and cooling and operation of electric appliances) and are all central to the way we live. They are, furthermore, highly interconnected and mutually dependent, and adjustments need to be considered in a holistic manner. Meaningful change will require policies and a political framework that overcomes the lock-in situations where the unsustainable option is: (1) the most rational (eg flying is typically cheaper than travelling by train); (2) the structurally convenient (eg obligatory parking spaces for personal cars in residential areas but no bicycle lanes); and (3) the traditionally supported (eg inexpensive meat promoted by supermarkets in their weekly advertising) (Lorek et al, 2021a).

Creating synergies between ecological sustainability and social progress will require major institutional changes that are likely to generate resistance from the beneficiaries of current socioeconomic and political arrangements. There is a substantial and growing body of scholarship on alternative economies (D'Alisia et al, 2014; Hill et al, 2022), but such innovations are typically applied on a very small and experimental basis. Public policies can play an important role in supporting these pioneering efforts. Vast, ambitious and perhaps very bold political undertakings require partnerships with social movements seeking environmental justice and radical change to the dominant ways in which relationships of consumption and production are structured (Lorek and Fuchs, 2013).

It is also important to look at the role cities might play as incubators of scalable and transferable social innovations (Kosovac and Pejic, Chapter 12 in this volume). The C40 network (www.c40.org), for example, is an international organization of mayors of nearly 100 world-leading cities collaborating to deliver the action needed to confront the climate crisis. The explicit aim is to create a future where everyone everywhere can thrive. Although urban modes of living are often – and not incorrectly – associated with energy- and material-intensive lifestyles, many (but certainly not all) sustainable solutions are being initiated in cities (Lee and van de Meene, 2012). The establishment of partnerships and alliances that span customary urban/rural divides will be essential for the renewal of SCP systems given the geographic expansiveness and interdependency that is a common feature of contemporary supply chains and provisioning arrangements (Koloffon Rosas and Pattberg, Chapter 13 in this volume).

Aside from the fiscal dimension, these interventions open up policy debates to stricter and more regulatory measures such as setting a cap on emissions on either an absolute or per capita basis. These objectives could be achieved by implementing annually renewable personal carbon budgets. Under such a policy, all consumers will be provided with a minimum emissions allowance (with additional increments for those with greater needs, for example, because of age or health status). Individual carbon budgets have been discussed in recent years in Ireland and France as well as quite prominently in the UK in the context of a personal carbon trading (PCT) scheme (Cohen, 2010). Research shows that PCT had a similar level of social acceptance as an alternative taxation policy and was publicly perceived as fair and effective because it allowed for the consideration of individual needs (Fawcett, 2010). On a voluntary level, living within capped emissions is already being tested by so-called carbon rationing action groups in organizational contexts and at the municipal level, for example, in the Finnish city of Lahti (Lorek et al, 2021b).

In conclusion, realizing SCP patterns reaches far beyond handling waste streams and changing certain production processes. Strong sustainable consumption requires fundamental and systemic changes in the critical features of how our economy and our societies are organized, with an emphasis on well-being for all based on significantly lower resource flows. The ways in which work is shared and managed is as important in this context as access to basic services and a fair distribution of financial and material resources.

References

Akenji, L., Bengtsson, M., Toivio, V., Lettenmeier, M., Fawcett, T., Parag, Y. et al (2021) *1.5-Degree Lifestyles: Towards A Fair Consumption Space for All*, Berlin: Hot Or Cool Institute.

Alfredsson, E., Bengtsson, M., Brown, H., Isenhour, C., Lorek, S., Stevis, D., and Vergragt, P. (2018) 'Why achieving the Paris Agreement requires reduced overall consumption and production', *Sustainability: Science, Practice and Policy*, 14(1): 1–5.

Ayres, R. and Simonis, U. (eds) (1994) *Industrial Metabolism: Restructuring for Sustainable Development*, Tokyo: United Nations University Press.

Bengtsson, M., Alfredsson, E., Cohen, M., Lorek, S. and Schroeder, P. (2018) 'Transforming systems of consumption and production for achieving the Sustainable Development Goals: Moving beyond efficiency', *Sustainability Science*, 13: 1533–47.

Bringezu, S. (2015) 'Possible target corridor for sustainable use of global material resources', *Resources*, 4(1): 25–54.

Brockway, P., Sorrell, S., Semieniuk, G., Kuperus Heun, M. and Court, V. (2021) 'Energy efficiency and economy-wide rebound effects: A review of the evidence and its implications', *Renewable and Sustainable Energy Reviews*, 141: 110781.

Cohen, M. (2010) 'Is the UK preparing for "war"? Military metaphors, personal carbon allowances, and consumption rationing in historical perspective', *Climatic Change*, 104(2): 199–222.

Cohen, M. (2017a) 'Workers and consumers of the world unite! Opportunities for hybrid cooperativism', in J. Michie, J. Blasi and C. Borzaga (eds) *Oxford Handbook of Cooperative and Mutual Business*, Oxford: Oxford University Press, pp 374-85.

Cohen, M. (2017b) *The Future of Consumer Society: Prospects for Sustainability in the New Economy*, Oxford: Oxford University Press.

Coote, A. (2021) 'Universal basic services and sustainable consumption', *Sustainability: Science, Practice and Policy*, 17(1): 32–46.

Creutzig, F., Fernandez, B., Haberl, H., Khosla, R., Mulugetta, Y. and Seto, K. (2016) 'Beyond technology: Demand-side solutions for climate change mitigation', *Annual Review of Environment and Resources*, 41: 173–98.

D'Alisa, G., Demaria, F. and Kallis, G. (eds) (2014) *Degrowth: A Vocabulary for a New Era*, New York: Routledge.

Defila, R. and Di Giulio, A. (2020) 'The concept of "consumption corridors" meets society: How an idea for fundamental changes in consumption is received', *Journal of Consumer Policy*, 43(2): 315–44.

Doherty, B., Haugh, H., Sahan, E., Wills, T. and Croft, S. (2020) *Creating the New Economy: Business Models that Put People and Planet First*, Leeds: White Rose University Press.

Dyrstad, J., Skonhoft, A., Christensen, M. and Ødegaard, E. (2019) 'Does economic growth eat up environmental improvements? Electricity production and fossil fuel emission in OECD countries 1980–2014', *Energy Policy*, 125: 103–9.

EEA (European Environment Agency) (2020) *The European Environment State and Outlook 2020*, Copenhagen: EEA.

Fawcett, T. (2010) 'Personal carbon trading: A policy ahead of its time?', *Energy Policy*, 38(11): 6868–76.

Fischer-Kowalski, M. and Haberl, H. (1997) 'Tons, joules, and money: Modes of production and their sustainability problems', *Society & Natural Resources*, 10(1): 61–85.

Fuchs, D., Sahakian, M., Gumbert, T., Di Giulio, A., Maniates, M., Lorek, S. and Graf, A. (2021) *Consumption Corridors: Living a Good Life within Sustainable Limits*, London: Routledge.

Gasper, D., Shah, A. and Tankha, S. (2019) 'The framing of sustainable consumption and production in SDG 12', *Global Policy*, 10: 83–95.

Gore, T. and Alestig, M. (2020) *Confronting Carbon Inequality in the European Union*, Oxford: Oxfam.

Haberl, H., Wiedenhofer, D., Virág, D., Kalt, G., Plank, B., Brockway, P. et al (2020) 'A systematic review of the evidence on decoupling of GDP, resource use and GHG emissions, part II: synthesizing the insights', *Environmental Research Letters*, 15(6): 065003.

Hanbury, H., Bader, C. and Moser, S. (2019) 'Reducing working hours as a means to foster low(er)-carbon lifestyles? An exploratory study on Swiss employees', *Sustainability*, 11(7): 2024.

Hickel, J. and Kallis, G. (2020) 'Is green growth possible?', *New Political Economy*, 25(4): 469–86.

Hill, S., Yagi, T. and Yamash'ta, S. (eds) (2022) *Creating an Economy of Care*, Singapore: Springer.

IEA (International Energy Agency) (2013) *World Energy Outlook, 2013*, Paris: IEA.

IEA (2021) *World Energy Outlook, 2021*, Paris: IEA.

IPCC (Intergovernmental Panel on Climate Change) (2021) *Climate Change, 2021: The Physical Science Base*, Geneva: IPCC.

Ivanova, D. and Wood, R. (2020) 'The unequal distribution of household carbon footprints in Europe and its link to sustainability', *Global Sustainability*, 3: e18.

Ivanova, D., Barrett, J., Wiedenhofer, D., Macura, B., Callaghan, M. and Creutzig, F. (2020) 'Quantifying the potential for climate change mitigation of consumption options', *Environmental Research Letters*, 15(9): 0930001.

Kallis, G., Kalush, M., O'Flynn, H., Rossiter, J. and Ashford, N. (2013) 'Friday off: Reducing working hours in Europe', *Sustainability*, 5(4): 1545–67.

Khanna, T., Baiocchi, G., Callaghan, M., Creutzig, F., Guias, H., Haddaway, N. et al (2021) 'A multi-country meta-analysis on the role of behavioural change in reducing energy consumption and CO_2 emissions in residential buildings', *Nature Energy*, 6(9): 925–32.

Korhonen, J., Honkasalo, A. and Seppälä, J. (2018) 'Circular economy: The concept and its limitations', *Ecological Economics*, 143: 37–46.

Lamb, W., Antal, M., Bohnenberger, K., Brand-Correa, L., Müller-Hansen, F. et al (2020) 'What are the social outcomes of climate policies? A systematic map and review of the ex-post literature', *Environmental Research Letters*, 15(11): 113006.

Lebel, L. (2004) 'Transitions to sustainability in production–consumption systems', *Journal of Industrial Ecology*, 9(1):1–3.

Lee, T. and van de Meene, S. (2012). 'Who teaches and who learns? Policy learning through the C40 cities climate network', *Policy Sciences*, 45(3): 199–220.

Le Quéré, C., Andrew, R., Friedlingstein, P., Sitch, S., Hauck, J., Pongratz, J. et al (2018) 'Global carbon budget 2018', *Earth System Science Data*, 10(4): 2141–94.

Lettenmeier, M. (2018) *A Sustainable Level of Material Footprint: Benchmark for Designing One-Planet Lifestyles*, Helsinki: Aalto University.

Lorek, S. and Fuchs, D. (2013) 'Strong sustainable consumption governance: Precondition for a degrowth path?', *Journal of Cleaner Production*, 38: 36–43.

Lorek, S., Weber, L., Kiss-Dobronyi, B., Gran, C., Barth, J. and Tomany, S. (2021a) 1.5 Degree Policy Mix: Demand-Side Solutions to Carbon-Neutrality in the EU: Introducing the Concept of Sufficiency, Transformation Policy Brief #5, Cologne: ZOE, Institute for Future-Fit Economies.

Lorek, S., Gran, C., Lavorel, C., Tomany, S. and Oswald, Y. (2021b) *Equitable 1.5-Degree Lifestyles How Socially Fair Policies Can Support the Implementation of the European Green Deal*, Policy Brief #1, Cologne: ZOE, Institute for Future-Fit Economies.

Martin, M., Alcaraz Sendra, O., Bastos, A., Bauer, N., Bertram, C., Blenckner, T. et al (2021) 'Ten new insights in climate science 2021: A horizon scan', *Global Sustainability*, 4(e25): 1–20.

Martínez-Alier, J., Pascual, U., Vivien, F. and Zaccai, E. (2010) 'Sustainable de-growth: Mapping the context, criticisms and future prospects of an emergent paradigm', *Ecological Economics*, 69(9): 1741–7.

Millward-Hopkins, J., Steinberger, J., Rao, N. and Oswald, Y. (2020) 'Providing decent living with minimum energy: A global scenario', *Global Environmental Change*, 65: 102168.

Newell, P., Daley, F. and Twena, M. (2021) *Changing Our Ways? Behaviour Change and the Climate Crisis: The Report of the Cambridge Sustainability Commission on Scaling Behaviour Change*, Cambridge: Cambridge University Press.

One Planet Network (2020) *Tax reduction and lower VAT for repairment of certain products*, Available from: www.oneplanetnetwork.org/knowledge-centre/policies/tax-reduction-and-lower-vat-repairment-certain-products [Accessed 20 August 2022].

One Planet Network (2021) 'Single-use carrier bag charge extension', Available from: www.oneplanetnetwork.org/knowledge-centre/polic ies/single-use-carrier-bag-charge-extension [Accessed 20 August 2022].

One Planet Network (2022a) 'Inspection plans for buildings promoting reuse', Available from: www.oneplanetnetwork.org/knowledge-cen tre/policies/inspection-plans-buildings-promoting-reuse [Accessed 20 August 2022].

One Planet Network (2022b) 'Environment Act: Powers on ecodesign and information', Available from: www.oneplanetnetwork.org/knowle dge-centre/policies/environment-act-powers-ecodesign-and-information [Accessed 20 August 2022].

OECD (Organisation for Economic Co-operation and Development) (2022) *The Short and Winding Road to 2030 Measuring Distance to the SDG Targets*, Paris: OECD.

O'Neill, D., Fanning, A., Lamb, W. and Steinberger, J. (2018) 'A good life for all within planetary boundaries', *Nature Sustainability*, 1(2): 88–95.

Paech, N. (2018) 'Postwachstumsökonomik' [Post-growth economics], in R. Kümmel, D. Lindenberger and N. Paech (eds) *Energie, Entropie, Kreativität*, Berlin: Springer Spektrum, pp 101–35.

Parrique, T., Barth, J., Briens, F., Kerschner, C., Kraus-Polk, A., Kuokkanen, A. and Spangenberg, J. (2019) *Decoupling Debunked: Evidence and Arguments against Green Growth as a Sole Strategy for Sustainability*, Copenhagen: European Environment Bureau.

Parry, I., Heine, M., Lis, E. and Li, S. (2014) *Getting Energy Prices Right: From Principle to Practice*, Washington, DC: International Monetary Fund.

Partzsch, L., Hartung, K., Lümmen, J. and Zickgraf, C. (2021) 'Water in your coffee? Accelerating SDG 6 through voluntary certification programs', *Journal of Cleaner Production*, 324: 129252.

Schandl, H., Hatfield-Dodds, S., Wiedmann, T., Geschke, A., Cai, Y., West, J. et al (2016) 'Decoupling global environmental pressure and economic growth: Scenarios for energy use, materials use and carbon emissions', *Journal of Cleaner Production*, 132: 45–56.

Schmidt-Bleek, F. (2008) 'Factor 10: The future of stuff', *Sustainability: Science, Practice and Policy*, 4(1): 1–4.

Schröder, P. and Raes, J. (2021) *Financing an Inclusive Circular Economy: De-risking Investments for Circular Business Models and the SDGs*, Chatham House Research Paper, London: Royal Institute of International Affairs.

Steinberger, J., Krausmann, F., Getzner, M., Schandl, H. and West, J. (2013) 'Development and dematerialization: An international study', *PloS One*, 8(10): E70385.

Steinberger, J., Lamb, W. and Sakai, M. (2020) 'Your money or your life? The carbon-development paradox', *Environmental Research Letters*, 15(4): 1–9.

Tudor, C. and Sova, R. (2021) 'On the impact of GDP per capita, carbon intensity and innovation for renewable energy consumption: Worldwide evidence', *Energies*, 14(19): 6254.

UN (United Nations) (2015) *Transforming Our World: The 2030 Agenda for Sustainable Development*, New York: United Nations General Assembly.

UN (2021) *High Level Political Forum Thematic Review Expert Group Meeting, 18–20 May*, New York: United Nations.

Vadén, T., Lähde, V., Majava, A., Järvensivu, P., Toivanen, T., Hakala, E. and Eronen, J. (2020) 'Decoupling for ecological sustainability: A categorisation and review of research literature', *Environmental Science & Policy*, 112: 236–44.

Weizsäcker, E., Lovins, A. and Lovins, H. (1998) *Factor Four: Doubling Wealth – Halving Resource Use: The New Report to the Club of Rome*, London: Earthscan.

Wiedmann, T., Lenzen, M., Keyßer, L. and Steinberger, J. (2020) 'Scientists' warning on affluence', *Nature Communications*, 11(1): 1–10.

Zero Waste Europe (2020) 'France's Law for Fighting Food Waste', Zero Waste Europe Factsheet, Available from: https://zerowasteeurope.eu/wp-content/uploads/2020/11/zwe_11_2020_factsheet_france_en.pdf [Accessed 20 August 2022].

Interview with Sylvia Lorek and Magnus Bengtsson: Contradictions between Economic Growth and Sustainable Consumption

Kathrin Lehmann and Laura Kräh

SDG 12 has been continuously criticized, especially when it comes to making actors fully commit to sustainable practices. Do you consider the targets of SDG 12 regarding consumption and production sufficient?

Bengtsson: The emphasis has always been on the production side, making things more efficient. To comply with the targets of the Paris Agreement, we need to make production systems less carbon polluting, but we also need to look at the demand side and work with lifestyle and consumption changes. However, SDG 12 is more about encouraging companies to report on sustainability and less about actually producing sustainably. Regarding consumers, SDG 12 only addresses the need for knowledge of sustainable living. But we live in a society where knowledge is not enough to shift individuals' behaviours. We need structural changes in pricing systems. Furthermore, the discourse, for example, on upgrading waste management, is questionable. Sustainability is not going to increase by improving such processes. We need to re-engineer how we produce and consume goods in order to minimize the generation of waste in the first place. On the bright side, there are also positive processes such as the discussion on target-setting for material consumption by the European Parliament.

Lorek: I'm happy that sustainable production and consumption was formulated as its own goal but, nevertheless, it is too deeply influenced by systemic flaws. They talk about efficiency and reduction, but there are no measurement criteria or benchmarks for resource consumption reduction. And as long as economic growth stays a central goal, the issue of rebound effects prevails and the chance to achieve sustainable consumption and production remains low.

Regarding this trade-off: would you suggest that SDG 12 should be prioritized over SDG 8?

Bengtsson: Yes. There is indeed a strong contradiction between the continuous pursuit of growth and the need for consumption reduction as a society, especially among high consumer groups. I see in SDG 12 an integrative function of easing the tensions between social objectives such as ending poverty and hunger and the need to stay within the ecological limits.

Lorek: For me, everything could be subsumed under SDG 12, together with SDG 10. If we have sustainable consumption and production, there would be no poverty, there would be no hunger, the oceans would be protected better ... It should be in the very centre, but it would have to be formulated completely differently.

How do you think the COVID-19 pandemic has affected the pathway of SDG 12 in terms of individual consumption?

Lorek: At first, the pandemic involuntarily threw people back to totally different lifestyles. They had the possibility to rethink what was necessary but as we see now, old habits come back. I don't really see a big change in mindsets, habits, consumption and in material footprints. The 'new normal' is not too different from the 'old normal' and definitely too far away from what would be needed for a sustainable lifestyle.

PART III

The Goals Relevant
to an Environmentally
Sound Implementation

12

Cities and the SDGs: A Spotlight on Urban Settlements

Anna Kosovac and Daniel Pejic

Sustainable Development Goal 11 has been a long time in the making. Incorporating elements of housing, transport, planning processes, disaster response, social inclusivity, and air and water quality, it is a goal that is broad in approach in recognition of the variety of global challenges confronting cities today. Five of the nine targets have a direct link to climate change and the environment generally. These are areas where cities have a significant current policy focus, as is reflected in recent studies (Kosovac et al, 2020b). Cities around the globe are increasingly facing sustainability challenges, including rapid urbanization leading to unsustainable housing. Mortality rates are rising in relation to poor air quality in urban areas (Liu et al, 2019), and this is linked to one in nine deaths globally every year (WHO, 2016. Furthermore, climate change effects such as more frequent extreme weather events (eg floods and bushfires) have resulted in calls for more action on committing to the Paris Agreement (C40 Cities, 2020) in line with many urban areas introducing cleaner energy systems in transport, electricity, housing and planning. With more than half of the world's population currently living in urban areas, a number that is projected to increase to 70 per cent by 2050 (UN Habitat, 2020), cities are feeling the direct impact of global challenges and are essential to addressing them.

This chapter considers the role of cities and academic advocates in promoting a dedicated urban focus in the lead-up to the announcement of the Sustainable Development Goals, why a separate city goal is important for global sustainability practices, and how cities are progressing on achieving the SDG 11 objectives. The chapter will also discuss how cities have been asserting themselves on the global stage and advocating for a seat at the table of multilateral discussions that are usually the domain of nation-states.

'Nations Talk, cities act' is an oft-cited statement overheard in global city-based discussions, often in the context of these climate change issues (Curtis, 2014). Much of this rhetoric stems from cities being recognized as the frontline of various global problems, from housing and well-being through to sustainability, a point that has only been broadly recognized in the past two decades. Given growing urban populations, it is no wonder that the United Nations has taken note and is acknowledging cities in its frameworks and meetings now more than ever (Kosovac et al, 2020a), and nothing expresses the need for a better quality of life for the urbanizing world than the inclusion of a cities focus in the SDGs.

Key idea: 'Urbanization' as a term does not refer solely to an increase in population in urban areas and cities, but rather to an *increase in the proportion* of those living in urban areas compared to non-urban areas.

Although the idea of cities sharing global power status with nation-states is considered anomalous, historical precedents suggest otherwise. This is not the first time that cities have exercised significant political power. For example, in (pre-unified) Italy, city-states such as Florence and Milan were independently sovereign. Similar sovereign arrangements were also present for the city of Athens in the classical period of ancient Greece. These city-states ruled their regions, as places where power rested and leaders presided. A modern example of a city-state exists in the case of Singapore, a city made up of over 5.7 million people (in 2020), with its own legislative system and currency. As such, the role of cities as global powerhouses is not new, and cities have taken a back seat only in recent history to make way for the predominance of nation-states, most notably with the adoption of the sovereignty-based Treaty of Westphalia in 1648. However, the current international order developed as a result of the establishment of the United Nations in 1945, which adopted an acute focus on nation-state representation with little regard for intra-statal issues, including those of cities. Not only have discussions in the 20th century centred on *inter*state relations, but this has also served to neglect urban issues occurring within state borders. The driver for this behaviour was largely the reluctance of the UN to overstep the recently asserted sovereignty claims of nation-states.

While states remain the essential interlocuters of the international system, over the past 30 years a shift in these relations has led to the higher visibility of cities on issues such as climate change, and sustainability (Acuto, 2016; Gordon and Johnson, 2018; Aust and Nijman, 2020; 2021). A recent analysis of UN frameworks and key documentation has highlighted the increasing recognition of the role of cities in environmental sustainability not only as sites of issues, but also as global actors (Kosovac et al, 2021a). This same study also found that cities were more than ever before considered formally in discussions and documentation (and this figure was rising over time), especially in the areas of development and the environment[1] (see Figure 12.1).

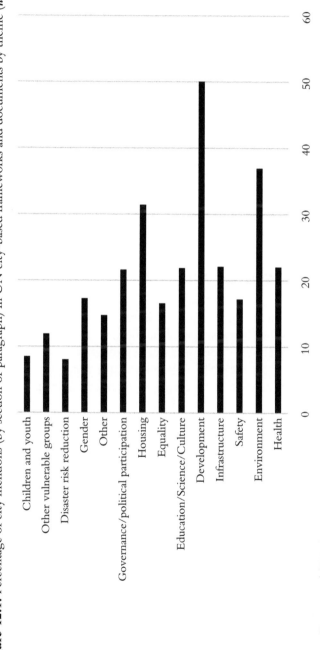

Figure 12.1: Percentage of city mentions (by section or paragraph) in UN city-based frameworks and documents by theme (*n* = 32)

Source: Kosovac et al (2020a).

While cities remain marginal participants in most multilateral forums, they are becoming more prominent in both the development and the implementation of global agendas.

City-specific environmental sustainability initiatives are widespread, as cities exist at the frontier of climate change impacts because of their often close proximity to coastal areas, increased risk of flooding and heat island effects, and as high producers of greenhouse gas emissions (70 per cent of global output: World Bank, 2021). Above all other issues, cities have been most vocal in their push for global environmental action on the climate crisis. The C40 Cities Climate Leadership Group represents one of the largest global city networks and its remit is based on influencing the global agenda on climate action through advocacy, research and promoting city-based sustainability planning. Similarly, the Global Covenant of Mayors for Climate & Energy is another high-profile urban network focused on environmental sustainability. As well the being the largest producers of GHG emissions, cities also generate the majority of the world's economic activity, 80 per cent of global GDP (World Bank, 2021). Hence, urban settlements play a central role in global environmental sustainability practices.

12.1 City networks as unifying agencies in global governance

Cities have worked to amplify their voices globally through membership of an increasing number of transnational city networks (see Table 12.1 for list). These networks represent a conglomerate of city members that work towards a common goal or goals (Acuto and Rayner, 2016). Some are based around a specific thematic issue, while others focus on several policy domains (Acuto and Leffel, 2020). Through networks, cities engage in diplomatic relations (called either city diplomacy or paradiplomacy), resulting in agreements, cooperation and partnerships across cultural, environmental, economic and political spheres (Acuto and Rayner, 2016; Acuto et al, 2021a; Kosovac and Pejic, 2021). City networks also produce a wide variety of outputs aimed at policy changes and knowledge mobilization, such as reports (62 per cent of total networks) and joint pilot programmes (32 per cent) (Acuto and Leffel, 2020).

A study of over 200 city networks found that almost 30 per cent have a focus on environmental issues (Acuto and Leffel, 2020), with the Rio Earth Summit in 1992 acting as the key instigator of this proliferation. Since then, two of the most prominent sustainability-focused transnational city networks have been Local Governments for Sustainability (ICLEI), a global network of more than 2,500 local and regional governments dedicated to sustainable urban development and C40 Cities, a network of almost 100 of the world's largest cities committed to urgent climate action. More recently

Table 12.1: Key global city networks with a focus on climate change and sustainability

City network	Founded	Type of network	No. of members
C40	**2005**	**Cities and local governments**	**97 cities and local governments**
Global Covenant of Mayors for Climate and Energy	2014	City networks and partners	12 networks and partners
Eurocities	1986	Cities and local governments	137 cities and local governments
Global Taskforce of Local and Regional Governments	2013	City networks	27 City networks
Global Parliament of Mayors	2016	Mayor or city	41 mayors/cities
Local Governments for Sustainability (ICLEI)	1990	Cities, local and regional governments	2,500+ local and regional governments
Metropolis	1994	Cities and local governments	141 local governments and cities
Regions4	2002	Regional governments	42 regional governments
United Cities and Local Governments (UCLG)	2004	Cities, local, regional and metropolitan governments	240,000 cities, regions and metropolitan governments
Resilient Cities Network	2011	Cities and local governments	97 cities and local governments

large networks of networks have emerged, such as the Global Covenant of Mayors for Climate & Energy (representing over 11,000 cities) and the Global Taskforce of Local and Regional Governments. Many other prominent city networks that are not specifically focused on climate and sustainable development have made these issues a key part of their work, such as United Cities and Local Governments (UCLG), Eurocities, Metropolis and the Global Parliament of Mayors.

Some of these city networks have also become institutional partners in multilateral systems governing sustainable development. For example, ICLEI's partnership agreements with the UN Environment Programme, UN-Habitat and the United Nations Office for Disaster Risk Reduction (UNDRR) or UCLG's chairing of the United Nations Advisory Committee of Local Authorities (UNACLA), a body that strengthens the partnership between the UN system and local authorities in the implementation of the

Habitat Agenda (discussed later). Since 1995, the Local Governments and Municipal Authorities Constituency (LGMA) has represented the views of local and regional governments in the UN Framework Convention on Climate Change Conference of Parties meetings. The LGMA now works on behalf of the Global Taskforce of Local and Regional Governments and advocated strongly for multilevel action on climate change at the COP 26 meeting in Glasgow in 20, a position which is reflected in the outcome document. The Global Taskforce of Local and Regional Governments has a range of partnerships across the UN system.

City network initiatives themselves are also often multistakeholder partnerships incorporating representatives or support from philanthropies, non-profit organizations and the private sector. C40, for example, has been underpinned by funding from Bloomberg Philanthropies but now has a broad range of funding partners including national governments, foundations and global brands such as IKEA and Arup. Many networks also work closely with academic institutions and think tanks to develop research and policy outputs to inform cities, national governments and international organizations (Kosovac and Pejic, 2021).

While the inclusion of local authorities in the global governance of sustainable development may be incremental, prominent advocacy from urban group, such as city networks have been an important factor in a broader 'urbanization' of the way sustainable development is thought about and discussed. The inclusion of a specific SDG on cities (discussed later) is a strong recognition of this. As city networks have grown in scale, they have also expanded the geographies of the cities included, with many more cities from the Global South participating in these initiatives (Acuto and Leffel, 2020). However, they are often still largely driven by major cities in the Global North who have the resources and capacity to participate meaningfully in these types of initiatives.

The centrality of urban settlements to the sustainability agenda has raised questions about whether the role of cities in global governance should be more substantial and further formalized. Barber (2013) famously advocated for city leaders to usurp national ones as the key international interlocuters and decision makers. His campaign led to the development of the Global Parliament of Mayors. It has been argued the often progressive politics of the city represents a dynamic and pragmatic alternative to the quagmire of international politics and geopolitical struggles (Lee, 2016). Other scholars have been more sober regarding the prospects of city leaders 'saving the planet' (Angelo and Wachsmuth, 2020). While urban settlements undoubtedly matter more than ever to global challenges such as sustainable development, city leaders are often not masters of their domain and lack the capacities, legal authority and competencies to implement many of the changes required to promote and foster sustainable development. This has made the multilevel governance of issues such as sustainability and climate change increasingly critical (Betsill and Bulkeley, 2006). City leaders have and continue to make an important contribution to the way sustainable development is governed within nations and globally, but this contribution needs to be made in partnership with other actors.

12.2 The rise of cities globally: a brief history of SDG 11

Before delving into SDG 11, the preceding actions and discussions that have led to the current rendition of the global framework to address urban issues should be explored (see Figure 12.3). Habitat I, held in 1976 in Vancouver, Canada, represents the first UN conference specifically focused on issues related to urban areas and human settlements. The ensuing Vancouver Action Plan recommended over 60 actions, which also prompted the formation of a UN arm, UN-Habitat, to further advance urban agendas. With urban issues institutionalized within a settlement-based organization such as UN-Habitat, there seemed to be little need to address these issues across the United Nations more broadly. The approach by and large was one of economic bolstering through funding of infrastructure in poor urban settlements. This method was spearheaded by the World Bank to produce (arguably limited) developmental outcomes at a time where urban modernization was accelerating globally (Parnell, 2016).

ICLEI, established in 1990, has been actively pursuing engagement with the UN at the global scale over the last 30 years (Gordon and Johnson, 2018) and, notably, was heavily involved in discussions prompting the formation of an urban-based dynamic in the UN's Agenda 21 at the 1992 Rio Summit. Furthermore, off the back of Agenda 21, there was an increase in city-based rhetoric within the UN, particularly recognizing cities as actors with a role to play in achieving global goals (see Figure 12.2).

Fuelled by these discussions, Habitat II followed in 1996, resulting in the Istanbul Declaration on Human Settlements and Habitat Agenda to promote safe housing in highly urbanized areas. The outcomes of Habitat II did not depart significantly from those of Habitat I, with an ongoing focus on cities in the Global South. However, a notable aspect was the progressive notion in both conferences to elevate civil society in shaping global urban debates around international development. Nevertheless, the discussion was limited in scope to issues of housing and settlements.

Over the following decade, C40, UCLG and urban scholars were actively voicing their concerns regarding broader city representation on the global stage and in frameworks (Parnell, 2016). As the year 2000 approached, these calls were heeded with the development of the Millennium Development Goals, wherein a goal specifically on cities was included. However, the remit of

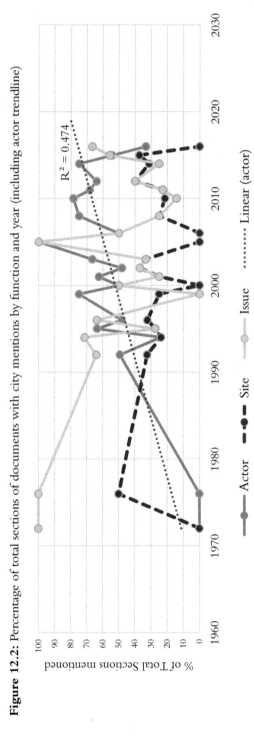

Figure 12.2: Percentage of total sections of documents with city mentions by function and year (including actor trendline)

Source: Acuto et al (2021a). Reproduced with permission.

the urban goal was very limited in scope. Focusing only on slum settlements, and pushing strongly for more public investment, the MDGs ignored many of the other prevailing issues facing cities such as clean water, sanitation, health, education and transport (Cohen, 2014). The target within the goal stated that the world must 'have achieved by 2020 a significant improvement in the lives of at least 100 million slum dwellers' (UN, 2015) (see Figure 12.3).

Not only was the target vague, but it also focused on factors that are spatial in nature, without considering broader social and economic issues within cities. It can also, arguably, promote the state-sponsored eviction of slum dwellers without providing alternative suitable accommodation to meet the MDG target (Meth, 2013).

Accordingly, the MDGs treated urbanization in ways that were outdated, focusing on housing (in particular, slums) as the key issue facing urban spaces and ignoring the context of wider-ranging underlying sociocultural concerns within cities (Rudd et al, 2015). There is no doubt that housing, when consciously designed to consider environmental practices, can have a significant impact on urban carbon emissions in how construction materials are sourced, used and maintained and the location of housing on peripheral environmental landscapes (Winston, 2009). However, the low incomes of slum dwellers do not allow for the incorporation of expensive sustainability construction solutions (Winston, 2009) but rather rely on slum upgrading (or demolition) options in the bid to improve unsafe building practices, overcrowding and access to sanitation (Doe et al, 2020). Therefore, adopting tunnel vision in redeveloping slum housing carries implications for sustainability issues that are inherent in city living and urban environments. Housing alone cannot solve sustainability issues such as water scarcity and flooding, or social–cultural problems, and cannot produce healthy ecosystems and air quality.

The focus on slum development in the MDGs was inadequate to address broader urban sustainability issues. UCLG, supported by other global organizations such as the Sustainable Development Solutions Network (SDSN), ICLEI and the Global Covenant of Mayors, pressurized the UN to develop a globally applicable urban-based goal. These calls were eventually heeded in 2015 with a city-specific SDG that incorporated the challenges of urban sustainable development (Arajarvi, 2018).

12.3 Overview of SDG 11

Planning for SDG 11 rested on a vision for urban areas in 15 years' time (in 2030). The vague language from the MDGs was replaced by more specific active goals with clear deadlines for meeting each.

SDG#11 unambiguously signals UN members' acceptance of some form of devolution in governance, the imperative of an integrated

Figure 12.3: Key time points for SDG 11

1976	HABITAT I A/CONF.70/15 UN Conference on Human Settlements *Vancouver, Canada*
1992	AGENDA 21 The Rio Declaration on Environment and Development *Rio de Janeiro, Brazil*
1996	HABITAT II A/CONF.70 UN Conference on Human Settlements *Istanbul, Turkey*
2000	UNITED NATIONS MILLENNIUM DECLARATION A/RES/55/2
2001	WORLD URBAN FORUM Conference on rapid urbanization
2015	**2030 AGENDA FOR SUSTAINABLE DEVELOPMENT** A/RES/70/1 SENDAI FRAMEWORK FOR DISASTER RISK REDUCTION A/RES/69/283 ADDIS ABABA ACTION AGENDA PARIS AGREEMENT
2016	HABITAT III A/CONF.226 United Nations Conference on Housing and Sustainable Urban Development *Quito, Equador*

vision of sustainable urban development that not exclude social, economic, or ecological imperatives and (implicitly) a collective acknowledgement that the spatial concentration of resources and flows that cities represent can act as a driver of sustainable development. (Parnell, 2016, 530)

SDG 11 adopted a multifaceted perspective on cities and their wide-ranging sustainability issues. The adoption of these principles, as described in the previous section, was the result of extensive and sustained campaigning by city networks, academics and cities themselves (Parnell, 2016). To formulate the goals, multiple workshops were run, incorporating governments, academia, philanthropy and the private sector, to develop sets of targets and indicators for the goal (Rudd et al, 2015). A wide range of urban issues were considered in the final iteration of the goal, including housing, transport, participatory approaches, heritage, disaster planning, pollution, green spaces and urban planning (see Box 12.1 for full list). This ambitious goal focuses on a global approach to urbanization and its impacts rather than a goal that is geared towards low- and middle-income nation-states. The topics explored in SDG 11 reflected discussions occurring across various UN conferences, and within declarations in the 30 years leading up to the goal, that cities and urban areas are often included in themes of development, environment and housing (see Figure 12.1). This is cognisant of a marked increase in populations living in urban areas without clean water, essential services, education and health services (UN-Habitat, 2020).

However, the focus on the housing aspect of the MDGs has not been abandoned in the SDGs, as it is still captured by target 11.1, with an expansion for the provision of safety to 'ensure access for all to adequate, safe and affordable housing and basic services and upgrade slums'. There is arguably a move away from eradicating slums to upgrading these settlements, recognizing justice framings of slum dwellers' 'right to the city' claims (Roy, 2005). However, the use of 'informal settlements' in the SDGs does not recognize the differentiated nature of informality, a term that is not solely related to the poor and has been shown to exist in the domain of the middle and elite classes (eg enclave urbanism) in a bevy of global cities (Roy and AlSayyad, 2004; Müller, 2017). Furthermore, informal settlements represent neighbourhood identities and highlight the adaptive capacity of these marginal groups that is wholly ignored as 'messy ... urban voids', underpinning the SDG eradication target (Lehmann, 2020). Although the goal has improved from its MDG predecessor, it nevertheless does not consider urban informality in line with issues of citizenship, migration and informal entrepreneurialism (Lehmann, 2020), a seeming area of improvement for future iterations. Further areas of interest were recognized in the aim for safe, accessible transport (target 11.2). Climate change adaptation in cities is addressed through targets 11.6 (Waste management), 11.7 (Green Spaces) and 11.b (including specific mention of the Sendai Framework for Disaster Risk Reduction). These goals have been heavily informed by discussions leading to the Paris Agreement and the Sendai Framework for Disaster Risk Reduction. The deliberations underpinning these agreements and frameworks focused on urban issues as key topic areas in negotiation, and this is further reiterated in SDG 11. Traditionally, cities have been managing climate actions using risk framework mechanisms, and have been criticized for their reductionist and project-specific approach (Sanchez Rodriguez et al, 2018; Wise et al, 2014). The SDGs adopt a broader approach to climate adaptation in urban areas, reducing the need for path dependency in planning while actively supporting the incorporation of participatory planning in urban planning (target 11.3), first discussed at the Habitat II conference in 1996. Furthermore, planning targets have also been captured to firmly include the oft-forgotten peri-urban areas, regional planning in urban development (target 11.a) and the preservation of cultural and natural heritage (target 11.4).

12.4 SDG 11 in action

SDG 11 has for the most part been embraced by mayors and city leaders globally (though admittedly more in the Global North), with its adoption in key city planning documents such as local government strategic priorities and organizational values (Wittmayer et al, 2018). Globally many local governments have integrated the SDGs into planning in a bid to localize the global nature of the goals. More than 125 city governments have also joined a global movement to conduct voluntary local reviews (VLRs), where they track and report their progress on SDGs in a similar manner to national governments (Ciambra, 2021). The lack of a high-level of prescriptiveness of the goals provides cities with ample opportunity to contextualize the goals for their own localization practices. However, it can then present a challenge for cities to acknowledge whether their approach works, a deterrent for cities pursuing goal localization (Andrea et al, 2020). Localizing the SDGs and conducting a comprehensive VLR is a complex process and requires effective urban governance (Fox and Macleod, 2021). While the VLR concept has had impressive take-up from large cities of the Global South, particularly in South America and Africa (Ciambra, 2021), the instigation of this initiative was driven by cities in Global North (New York City and several Japanese municipalities), and it remains a much greater challenge for small and middle-sized cities, particularly those in the Global South, to implement this kind of sustainable development initiative.

The measurement of goal progression is an area of increasing interest for cities needing to track and monitor SDG targets. The approach, used correctly, can also provide for effective city-to-city peer learning (Leavesley, 2021), as well as informing reflexive practice to drive SDG goal attainment.

Behind each target are indicators by which to track sustainable development progress (see Box 12.1). Indicators are a way of monitoring and uniformly testing progression of the goals across differing urban landscapes (Hiremath et al, 2013). Although the indicators are clearly defined, there have been various critiques aimed at how they are positioned and utilized in practice (Barnett and Parnell, 2016; Klopp and Petretta, 2017; du Plessis and Aust, 2018). For many cities, a constraint exists in their capacity to collect and track required data by which measure indicators (Simon et al, 2016). In a study of five cities' abilities to report on the indicators, Simon et al found that three of the draft indicators (11.3.2, 11.3 and 11.b.) were relatively easy for local governments to report on, while others strained already overwhelmed city departments, becoming reporting burdens and thus limiting cities' active participation in tracking programs (Simon et al, 2016).

Other studies have incorporated wider city samples, for example the SDG Cities Challenge Project run by the Melbourne Centre for Cities at the University of Melbourne. This project looked at 14 cities in Asia-Pacific to determine the challenges of localizing SDG 11. A difficulty faced universally across these cities was inadequate funding provisions and limited staff resourcing dedicated to tracking the SDGs (Leavesley, 2021; Leavesley et al, 2022) This lack of funding was also a primary factor in the ability of cities to conduct city-to-city learning on key urban issues to encourage goal progression (Kosovac et al, 2021).

As previously mentioned, there has been wide-ranging criticism of the limited specificity regarding SDG 11 implementation, resulting in ambiguity for city planning departments. This can lead to city governments questioning whether the approach they have taken is appropriate for the goal, as local-level tactics are not clearly stipulated in the goals, primarily to increase universality (Leavesley, 2021). Comparative analysis across cities on goal progress can also be problematic in understanding improvement, as measurement procedures and approaches can vary widely, largely as a result of differing governance structures of cities (Cottineau et al, 2017). The challenges are further exacerbated for cities of the Global South, which tend to have fewer resources to actively measure and track progress.

Conversely, the ambiguity of the goals and indicators lends itself well to allowing cities to establish their own version of the goals that are context specific, which also reduces the chance of path dependence (the ability to stray off a planned trajectory and be adaptive to changing situations and values) (Hartley, 2019). This allows the strategic planning to remain flexible, and enables cities to take a reflexive approach in their uptake, an important element in sustainability planning.

Box 12.1: Indicators for Sustainable Development Goal 11

Agenda 2030 Sustainable Development Goal 11: Targets and indicators

Target 11.1: By 2030, ensure access for all to adequate, safe and affordable housing and basic services and upgrade slums

Indicator 11.1.1: Proportion of urban population living in slums, informal settlements or inadequate housing

Target 11.2: By 2030, provide access to safe, affordable, accessible and sustainable transport systems for all, improving road safety, notably by expanding public transport, with special attention to the needs of those in vulnerable situations, women, children, persons with disabilities and older persons

Indicator 11.2.1: Proportion of population that has convenient access to public transport, by sex, age and persons with disabilities

Target 11.3: By 2030, enhance inclusive and sustainable urbanization and capacity for participatory, integrated and sustainable human settlement planning and management in all countries

Indicator 11.3.1: Ratio of land consumption rate to population growth rate

Indicator 11.3.2: Proportion of cities with a direct participation structure of civil society in urban planning and management that operate regularly and democratically

Target 11.4: Strengthen efforts to protect and safeguard the world's cultural and natural heritage

Indicator 11.4.1: Total per capita expenditure on the preservation, protection and conservation of all cultural and natural heritage, by source of funding (public, private), type of heritage (cultural, natural) and level of government (national, regional and local/municipal)

Target 11.5: By 2030, significantly reduce the number of deaths and the number of people affected and substantially decrease the direct economic losses relative to global gross domestic product caused by disasters, including water-related disasters, with a focus on protecting the poor and people in vulnerable situations

Indicator 11.5.1: Number of deaths, missing persons and directly affected persons attributed to disasters per 100,000 population

Indicator 11.5.2: Direct economic loss in relation to global GDP, damage to critical infrastructure and number of disruptions to basic services, attributed to disasters

Target 11.6: By 2030, reduce the adverse per capita environmental impact of cities, including by paying special attention to air quality and municipal and other waste management

Indicator 11.6.1: Proportion of municipal solid waste collected and managed in controlled facilities out of total municipal waste generated, by cities

Indicator 11.6.2: Annual mean levels of fine particulate matter (eg PM2.5 and PM10) in cities (population weighted)

Target 11.7: By 2030, provide universal access to safe, inclusive and accessible, green and public spaces, in particular for women and children, older persons and persons with disabilities

Indicator 11.7.1: Average share of the built-up area of cities that is open space for public use for all, by sex, age and persons with disabilities

Indicator 11.7.2: Proportion of persons victim of physical or sexual harassment, by sex, age, disability status and place of occurrence, in the previous 12 months

Target 11.a: Support positive economic, social and environmental links between urban, peri-urban and rural areas by strengthening national and regional development planning

Indicator 11.a.1: Number of countries that have national urban policies or regional development plans that (a) respond to population dynamics; (b) ensure balanced territorial development; and (c) increase local fiscal space

Target 11.b: By 2020, substantially increase the number of cities and human settlements adopting and implementing integrated policies and plans towards inclusion, resource efficiency, mitigation and adaptation to climate change, resilience to disasters, and develop and implement, in line with the Sendai Framework for Disaster Risk Reduction 2015–2030, holistic disaster risk management at all levels

Indicator 11.b.1: Number of countries that adopt and implement national disaster risk reduction strategies in line with the Sendai Framework for Disaster Risk Reduction 2015–2030

Indicator 11.b.2: Proportion of local governments that adopt and implement local disaster risk reduction strategies in line with national disaster risk reduction strategies

Target 11.c: Support least developed countries, including through financial and technical assistance, in building sustainable and resilient buildings utilizing local materials

Indicator 11.c.1: No indicator is currently listed under 11.c. See E/CN.3/2020/2, paragraph 23. (UN, 2017)

Agenda 2030 Sustainable Development Goal 11: Targets and Indicators

Target 11.1: By 2030, ensure access for all to adequate, safe and affordable housing and basic services and upgrade slums

Indicator 11.1.1: Proportion of urban population living in slums, informal settlements or inadequate housing

Target 11.2: By 2030, provide access to safe, affordable, accessible and sustainable transport systems for all, improving road safety, notably by expanding public transport, with special attention to the needs of those in vulnerable situations, women, children, persons with disabilities and older persons

Indicator 11.2.1: Proportion of population that has convenient access to public transport, by sex, age and persons with disabilities

Target 11.3: By 2030, enhance inclusive and sustainable urbanization and capacity for participatory, integrated and sustainable human settlement planning and management in all countries

Indicator 11.3.1: Ratio of land consumption rate to population growth rateIndicator 11.3.2: Proportion of cities with a direct participation structure of civil society in urban planning and management that operate regularly and democratically

Target 11.4: Strengthen efforts to protect and safeguard the world's cultural and natural heritage

Indicator 11.4.1: Total per capita expenditure on the preservation, protection and conservation of all cultural and natural heritage, by source of funding (public, private), type of heritage (cultural, natural) and level of government (national, regional, and local/municipal)

Target 11.5: By 2030, significantly reduce the number of deaths and the number of people affected and substantially decrease the direct economic losses relative to global gross domestic product caused by disasters, including water-related disasters, with a focus on protecting the poor and people in vulnerable situations

Indicator 11.5.1: Number of deaths, missing persons and directly affected persons attributed to disasters per 100,000 population Indicator 11.5.2: Direct economic loss in relation to global GDP, damage to critical infrastructure and number of disruptions to basic services, attributed to disasters

Target 11.6: By 2030, reduce the adverse per capita environmental impact of cities, including by paying special attention to air quality and municipal and other waste management

Indicator 11.6.1: Proportion of municipal solid waste collected and managed in controlled facilities out of total municipal waste generated, by cities Indicator 11.6.2: Annual mean levels of fine particulate matter (eg PM2.5 and PM10) in cities (population weighted)

Target 11.7: By 2030, provide universal access to safe, inclusive and accessible, green and public spaces, in particular for women and children, older persons and persons with disabilities

Indicator 11.7.1: Average share of the built-up area of cities that is open space for public use for all, by sex, age and persons with disabilitiesIndicator 11.7.2: Proportion of persons victim of physical or sexual harassment, by sex, age, disability status and place of occurrence, in the previous 12 months

Target 11.a: Support positive economic, social and environmental links between urban, peri-urban and rural areas by strengthening national and regional development planning

Indicator 11.a.1: Number of countries that have national urban policies or regional development plans that (a) respond to population dynamics; (b) ensure balanced territorial development; and (c) increase local fiscal space

Target 11.b: By 2020, substantially increase the number of cities and human settlements adopting and implementing integrated policies and plans towards

inclusion, resource efficiency, mitigation and adaptation to climate change, resilience to disasters, and develop and implement, in line with the Sendai Framework for Disaster Risk Reduction 2015–2030, holistic disaster risk management at all levels

Indicator 11.b.1: Number of countries that adopt and implement national disaster risk reduction strategies in line with the Sendai Framework for Disaster Risk Reduction 2015–2030 Indicator 11.b.2: Proportion of local governments that adopt and implement local disaster risk reduction strategies in line with national disaster risk reduction strategies

Target 11.c: Support least developed countries, including through financial and technical assistance, in building sustainable and resilient buildings utilizing local materials

Indicator 11.c.1: No indicator is currently listed under 11.c. See E/CN.3/2020/2, paragraph 23.Source: UN (2017).

12.5 The story so far

Many cities have taken it on themselves to report on their own strategies around SDGs and to encourage city-to-city learning. Despite this, it can be difficult to compare cities because of the inconsistent measurement practices and differences in baseline starting points (du Plessis and Aust, 2018; Pipa and Bouchet, 2020). Despite the previously described weaknesses in the indicator process, there are nevertheless wide-ranging attempts to report and compare SDG 11 progress across different nation-states. Data is being collected by UN-Habitat, the Institute for Health Metrics and Evaluation, WHO, United Nations Children's Fund (UNICEF) and the Organisation for Economic Co-operation and Development to actively track the progression of goals.

A key aspect to note is that these data are collated on the basis of nation-states, not cities, which increases the difficulty of establishing a city-level comparison. The following figures highlight some of the reporting on progress for different goal indicators. The nation-states included in the diagrams were chosen on the basis of either having the highest shift in progress across time or representing a high area of concern for that country. Voluntary national or local reviews are by nature voluntary. While the 193 countries that signed the 2030 Agenda for Sustainable Development committed to completed at least two voluntary national reviews (VNRs), five (Haiti, Myanmar, South Sudan, Yemen and the US) have yet to complete one report. The US is a significant outlier here as a high-income country where several cities, including New York City, Los Angeles and Pittsburgh have completed VLRs. While there has been strong participation globally in the VNR process, the voluntary nature of reporting and tracking, and the absence of consequences for inadequate performance, coupled with challenges in data collection and validation, are all hindrances to SDG 11, and indeed to all the SDGs, in driving global progress.

Figure 12.4: Tracking progress on indicator 11.1.1 (percentage of urban population living in slums)

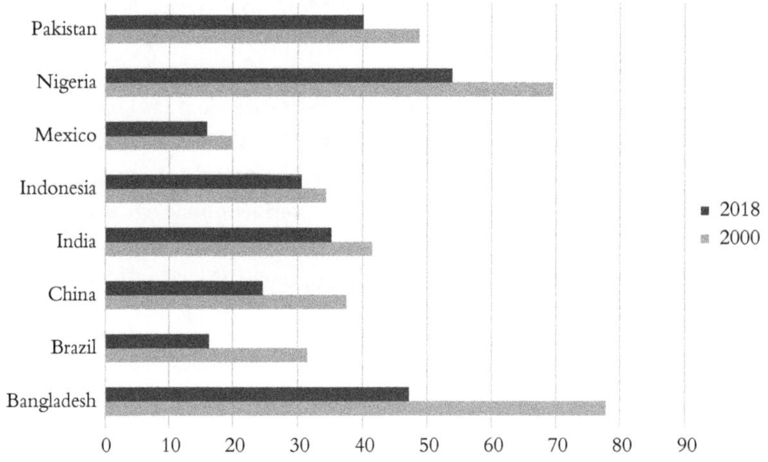

Source: UN-Habitat (2021).

Figures 12.4, 12.5 and 12.6 show one of the approaches taken to track SDG 11 progress across nation-states. As shown, it provides a broad outlook and comparability across countries themselves rather than cities, highlighting the persistent nation-state-centric nature of the endeavour. While Figures 12.4 and 12.6 highlight a goal that is specifically tracked within urban areas, Figure 12.5 does not make clear whether or not the data are based in urban spaces. The definition of *urban* also requires interpretation. There has been a widespread debate in urban studies on what the terms *urban* and *city* mean, and in particular, the boundary constraints, spatially or otherwise (see eg Allen et al, 1999; Marcotullio and Solecki, 2013; Wachsmuth, 2014). This ongoing conflict over definition presents a further issue with regard to what is being measured and where, resulting in difficulty in comparing across cities/urban areas. Despite the lack of comparable data, many cities were adamant about the importance of city-to-city benchmarking, tracking and learning (Leavesley, 2021).

As mentioned, the SDG Cities Challenge takes an approach that tracks cities themselves rather than nation-states, relying on local governments to actively measure and share their approaches with other participants. Not only does this create a strong shared-learning platform, but it also encourages further city-to-city relations, building on existing city diplomatic actions (Acuto et al, 2021a; Kosovac et al, 2021). The SDG Cities Challenge undertaken in 2019 and 2020 relied on voluntary local reporting of cities within the challenge, which looked at eight cities across the Asia-Pacific region (Leavesley, 2021). A further iteration of this program has expanded the remit to a range of cities in the US. Another example of a program that

Figure 12.5: Tracking progress on indicator 11.6.2 (air quality) (Annual mean concentration of particulate matter of less than 2.5 microns in diameter [PM2.5] percentage change between 2000 and 2019)

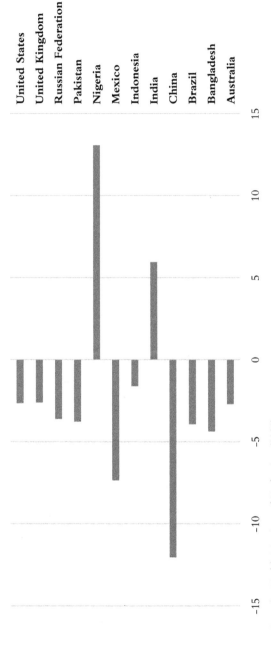

Annual mean concentration of particulate matter of less than 2.5 microns in diameter (PM2.5)

Source: Institute for Health Metrics and Evaluation (2021).

Figure 12.6: Percentage change on access to safe water for tracking Goals 6.1 (Safe water access) and 11.1 (Access to basic services) (percentage change between 2000 and 2017)

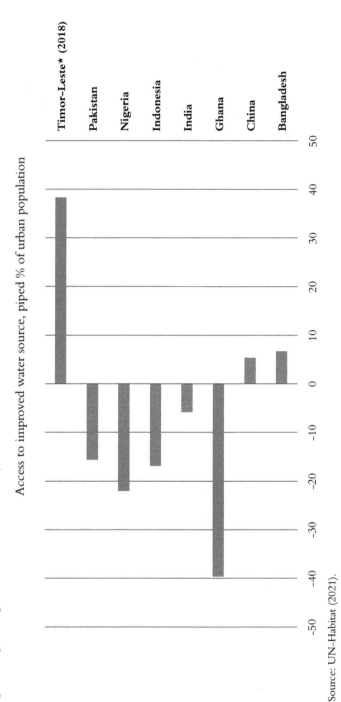

Access to improved water source, piped % of urban population

Source: UN-Habitat (2021).

supports knowledge sharing on SDG localization is the SDG Leadership Cities initiative run by the Brookings Institution (see also Ortiz Moya et al, 2021). These types of city-centric programmes are flourishing to further embed and operationalize SDG 11 in urban areas.

12.6 Conclusion

Issues of sustainability are inherently complex, involving a range of disciplines and sectors in planning. This is reflected in how cities have embraced the goals, with many incorporating public–private and civil society partnerships to encourage a shared responsibility approach to goal delivery. The boundary work encompassed by these relations highlights the importance not only of engaging local governments in the achievement of the goals but also of private business, other levels of government and civil societies playing a key role in ensuring strong multistakeholder partnerships that share knowledge, expertise, technology and financial resources (Kosovac and Pejic, 2021). This is captured by the very last goal in the SDGs, SDG 17, to strengthen the means of implementation and revitalize the Global Partnership for Sustainable Development. By and large, a strong acknowledgement exists in sustainability governance that interdisciplinary approaches are key to ensuring continued and long-lasting environmental and social outcomes (Halffman et al, 2019).

However, the legal and budgetary limitations of cities are also worth considering. The power to enact new laws, policies and fiscal strategies related to environmental action is often determined by national governments and varies distinctly between countries. Although some cities do not have direct policy influence over elements such as energy regulation, they nevertheless can still be key instigators of on-the-ground action. Cities therefore have the ability to implement climate change and sustainability measures directly, as their remit often covers areas of responsibility such as waste management, housing, water management and urban planning, to name a few. This results in a direct path to action in areas where nation-states can struggle to implement quickly and effectively at the ground level. Urban areas are key sites for sustainability issues, but also feature increasingly as actors in the implementation of global sustainability practices (Acuto et al, 2021b). Cities therefore do not just act as bystanders or implementers of global agendas dictated by nation-states, but are increasingly powerhouses of economic and social activity. SDG 11 speaks directly to the importance of cities in sustainability and to their wide-ranging nature. Given that the capacity to collect local data is still a barrier to effective SDG 11 reporting, the introduction of programs such as the voluntary local reviews is promising, even though structural inequalities mean that some cities are better placed to be part of such initiatives than others. Without some form of equalizer to aid cities in their reporting functions, it will be difficult to track progress meaningfully. As described in this chapter, cities and urban-focused academics advocated strongly for a dedicated urban SDG, ultimately leading to the 2015 announcement. SDG 11 represents a starting point. It is important not only for its measures but also for its symbolic power in recognizing cities as actors and essential sites for advancing sustainability practices.

Note

1   The term 'environment' in Kosovac et al. (2020a) referred to references to climate, pollution, forest, air quality, desertification, biodiversity, climate change, coastal, ecosystems, pollutants and environmental management.

References

Acuto, M. (2016) 'Give cities a seat at the top table', *Nature*, 537(7622): 611–13.

Acuto, M. and Rayner, S. (2016) 'City networks: Breaking gridlocks or forging (new) lock-ins?', *International Affairs*, 92(5): 1147–66.

Acuto, M. and Leffel, B. (2020) 'Understanding the global ecosystem of city networks', *Urban Studies*, 58(9): 1758–74.

Acuto, M., Kosovac, A. and Hartley, K. (2021a) 'City diplomacy: Another generational shift?', *Diplomatica*, 3(1): 137–46.

Acuto, M., Kosovac, A., Pejic, D. and Jones, T. L. (2021b) 'The city as actor in UN frameworks: Formalizing "urban agency" in the international system?', *Territory, Politics, Governance*, 11(3): 519–36, doi: 10.1080/21622671.2020.1860810.

Allen, J., Massey, D. and Pile, S. (1999) *City Worlds*, New York: Routledge.

Andrea, C., Fernandez, A., Calvete, A., Miraglia, M. and Herrera-Favela, L. (2020) *Guidelines for Voluntary Local Reviews*, vol. 1: *A Comparative Analysis of Existing VLRs*, UN–Habitat, Available from: https://unhabitat.org/gui delines-for-voluntary-local-reviews-volume-1-a-comparative-analysis-of-existing-vlrs [Accessed 23 June 2023].

Angelo, H. and Wachsmuth, D. (2020) 'Why does everyone think cities can save the planet?', *Urban Studies*, 57(11): 01–21.

Arajarvi, N. (2018) 'Including cities in the 2030 Agenda A review of the post-2015 process', in H.P. Aust and A. du Plessis (eds), *The Globalisation of Urban Governance*, New York: Routledge, pp 14–35.

Aust, H.P. and Nijman, J.E. (2020) 'Planetary boundaries intra muros: Cities and the Anthropocene', in D. French and L.J. Kotze (eds), *Research Handbook on Law, Governance and Planetary Boundaries*, Cheltenham: Edward Elgar, pp 103–24.

Aust, H.P. and Nijman, J.E. (2021) 'The emerging roles of cities in international law Introductory remarks on practice, scholarship and the handbook', in H.P. Aust and J.E. Nijman (eds), *Research Handbook on International Law and Cities*, Cheltenham: Edward Elgar, pp 1–16.

Barber, B.R. (2013) *If Majors Ruled the Word: Dysfunctional Nations, Rising Cities*, New Haven, CT: Yale University Press.

Barnett, C. and Parnell, S. (2016) 'Ideas, implementation and indicators: Epistemologies of the post-2015 urban agenda', *Environment and Urbanization*, 28(1): 87–98, doi: 10.1177/0956247815621473.

Betsill, M.M. and Bulkeley, H. (2006) 'Cities and the multilevel governance of global climate change', *Global Governance*, 12(2): 141–59.

C40 Cities (2020) *C40 Annual Report 2020*, Available from: www.c40.org/wp-content/uploads/2021/11/C40_Annual_Report_2020_vMay2021_lightfile.pdf [Accessed 20 August 2022].

Ciambra, A. (2021) *Guidelines for Voluntary Local Reviews: Towards a New Generation of VLRs: Exploring the Local–National Link*, Nairobi: UN-Habitat.

Cohen, M. (2014) 'The city is missing in the Millennium Development Goals', *Journal of Human Development and Capabilities*, 15(2–3): 261–74.

Cottineau, C., Hatna, E., Arcaute, E. and Batty, M. (2017) 'Diverse cities or the systematic paradox of urban scaling laws', *Computers, Environment and Urban Systems*, 63: 80–94.

Curtis, S. (2014) *The Power of Cities in International Relations*, New York: Routledge.

Doe, B., Peprah, C. and Chidziwisano, J. R. (2020) 'Sustainability of slum upgrading interventions: Perception of low-income households in Malawi and Ghana', *Cities*, 107: 102946.

du Plessis, A. and Aust, H.P. (2018) 'Summary of obserations and pointers for future research', in H.P. Aust and A. du Plessis (eds), *The Globalisation of Urban Governance*, New York: Routledge.

Gordon, D.J. and Johnson, C.A. (2018) 'City-networks, global climate governance, and the road to 1.5 °C', *Current Opinion in Environmental Sustainability*, 30: 35–41.

Halffman, W., Turnhout, E. and Tuinstra, W. (2019) *Connecting Science, Policy and Society*, Cambridge: Cambridge University Press.

Hartley, K. (2019) *Global Goals, Global Cities: Achieving the SDGs through Collective Local Action*, Chicago: Chicago Council on Global Affairs.

Hiremath, R.B., Balachandra, P., Kumar, B., Bansode, S.S. and Murali, J. (2013) 'Indicator-based urban sustainability A review', *Energy for Sustainable Development*, 17(6): 555–63.

Institute for Health Metrics and Evaluation (2021) *Health-Related SDGs*, https://vizhub.healthdata.org/sdg.

Klopp, J.M. and Petretta, D.L. (2017) 'The urban Sustainable Development Goal: Indicators, complexity and the politics of measuring cities', *Cities*, 63: 92–7.

Kosovac, A. and Pejic, D. (2021) 'What's next? New forms of city diplomacy and emerging global urban governance', in A. Fernández de Losada and M. Galceran-Vercher (eds) *Cities in Global Governance: From Multilateralism to Multistakeholderism?*, Barcelona: CIDOB, pp 87–96.

Kosovac, A., Acuto, M. and Jones, T. L. (2020a) 'Acknowledging urbanization: a survey of the role of cities in UN frameworks', *Global Policy*, 11(2): 293–304.

Kosovac, A., Hartley, K., Acuto, M. and Gunning, D. (2020b) *Conducting City Diplomacy: A Survey of International Engagement in 47 Cities*, Chicago: Chicago Council on Global Affairs.

Kosovac, A., Hartley, K., Acuto, M. and Gunning, D. (2021) 'City leaders go abroad: A survey of city diplomacy in 47 Cities', *Urban Policy and Research*, 39(2): 127–42, doi: 10.1080/08111146.2021.1886071.

Leavesley, A. (2021) *Sustainable Transitions in Cities: Local Transformation in an Urbanising World*, Parkville: University of Melbourne.

Leavesley, A., Trundle, A. and Oke, C. (2022) 'Cities and the SDGs: Realities and possibilities of local engagement in global frameworks', *Ambio*, 51: 1416–32, 10.1007/s13280-022-01714-2.

Lee, T. (2016) *Global Cities and Climate Change: The Translocal Relations of Environmental Governance*, London: Routledge.

Lehmann, S. (2020) 'The self-organising city and its modus operandi', in A. Di Raimo, S. Lehmann and A. Melis (eds), *Informality through Sustainability*, London: Routledge.

Liu, C., Chen, R., Sera, F., Vicedo-Cabrera, A.M., Guo, Y., Tong, S. et al (2019) 'Ambient particulate air pollution and daily mortality in 652 cities', *New England Journal of Medicine*, 381(8): 705–15.

Marcotullio, P.J. and Solecki, W. (2013) 'What is a city? An essential definition for sustainability', in F. DeClerck, J.C. Ingram and C. Rumbaitis del Rio (eds), *Integrating Ecology into Poverty Alleviation and International Development Efforts: A Practical Guide*, New York: Springer, pp 11–25.

Meth, P. (2013) 'Millennium Development Goals and urban informal settlements: unintended consequences', *International Development Planning Review*, 35(1): v–xiii.

Müller, F. I. (2017) 'Urban informality as a signifier: Performing urban reordering in suburban Rio de Janeiro', *International Sociology*, 32(4): 493–511.

Ortiz-Moya, F., Saraff Marcos, E., Kataoka, Y. and Fujino, J. (2021) *State of the Voluntary Local Reviews 2021: From Reporting to Action*, Institute for Global Environmental Strategies, Available from: www.iges.or.jp/en/pub/vlrs-2021/en [Accessed 20 August 2022].

Parnell, S. (2016) 'Defining a global urban development agenda', *World Development*, 78: 529–40.

Pipa, T. and Bouchet, M. (2020) *Next Generation Urban Planning*, Washington, DC: Brookings Institute.

Roy, A. (2005) 'Urban informality: Toward an epistemology of planning', *Journal of the American Planning Association*, 71(2): 147–58.

Roy, A. and AlSayyad, N. (2004) *Urban Informality: Translational Perspectives from the Middle East, Latin America, and South Asia*, Lanham, MD: Lexington Books.

Rudd, A., Birch, E. L., Revi, A., Rosenzweig, C., Arikan, Y., Jit, B. et al (2015) *Second Urban Sustainable Development Goal Campaign Consultation on Targets and Indicators: Bangalore Outcome Document*, Bangalore: Sustainable Development Solutions Network, World Urban Campaign, UN-Habitat, United Cities and Local Governments, Slum Dwellers International, Cities Alliance et al.

Rudd, A., Simon, D., Cardama, M., Birch, E. L. and Revi, A. (2018) 'The UN, the urban Sustainable Development Goal, and the new urban agenda', in T. Elmqvist, X. Bai, N. Frantzeskaki and C. Griffith (eds), *The Urban Planet: Knowledge towards Sustainable Cities*, Cambridge: Cambridge University Press, pp 180–96.

Sanchez Rodriguez, R., Ürge-Vorsatz, D. and Barau, A. S. (2018) 'Sustainable Development Goals and climate change adaptation in cities', *Nature Climate Change*, 8(3): 181–3.

Simon, D., Arfvidsson, H., Anand, G., Bazaz, A., Fenna, G., Foster, K. et al (2016) 'Developing and testing the urban Sustainable Development Goal's targets and indicators A five-city study', *Environment and Urbanization*, 28(1): 49–63.

UN (United Nations) (2015) *The Millennium Development Goals Report*, New York: UN.

UN (2017) 'Annex: Global Indicator Framework for the Sustainable Development Goals and targets of the 2030 Agenda for Sustainable Development', A/RES/71/313.

UN-Habitat (United Nations Human Settlements Programme) (2021) 'Sustainable Development Report Data explorer', Available from: https://dashboards.sdgindex.org/explorer [Accessed 20 August 2022].

UN-Habitat (2020) *World Cities Report 2020: The Value of Sustainable Urbanization*, Nairobi: UN-Habitat.

Wachsmuth, D. (2014) 'City as ideology: Reconciling the explosion of the city form with the tenacity of the city concept', *Environment and Planning D: Society and Space*, 32(1): 75–90.

WHO (World Health Organization) (2016) *Preventing Disease through Healthy Environments: A Global Assessment of the Burden of Disease from Environmental Risks*, Geneva.

Winston, N. (2009) 'Urban regeneration for sustainable development: The role of sustainable housing?', *European Planning Studies*, 17(12): 1781–96.

Wise, R.M., Fazey, I., Stafford Smith, M., Park, S.E., Eakin, H.C., Archer Van Garderen, E.R.M. and Campbell, B. (2014) 'Reconceptualising adaptation to climate change as part of pathways of change and response', *Global Environmental Change*, 28: 325–36.

Wittmayer, J.M., van Steenbergen, F., Frantzeskaki, N. and Bach, M. (2018) 'Transition management: Guiding principles and applications', in N. Frantzeskaki, K. Hölscher, M. Bach and F. Avelino (eds), *Co-creating Sustainable Urban Futures. Future City*, vol 11, Cham: Springer, pp 81–101.

World Bank (2021) 'Urban development: Overview', Available from: www. worldbank.org/en/topic/urbandevelopment/overview [Accessed 20 August 2022].

Interview with Daniel Pejic: Cities and City Networks: Bottom-Up Pioneers in a Top-Down World

Myrodis Athanassiou

How would you assess the impact of SDG 11 on how cities approach sustainable development?

Pejic: The SDGs have been effective as a coalescing framework for bringing together diverse actors within cities towards more sustainable development. They are a useful organizing principle: broad enough to get people on board but not too controversial. Cities have also become quite savvy at working internationally, speaking the language of sustainable development and using this to coordinate stakeholders and look for new avenues for funding. Here, city networks offer exposure. There has been a focus on this idea of a seat at the table of multilateral discussions. However, cities often perceive the multilateral system to be flawed and ineffective, and such a view has been an impetus for the development of major city networks.

City networks have been criticized as being somewhat skewed towards the involvement of European and North American cities. Would you agree?

Pejic: Indeed, the networks that became influential mostly come from the Global North, often funded by American philanthropic organizations, multilateral organizations and the EU. That is changing. More than ever, city networks try to diversify the geographies of involved cities. However, participating in and getting something out of city networks requires resourcing we're not dealing with an equal playing field while capacity issues remain. Many cities in the global

South have established their own networks and are leaders within networks based in the Global North. Yet, these are usually prominent, large megacities participating across different networks and with the funding and the institutional capacity to benefit from it. If you're a city that struggles to find the resources to manage the everyday in an urban context, investing significant resources in international activity can be hard to justify on the domestic level.

How did cities' roles change regarding local issues and international demands?

Pejic: When we look at urban governance more broadly, we see a shift away from the idea of urban managerialism where local governments were mainly in charge of providing services. Under austerity and neoliberalism, we've seen a move to urban entrepreneurialism, where the role of city leaders is to bring together various actors and to try to spur investment and make sure the private sector is delivering. City leaders oftentimes aren't masters of their own domain and require partnerships. They have to work with a range of stakeholders within the city as well as with higher government levels, and they have to work internationally to solve the increasingly globalized challenges of everyday urban governance. The flipside of that is the international system, which should attend more to what's happening in cities. There's agency in the sense that city leaders are promoting and asserting themselves in those discussions. Cities are more recognized as global actors with agency than ever before. The nature of modern urban challenges has forced cities to move from being service providers at the local level to international actors that can play a role in global efforts at sustainability.

13

Partnerships for SDGs: Facilitating a Biodiversity–Climate Nexus?

Montserrat Koloffon Rosas and Philipp Pattberg

In 2018, 15-year-old Greta Thunberg began a school strike for climate, and was soon joined by many more school children around the world. In 2019 Thunberg addressed the United Nations Climate Action Summit with an emotional speech, condemning world leaders for their betrayal of younger generations through their inertia over climate action. This was not the first time international delegates had applauded a dire warning and call to action from a teenager. In 1992, 12-year-old Severn Cullis-Suzuki delivered a similarly powerful speech at the United Nations' Earth Summit in Rio. 'All this is happening before our eyes and yet we act as if we have all the time we want and all the solutions' (Suzuki, 1992), she said after listing hunger, biodiversity loss, global warming and air pollution as some of the issues requiring most attention, the very same challenges still facing the world three decades later. So, what happened in global sustainability governance during the three decades between Severn's and Greta's memorable activist interventions? And is there a better plan today to finally halt the problems that, like climate change and biodiversity loss, have already concerned three generations of young people?

Since the Rio Earth Summit, the UN has led global governance of sustainable development through three different international policy agendas: Agenda 21, from 1992 to 2000; the Millennium Development Goals, from 2000 to 2015; and the current 2030 Agenda, from 2015 to 2030. With three consecutive global agendas addressing sustainable development, it might seem that the UN has found a clever way of extending the deadline to achieve the goals by renaming and relaunching the agendas for a new period. Yet the evolution of these agendas has been more than a mere rebranding exercise, with each taking a more mature perspective than its predecessor towards sustainability on the one hand and significant structural changes regarding implementation on the other.

In terms of perceptions of sustainable development, between 1992 and 2022, the governance of sustainable development shifted from clearly distinguishing between the different areas of sustainability (economic, environmental and social) and identifying some loose objectives and activities for each area, to an agenda listing 17 specific and measurable goals, while explicitly recognizing their interlinked nature across the different issue areas of sustainability. In other words, the 2030 Agenda aims to break down the silos that create the illusion that global issues can be solved in isolation (UN, 2018). In terms of implementation, these agendas have shifted from declaring that the successful implementation of the agenda is 'first and foremost the responsibility of Governments' (UN, 1992), to stating that 'all countries and all stakeholders, acting in collaborative partnership, will implement this plan' (UN, 2015). While many different means of implementation are available – such as regulations of various kinds, taxes and direct government budget spending (see eg Elder and Bartalini, 2019) it is in this all-hands-on-deck context that multistakeholder partnerships have gained their reputation as a promising governance mechanism to implement the SDGs.

In sum, since Severn delivered her speech in 1992, the UN has taken a much more integral *and* inclusive approach to governance of sustainable development with the 17 Sustainable Development Goals and the endorsement of multistakeholder partnerships (MSPs) for implementation. Consequently, a number of studies have been dedicated to the linkages and connections between the SDGs (eg Nilsson et al, 2016; Schleicher et al, 2018; Zelinka and Amadei, 2019). Similarly, research on MSPs has shown that MSPs indeed 'create bridges on a transnational scale among the public sector ... the private sector, and civil society' (Benner et al, 2004: 196). The UN 2030 Agenda, however, makes an implicit assumption about MSPs as a central implementation mechanism, namely that, next to the bridges they create among sectors, MSPs are *also* able and effective at creating bridges between the 17 different SDGs to devise and implement solutions that benefit multiple issue areas simultaneously. The question of whether this optimistic expectation is warranted has, however, received little scholarly attention.

The aim of this chapter, consequently, is to start addressing this question by exploring the extent to which the theoretical expectation of SDG interlinkages is reflected in the focus of partnerships working on the implementation of SDGs. For this purpose, it focuses on two closely interlinked SDGs: SDG 13 (Climate action) and SDG 15 (Life on land). The risk that partnerships follow the tendency not to address environmental challenges in their full complexity but rather to do it selectively by establishing niches of collaboration where the interests of like-minded partners ally, has been raised before (Andonova, 2010). Nevertheless, as Hale and Mauzerall (2004) point out, according to MSP advocates, framing projects as partnerships has the advantage of connecting disparate activities to a multilateral process. Climate change and biodiversity loss share root causes (Deprez et al, 2021), and the recognition that these two challenges need to be addressed together has become popular in academic and practitioner circles. The chapter explores the extent to which the theoretical understanding of interlinkages between these two SDGs is reflected in the focus of MSPs registered on UNDESA's Partnership Platform. Furthermore, it looks into the practices MSPs implement when they act as *nexus facilitators*.

This chapter begins by discussing the two characteristics that define the 2030 Agenda: its *inclusive* and *integral* approach. On inclusiveness, it briefly reviews the evolution of MSPs in sustainable development governance. With reference to the integral approach, it focuses on the nexus concept by discussing the biodiversity–climate nexus and introduces the idea of MSPs as nexus facilitators. After defining the relevant concepts, it presents the authors' research design for a first empirical exploration of the MSP landscape in the biodiversity–climate governance nexus. Finally, the chapter presents and discusses the authors' findings on the MSPs landscape and their practices, and concludes by outlining avenues for future research.

13.1 The evolution of MSPs in the broader sustainable development agenda

The idea of partnerships for sustainability as a main vehicle for successful implementation did not emerge with the 2030 Agenda. Agenda 21 had already called for a 'Global Partnership for Sustainable Development', and specifically for 'partnerships among the public, private, and community sectors' (UN, 1992). In 1992, however, partnerships were mainly expected to ensure 'the participation of vulnerable groups (such as women, workers, farmers, etc.) and major stakeholders (businesses, NGOs, etc.) to the decision making process of sustainable development' (Biermann et al, 2007: 1). While the role of partnerships was marginal at first, their recognition set the stage for what would be formalized ten years later as the so-called Type II partnerships as an outcome of the 2002 Rio+10 Summit (World Summit on Sustainable Development) in Johannesburg (Pattberg et al, 2012). With this, their role expanded from helping to fill the participation gap to prospects to address three functional deficits in global governance: regulation, participation and implementation (Haas, 2004). Interestingly, while MSPs have increasingly gained support, research has shown that, in reality, they have limited power to cover these gaps (Biermann et al, 2007).

MSPs have become mainstream nowadays, and the term 'partnership' carries largely positive connotations (Bäckstrand, 2006). However, the introduction of this form of collaboration for implementation did not come without resistance. The criticism ranged everywhere from 'EU delegations and environmental NGOs ... worried that partnerships could become an instrument to repudiate international environmental agreements' to 'delegations from the South [which] started to perceive partnerships as a threat to their sovereignty' (Mert and Chan, 2012: 25), as well as scholars pointing out democratic, accountability and legitimacy deficits (Bäckstrand, 2006). These reactions – next to some more pessimistic ones suggesting that MSPs are leading to the privatization of governance structures – questioned whether this modus operandi would promote effective and legitimate global governance (Benner et al, 2004). Partly given to these negative partnership estimates, initially, Type II partnerships were referred to as such in order 'to distinguish them from the politically negotiated agreements and commitments that were considered the first outcome of the summit' (Hale and Mauzerall, 2004: 221). In their early years, this difference was intentionally quite pronounced; the partnerships were conceived as a complementary approach to contribute to the implementation of the sustainable development commitments, and were not specifically intended as substitutes for intergovernmental commitments (Hale and Mauzerall, 2004). However, due to their rising popularity with the later agendas for sustainable development, their complementary role in filling the implementation gap shifted to a much more central one as an implementation mechanism.

The Millennium Development Goals were the first global agenda to dedicate one of its eight goals exclusively to partnerships, and the 2030 Agenda has done the same with SDG 17. Unlike the other 16 goals, SDG 17 is not dedicated to any particular sustainable development issue area, but instead 'serves as a convener and facilitator for all the other goals' (Schnurbein, 2020: 1). SDG 17 addresses several shortcomings of MDG 8 in its design, monitoring and review[1] (ECOSOC, 2015). The main difference is that, while the principle of common but differentiated responsibilities is recognized, the SDGs take a universally inclusive approach, no longer differentiating between the developed and developing world. The SDGs put an end to the tensions of previous agendas by addressing not only all member states but all actors equally, including the state, the market and civil society (Schnurbein, 2020). The variety of actors are expected to get involved and to support the achievement of the SDGs mainly through MSPs (UN, 2015).

From an academic perspective, Boas and colleagues (2016) propose partnerships as a possible avenue for institutionalizing the nexus approach in the context of SDG implementation. Something to consider, however, is that transnational MSPs have been defined as 'institutionalized transboundary interactions between public and private actors, which aim at the provision of collective goods' (Schäferhoff et al, 2009: 455). This definition features four characteristics: (1) institutionalized interactions; (2) transboundary scope; (3) public and private actors; and (4) a goal related to collective goods' provision – none of which refers to the competence of bridging issue areas (ie the capacity to facilitate a *governance nexus* approach). So, can MSPs be expected to act as nexus facilitators? Can MSPs bring about effective and efficient transformations in global sustainability governance? In general, evidence for partnerships' actual role and relevance in this field is scarce and inconclusive (Pattberg et al, 2012). While the effectiveness question has received some academic attention, together with assessments of MSPs' accountability and legitimacy (Bäckstrand, 2006), the efficiency question (ie synergistic effects that emerge from an effective nexus approach) remains largely unexplored. The next section introduces the biodiversity–climate nexus, which serves as an empirical starting point to explore MSPs' performance as nexus facilitators.

13.2 The biodiversity–climate nexus

A nexus generally refers to a series of connections linking two or more things and is now commonly used in relation to governance when several issue areas are intrinsically interlinked (Boas et al, 2016). The nexus concept had not yet been coined to refer to this type of integral approach, until Boas et al (2016: 451) reported that:

> the argument for considering connections between the economic, social, and environmental dimensions of sustainable development dates back more than 20 years, at least to the 1992 ... UN Conference on Environment and Development in Rio de Janeiro.

In the early stages, the conversation about an integral approach started by considering the three dimensions of sustainable development but not more specific issue areas. At the Rio Earth Summit, issue areas including climate and biodiversity received attention separately, as reflected by one of the main outcomes of the summit: the Rio Conventions. These three international environmental treaties governing climate, biodiversity and desertification[2] were negotiated simultaneously but were signed as independent outputs. When it comes to the UN Framework Convention on Climate Change and the UN Convention on Biological Diversity, 'surprisingly the two treaties do not refer explicitly, but only implicitly, to each other' (Maljean-Dubois and Wemaere, 2017: 4).

The nexus concept first appeared at the World Economic Forum in 2008 (Cairns and Krzywoszynska, 2016) and seriously entered the policy arena in 2011 at the Bonn Nexus Conference. There, the focus shifted from the integration of sustainable development dimensions to the integration of issue areas, with the water–energy–food nexus as the main linkage of concern (Keskinen et al, 2016; Sharmina et al, 2016).

Around the same time, solutions approaching climate and biodiversity as a nexus, referred to as nature-based solutions (NBSs) were 'initiated, guided and promoted by influential inter-governmental institutions' such as the World Bank, the International Union for Conservation of Nature and the European Commission (Davies et al, 2021: 1). The nexus approach was further encouraged in 2012, during the UN Conference on Sustainable Development (Rio+20), where it was recognized that the fragmentation of sectors could no longer be maintained (Boas et al, 2016). With this precedent, it is not a surprise that the 2030 Agenda was 'fully cognizant of this problem' and ended up including 'strong language that is meant to address the nexus problem' (Boas et al, 2016: 450), including explicit recognition of both the UNFCCC and the CBD.

The rationale behind a nexus approach is that it can support a transition to sustainability by focusing on system efficiency rather than on the productivity of isolated sectors (Hoff, 2011). Approaching two (or more) issues as a nexus (ie broadening the scope of the system being governed) implies aiming 'to find development paths that consider synergies and trade-offs among sectors' thus enabling 'sectoral policy-making cross-cutting dimensions ... such as sustainable consumption and production' (UNDESA, 2014: 1). An integrated approach addressing shared root causes from the beginning promises to be more efficient and in many cases much more cost effective, 'as investments made to achieve a given goal influence the approach, resourcing, and effectiveness of the delivery of others' (UNDESA, 2014: 2). This last point is crucial, as it recognizes both global warming and biodiversity loss not as the problems themselves but as symptoms of other unsustainable patterns. Once the shared causes are identified, it is easy to notice how even more SDGs are interlinked. In line with UNDESA's understanding of a nexus, Deprez and colleagues (2021: 5) point out that climate change and biodiversity loss 'share root causes which are linked to unsustainable production and consumption (eg in agri-food systems [SDG 2] and energy production [SDG 7]), resulting in damaging land-use changes (eg deforestation and land degradation [SDG 15])'.[3] Their study identifies, besides the shared root causes, three more ways in which the two issue areas are linked: (1) climate change hurts biodiversity; (2) biodiversity is essential to climate mitigation and adaptation; and (3) some climate mitigation solutions hurt biodiversity, in turn potentially compromising the world's ability to reach net zero emissions.

Governing sustainable development through a nexus approach has been receiving increasing attention, and 'the need to address [climate change and biodiversity] together has recently gained prominence in the scientific and political mainstream' (Deprez et al, 2021: 9). Examples of studies taking a nexus approach include Bellard et al (2012), Eitelberg et al (2016) and Ozturk (2016), and examples of political action include the Beijing

Call for Biodiversity Conservation and Climate Change, the UK COP26 Presidency's Nature Campaign, and – despite a joint liaison group (JLG) between the three Rio Conventions having been established since 2001 – the first ever joint collaboration between the Intergovernmental Panel on Climate Change and the Intergovernmental Science-Policy Platform on Biodiversity and Ecosystem Services. In this co-sponsored workshop, the two intergovernmental scientific bodies explicitly recognized the two issue areas to be 'intertwined through mechanistic links and feedbacks' (Pörtner et al, 2021: 14).

In the view of Deprez et al (2021: 16), while these are commendable and well intentioned efforts:

> the most key and problematic blind spot in the current global discussions on linking climate and biodiversity is that ... many [efforts] still seem incomplete [given that] the framing is often primarily on maximizing synergies ... while de-emphasizing or even not acknowledging the existence of potentially severe trade-offs between the two issues.

An example of trade-offs is the impact of carbon dioxide removal (CDR) on biodiversity. Modelling emission pathways to reach net zero CO_2 emissions by mid-century and thus limit temperature rise to 1.5 °C by 2100 shows that we have already reached a point where, to reach the 1.5 °C by 2100 goal, at least some level of CDR (or negative emissions) must be deployed. Bioenergy with carbon capture and storage (BECCS) is a popular form of CDR. However, given the large areas of land that would be required to operate this form of CDR on an industrial scale, if relied on too much,[4] BECCS is set to severely affect biodiversity conservation (and food security) goals (IPCC, 2018; Roe et al, 2019; Deprez et al, 2021). Despite the shift in conversation towards recognizing the risks from trade-offs, and acknowledging nature-based solutions, the shift has not yet been reflected in practice, as only 3 per cent of global climate finance is earmarked for these type of NBSs (COP26 UK Presidency, 2022).

This means that, for effective and efficient SDG implementation, MSPs focusing on the implementation of both SDG 13 and SDG 15 should not only be contributing to the implementation of NBSs but also contribute to 'refine the concept of climate ambition by integrating biodiversity, to select those Paris compatible emission reduction pathways most aligned with reaching biodiversity goals and other SDGs' (Deprez et al, 2021: 18), thus avoiding the risk of the trade-offs between both issues. This chapter's contribution is a first empirical exploration on whether existing MSPs seem to be tackling SDG implementation with a nexus approach and how. The next section discusses the concept of MSPs as nexus facilitators in more depth.

13.3 MSPs as nexus facilitators

MSPs emerged on the scene of sustainable development governance as a response to the lack of effectiveness of more traditional governance approaches (Pattberg et al, 2012). It would therefore be reasonable to attribute their growing popularity to their degree of effectiveness, broader accountability and legitimacy. In reality, however, there is no conclusive evidence of their positive performance (Pattberg and Widerberg, 2016). One of the few empirical evaluations finds, for example, that MSPs with high empirical legitimacy and an appropriate institutional design are best able to fulfil complex tasks in contexts of limited statehood [while] projects that lack legitimacy are prone to fail (Beisheim et al, 2014). When it comes to MSPs, as Schäferhoff et al (2009: 457) point out, it should be noted that 'measuring [MSP] effectiveness in a comparative perspective is difficult, because [MSPs] carry out various [different] functions', and therefore 'any assessment on the effectiveness of [MSPs] should relate to their functions'. They distinguish between the effectiveness of MSPs concerning policy formulation (output level) and policy implementation (outcome and/or impact level).

This section conceptualizes nexus facilitation as a specific implementation set-up that brings together different types of actors with the explicit intention of addressing two or more sustainable development goals simultaneously, thereby generating the potential to bring about synergistic effects on the output, outcome and/or impact level. For example, one MSP could be generating knowledge about the shared root causes of biodiversity loss and climate change, another could be formulating regulations accordingly and yet another could focus on technological innovation to comply with the regulations on the ground. MSPs that successfully generate synergies in SDG implementation through a nexus approach are arguably not only effective, but also more efficient compared to mechanisms implementing SDGs in silos. An important caveat here is that the nexus facilitation set-up is likely to aid but not guarantee the emergence of synergies. A synergistic implementation depends on the activities carried out by the MSPs, which can be categorized along three different dimensions: the *partner*, the *parlance* and the *practice* dimensions.[5] The partner dimension focuses on the participants of the MSP; for example, an MSP could bring together practitioners from the UNFCCC and the CBD together for collaboration. The parlance dimension focuses on discourses, specifically how SDGs link through different narratives. An example here would be to find evidence of MSPs on climate and biodiversity achieving a common understanding to avoid risks like BECCS. The practice dimension focuses on the goals of the MSP itself, and the concrete policy instruments it employs. For example, two MSPs working on the biodiversity–climate nexus could have entirely different goals; one MSP could have the goal to reduce unsustainable agricultural practices by creating regulations, while another could aim to reforest a specific area of land.

It is likely that certain configurations of the three dimensions, or pathways (independent variable), are more likely to lead to synergistic effects in SDG implementation (dependent variable). Determining these pathways is outside of the scope of this chapter,[6] but, as a first step towards this purpose, the MSP landscape in the biodiversity–climate governance nexus can be explored. The connections of SDGs through MSPs and how MSPs operate can be uncovered, and whether this seems to be in line with the expectations for them to act as nexus facilitators determined. The following section presents the authors' data collection and sample delimitation process, and discusses the MSP mapping, as well as the activities carried out by the MSPs in their sample.

13.4 Analysing the biodiversity–climate governance landscape

The Partnership Platform is 'a global registry of voluntary commitments and multistakeholder partnerships made by stakeholders in support of the implementation of the Sustainable Development Goals' (UNDESA, 2022). There is no exhaustive directory of MSPs working on SDG implementation; however, UNDESA's Partnership Platform serves as the official UN registry of MSPs working on SDG implementation. As of January 2022, there were around 6,000 entries[7] on the platform, of which 2,241 declare SDGs 13 and/or 15[8] to be in their focus. According to an independent study commissioned by UNDESA in 2019, 'initiatives are widely distributed across the goals' on the platform, but they are 'skewed to an extent by the approx. 1400 commitments and initiatives added from the Ocean Conference in 2017' (Clough et al, 2019: 4). This trend is still noticeable in the current distribution, with most entries including SDG 14 (Life below water) as one of the SDGs on which their work focuses. The SDGs in focus are self-reported by the entries, potentially reflecting the partnership's aspiration rather than an empirical nexus. This bias can be avoided by relying on text analysis of the partnerships' mission statements based on pre-coded dictionaries for each SDG. For this chapter's purpose, however, the analysis is limited to self-reported SDGs on the platform as a first exploratory step.

To obtain a picture of the constellation of SDGs connected by MSPs working mainly on climate and biodiversity, the authors' sample includes only entries reporting that they focus on a maximum of four SDGs, from which two must be SDGs 13 and 15. There are 66 entries on the Partnership Platform fulfilling this criterion.[9] From the 66 observations in the sample, only six (ie about 9 per cent) focus exclusively on SDGs 13 and 15 (see Figure 13.1). The other 60 report working on those two SDGs in

Figure 13.1: Number of partnerships registered on the Partnership Platform per nexus

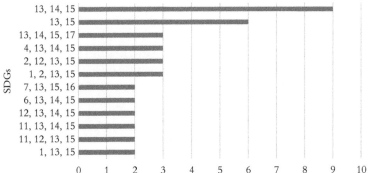

combination with at least one additional SDG. All the SDGs are included by at least one MSP in the sample, except SDG 10 (Reduced inequalities), which does not feature in any of the SDG combinations (see Figure 13.3). This, together with SDG 9 (Industry, innovation and infrastructure), which is featured in only one entry in the sample, correlates with their being the two least featured SDGs among all entries. The most reported nexus in the sample (nine partnerships) includes the two SDGs in focus together with SDG 14. This SDG combination (which could be called the aquaterrestial biodiversity–climate nexus) fits the biodiversity–climate nexus, suggesting that working on biodiversity in general (both on land and below water) may be more popular than focusing on biodiversity on land only. Of course, the popularity of this specific combination could also be attributed to the high frequency of registered partnerships including SDG 14 within their focus areas. Including the six partnerships focusing on the aquaterrestial biodiversity–climate nexus as part of those working exclusively on the climate–biodiversity nexus, amounts to about 22 per cent of the partnerships in the sample.

Additionally, four popular SDG combinations or nexuses were identified, each with three observations reporting to work on them. Figure 13.2 shows the constellation of SDGs according to the most popular connections generated by partnerships, with SDGs as nodes (SDGs 13 and 15 are shown as a single node, as they appear in all SDG combinations) and partnerships as the links connecting them (colours in Figure 13.2 correspond to those of Figure 13.1). The dark blue, magenta, yellow and orange links are triangular shapes, and each triangle is one MSP (instead of each single straight line). The six partnerships focusing exclusively on SDGs 13 and 15 appear in the centre of the figure in green, and the nine partnerships working on the

Figure 13.2: The biodiversity–climate nexus TIE fighter

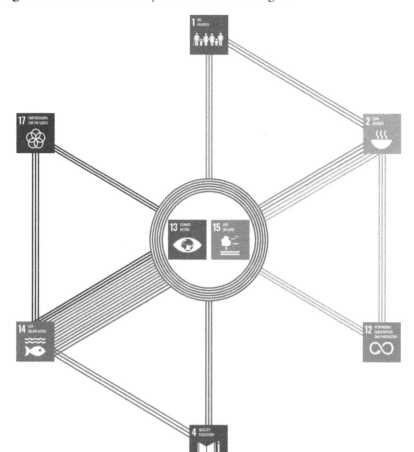

aquaterrestial biodiversity–climate nexus are shown in light blue. The four remaining SDG combinations (the four triangles) are the four next most popular SDG combinations. Two of these SDG combinations are instances of the aquaterrestial biodiversity–climate nexus including one more SDG, which means that this specific combination comes up in about 22 per cent of the partnerships of the sample (nine times on its own, and six times in combination with an additional SDG). In one case (dark blue triangle) this fourth SDG is SDG 17 (Partnerships for the goals), and in the other (magenta triangle) it is SDG 4 (Quality education). Both of the other two popular nexuses include SDG 2 (Zero hunger) and, since they do not include SDG 14, they are likely to be partnerships focusing on issues related to land use, agriculture and food production. One of these two popular combinations (yellow triangle) includes SDG 12 (Responsible consumption

Figure 13.3: Network of SDGs working primarily on SDGs 13 and 15 connected by MSPs

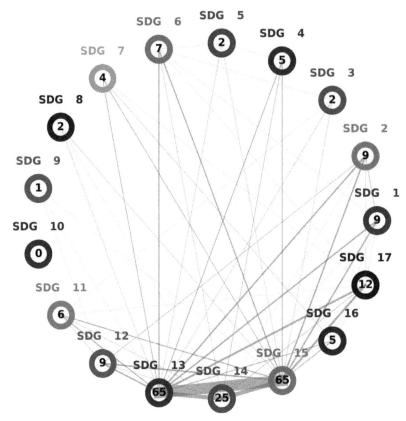

and production) as their fourth SDG, while the other one (orange triangle) includes SDG 1 (No poverty).

Finding that the data sample includes several entries declaring their work to focus on various SDGs simultaneously is a hopeful sign for the expectation that MSPs act as nexus facilitators. Furthermore, that the most popular SDG combinations in the sample focus exclusively on the (aquaterrestial) biodiversity–climate nexus suggests that there is a recognition by MSPs of the interconnected nature of these two issue areas. These findings seem to support the expectation of MSPs having the ability to bridge SDGs. At this point, the critical reader could argue that, on the basis of the very limited progress SDGs are making and the large number of MSPs registered on the Partnership Platform, the thousands of MSPs might not be very effective. This would be the case assuming that all entries registered on the platform were indeed MSPs. However, since registration on the platform is open and voluntary, the entries should be further scrutinized to determine whether their configuration is actually that of an MSP[10] and to understand the type of activities they are carrying out. In this process, significant weaknesses were detected. First, focusing on the sample subset of the most popular nexuses (ie those featured in Figure 13.2), only 19 (61 per cent) of the 31 observations include a (working) website, and even then some websites are not for the partnership itself. Out of the 19 observations including a website, seven are not engaged in transboundary projects; this, strictly speaking, does not fit the authors' working definition for MSPs. This leaves a reduced number of just 12 observations (see Table 13.1) that both fit the authors' requirements and provide a source of information (website) about them and their activities. This points to a different conclusion, that despite the vast amount of attention MSPs receive, the number of active and formal MSPs is actually quite low.

Different methods are required to conduct a comprehensive evaluation of the activities carried out by the MSPs. Some examples include network

Table 13.1: Activities mentioned in descriptions of observations in subsample

Name of entry on UNDESA'S Partnership Platform	Activities		
	Partner dimension	*Parlance dimension*	*Practice dimension*
Conservation and wise use of mangroves and coral reefs in Latin America and the Caribbean	✓	✓	✓
Mangrove restoration potential map	×	✓	✓
Over 2.5 million ha of forest landscape will be restored by countries in the Caucasus and Central Asia under the Bonn Challenge by 2030	✓	×	✓
Carbon farming school	✓	×	✓
WildAREness	×	×	✓
Hyundai Green Zone	✓	✓	✓
Conservation finance product development	×	×	✓
The lifetime carbon neutral commitment: taking responsibility for the past and future carbon emissions. VELUX–WWF partnership to capture the VELUX Group's historical carbon emissions through forest conservation	✓	×	✓
Watertrek	✓	✓	✓
No-till sunflower oil production		✓	✓
Platform for agricultural risk management	✓	✓	✓
Supporting the revision and developing of national legislation for climate change adaptation and mitigation	✓	×	✓

analysis for the partner dimension, quantitative or qualitative content analysis for the parlance and practice dimensions, and interviews for the latter. The more refined understanding of the MSP activities achieved through such methods is indispensable for an analysis of the potential pathways for synergies. This line of research will determine the added value of MSPs in terms of efficiency, and is currently being pursued by the Transformative Partnerships 2030 project. This chapter conducts a preliminary exploration of the descriptions of the 12 observations in the subsample to give an overview of the types of activities they perform, according to the three categories in the conceptual framework. This exercise is insufficient for evaluating, for example, the discourses (parlance dimension) generated through the MSPs' work. However, it is reported whenever a partnership's descriptions indicate the potential facilitation of dialogues contributing to a common framing of the narrative around problems and solutions.

The authors' first finding is that surprisingly, within this subsample, the self-reported SDGs in focus are used homogeneously by MSPs declaring that they work on the same SDG combinations. This is reflected by the similarities among the MSPs working on identical SDG nexuses, for example, the three MSPs focusing exclusively on the biodiversity–climate nexus (green), are all conducting reforestation-related activities, while the two MSPs focusing on the aquaterrestial biodiversity–climate nexus (light blue) conduct work related to mangroves. Among the partnerships focusing on four SDGs, the three MSPs focusing on the aquaterrestial biodiversity–climate nexus *and* partnerships (dark blue) are led by the private sector, and the two including the SDGs on poverty and hunger (orange) have activities related to agriculture. Finally, the only observation in the subsample working on the aquaterrestial biodiversity–climate nexus *and* education (magenta) reports working on education about the nexus, explicitly mentioning mangroves. Interestingly, several of these activities could be considered as nature-based solutions, but this term is not used in any of the descriptions in the subsample. A relevant question is whether the observed homogeneous understanding of reported SDGs holds for MSPs focusing on other issue areas or working on more (than four) SDGs simultaneously.

Next, from their descriptions, it can be observed that for all but one MSP (WildAREness), it is possible to find traits that point to activities fitting at least two of the dimensions of the authors' conceptual framework, with almost half of them having a fitting trait for the three dimensions. One of the most central questions for future research is to determine whether covering more dimensions is advantageous for a more effective and efficient implementation. More in-depth methodologies at a later stage will allow for a more rigorous categorization of activities between dimensions that allows the different pathways for synergistic SDG implementation to be studied.

Finally, most MSPs in the subsample are relatively small. The absence of more well-established partnerships working on the biodiversity–climate nexus[11] is noteworthy and raises different possible hypotheses. One interpretation could be that (big) actors do not see the Partnership Platform as a valuable resource and therefore find the registration process unnecessary. Another is that actors do not regard the platform as a valuable resource because it brings no added value for them (eg for the MSPs, the registration could be but one more bureaucratic requirement draining their limited human resources). Another possibility could be that the platform has not received enough visibility, that is, the low representation is a result of weak orchestration efforts from the UN. Lastly, while unlikely, the Partnership Platform may accurately capture the existing MSP landscape in the biodiversity–climate governance nexus, which would mean that this community happens to be rather inactive, with just a few small initiatives actively working at it.

13.5 Conclusion

This chapter starts from two observations. First, the 2030 Agenda explicitly recognizes the interlinkages between the SDGs and calls for an integrated approach to implementation. Second, SDG 17 calls for MSPs as a key vehicle for delivering sustainable development globally. This means that it is implicitly assumed that MSPs are both able and effective at bringing SDGs closer together to devise and implement solutions that benefit multiple issue areas simultaneously. Whether this is the case empirically is a question that has received little scholarly attention to date. Consequently, this chapter has begun to address this question. It focuses empirically on two environmental SDGs that are widely accepted as being closely related, SDG 13 (Climate action) and SDG 15 (Life on land).

Given the early stage of this line of research, the authors have gone back to basics and reviewed the evolution trajectory that MSPs followed to become such a central tool in the implementation of SDGs. Similarly, they introduce the concept of nexus as an approach to sustainability governance. In terms of their empirical focus on climate change and biodiversity, they provide an overview of the existing understanding on how the two issue areas are related. In spite of the unquestioned close relationship between these topics, they discuss the limited response both in research and in practice to address them as a nexus. Their main conceptual contribution is developing the idea that MSPs can act as nexus facilitators and their conceptual categorization of institutional linkages of SDGs along the three dimensions of partner, parlance and practice.

To make a first empirical assessment of the MSP biodiversity–climate governance landscape, the authors analysed a sample of 66 MSPs declaring that they work mainly[12] on the biodiversity–climate nexus, as registered on UNDESA's Partnership Platform. Preliminary assessment does not allowed conclusions about what MSPs are doing in much detail or their overall impact, but some interesting observations have arisen from this exploratory exercise. First, the climate–biodiversity nexus is the most prevalent among the most popular SDG combinations in their dataset, with most observations (c. 22 per cent) focusing exclusively on biodiversity- and climate-related SDGs (including SDG 14). Second, there are four additional popular nexuses (each with at least three observations reporting that they work on it) that, next to biodiversity and climate change, declare at least one more SDG in their thematic focus (SDGs 1, 2, 4, 12 and 17). That the authors have found several partnerships declaring their work focuses on various SDGs simultaneously is a hopeful sign for the expectation of MSPs acting as nexus facilitators. Furthermore, that the most popular SDG combinations in the sample focus exclusively on the (aquaterrestial) biodiversity–climate nexus suggests that there is a recognition by MSPs of the interconnected nature of these two issue areas, which seems to support the expectation that MSPs can bridge SDGs. Similarly, the descriptions of the majority of MSPs in the sample show traits pointing to activities fitting at least two of the dimensions in the conceptual framework. At a later stage, this information will allow the added value of MSPs to be evaluated in relation to a synergistic implementation of SDGs. On a less positive note, some worrying limitations with the available data were also found, and possible explanations given for the Partnership Platform's shortcomings.

In sum, this chapter has provided a first exploration in an understudied field, and is meant as a starting point for further research in this area. Future research would ideally expand the scope of SDGs and evaluate whether the observations made in the biodiversity–climate nexus hold along other issue areas, as well as MSPs with a focus on more SDGs than in the sample. Furthermore, the debate on data limitations has the potential to be an entire line of research of its own and to contribute to improve the orchestration, as well as the monitoring, reporting and validating (MRV) practices around SDGs. Lastly, the efficiency question should be investigated in more depth. For this, case studies of MSPs would bring clarity on whether or not a nexus approach has the potential to lead to the urgently needed transformative changes to make progress towards achieving the 2030 Agenda in the years to come.

Notes

[1]   The shortcomings were: (1) Goal 8 did not cover all aspects of the global partnership for development, as set out in the Millennium Declaration, the Monterrey Consensus and the Johannesburg Plan of Implementation; (2) there was a lack of quantitative and time-bound

targets, as well as inconsistencies between goals, targets and indicators; (3) structurally, MDG 8 stood apart from the rest of the MDGs; (4) MDG 8 was often misinterpreted as focusing solely on aid commitments.

[2] UNFCCC, CBD and UNCCD.

[3] Acknowledging this much is already a big step ahead, but of course, in the spirit of formal root cause analysis (Reppening et al, 2017), one should further ask: what is the cause of unsustainable production and consumption patterns?

[4] Deprez et al (2021) point out that 'across the IPCC's Report's four illustrative emission pathways, the amount of CDR deployed varies widely'. Indeed, the difference between scenarios P2 and P4 amounts to planting bioenergy crops on 7 per cent of global agricultural land by 2050 (an area the size of Nigeria) or on 33 per cent of global agricultural land (an area the size of Australia), respectively. They declare that, 'in order to reach both climate *and* biodiversity goals, it is essential to avoid at all costs taking such widespread CDR and BECCS deployment as in P4'.

[5] A broader overview of this conceptualization is forthcoming in the framework of the Transformative Partnerships 2030 research project (www.transform2030.se).

[6] This line of research is currently being pursued in the framework of the Transformative Partnerships 2030 research project.

[7] The number of entries in the platform is currently unreliable, as the authors found several duplicates during the coding process. This problem does not seem to have been corrected on the platform as of January 2022.

[8] Unfortunately, the platform does not report disaggregated data, which makes it hard to tell precisely how many partnerships feature in both SDGs 13 and 15. A dataset that will allow these figures to be determined is currently being coded for the Transformative Partnerships 2030 project (www.transform2030.se).

[9] The full sample consists of 143 observations, from which 77 are duplicate data points. This means that some partnerships are registered more than once or even twice.

[10] This has now been done for the full set of entries on the Partnership Platform in the framework of the Transformative Partnerships 2030 project (see www.transform2030.se/output/data).

[11] Such as the Forest Carbon Partnership Facility, the Global Partnership on Forest Landscape Restoration and the Roundtable on Sustainable Biomaterials.

[12] This has been operationalized as MSPs declaring their activities to focus on SDGs 13 and 15, and a maximum of two additional SDGs.

References

Andonova, L.B. (2010) 'Public–private partnerships for the earth: politics and patterns of hybrid authority in the multilateral system', *Global Environmental Politics*, 10(2): 25–53.

Bäckstrand, K. (2006) 'Multi-stakeholder partnerships for sustainable development: rethinking legitimacy, accountability and effectiveness', *European Environment*, 16(5): 290–306.

Beisheim, M., Liese, A., Janetschek, H. and Sarre, J. (2014) 'Transnational partnerships: conditions for successful service provision in areas of limited statehood', *Governance*, 27(4): 655–73.

Bellard, C., Bertelsmeier, C., Leadley, P., Thuiller, W. and Courchamp, F. (2012) 'Impacts of climate change on the future of biodiversity', *Ecology Letters*, 15(4): 365–77.

Benner, T., Reinicke, W.H. and Witte, J.M. (2004) 'Multisectoral networks in global governance: towards a pluralistic system of accountability', *Government and Opposition*, 39(2): 191–210.

Biermann, F., Chan, M., Mert, A. and Pattberg, P. (2007) 'Multi-stakeholder partnerships for sustainable development: does the promise hold?', in P. Glasbergern, F. Biermann and A.P.J. Mol (eds) *Partnerships, Governance and Sustainable Development*, Cheltenham: Edward Elgar, pp. 230–69.

Boas, I., Biermann, F. and Kanie, N. (2016) 'Cross-sectoral strategies in global sustainability governance: towards a nexus approach', *International Environmental Agreements: Politics, Law and Economics*, 16(3): 449–64.

Cairns, R., and Krzywoszynska, A. (2016) 'Anatomy of a buzzword: the emergence of "the water–energy–food nexus" in UK natural resource debates', *Environmental Science and Policy*, 64: 164–70.

Clough, E., Long, G. and Rietig, K. (2019) *A Study of Partnerships and Initiatives Registered on the UN SDG Partnerships Platform*, UNDESA, Available from: https://sustainabledevelopment.un.org/content/docume nts/21909Deliverable_SDG_Partnerships_platform_Report.pdf [Accessed 23 June 2023].

COP26 UK Presidency (2022) 'Protecting and restoring nature for the benefit of people and climate', UN Climate Change Conference (COP26), Glasgow 2021, Available from: https://ukcop26.org/nature [Accessed 20 August 2022].

Davies, C., Chen W.Y., Sanesi, G. and Lafortezza, R. (2021) 'The European Union roadmap for implementing nature-based solutions: a review', *Environmental Science and Policy*, 121: 49–67.

Deprez, A., Rankovic, A., Landry, J., Treyer, S., Vallejo, L. and Waisman, H. (2021) 'Aligning high climate and biodiversity ambitions in 2021 and beyond: why, what, and how?', Available from: https://policycommons. net/artifacts/1761520/aligning-high-climate-and-biodiversity-ambitions-in-2021-and-beyond/2493167 [Accessed 20 August 2022].

ECOSOC (Economic and Social Council) (2015) 'A revitalized global partnership for sustainable development and adjusting development cooperation for implementing the SDGs', Policy Brief 12, New York.

Eitelberg, D.A., van Vliet, J., Doelman, J.C., Stehfest, E. and Verburg, P.H. (2016) 'Demand for biodiversity protection and carbon storage as drivers of global land change scenarios', *Global Environmental Change*, 40: 101–11.

Elder, M. and Bartalini, A. (2019) *Assessment of the G20 Countries' Concrete SDG Implementation Efforts: Policies and Budgets Reported in their 2016–2018 Voluntary National Reviews*, Institute for Global Environmental Strategies, Available from: https://iges.or.jp/en/pub/assessment-g20-countries'-concr ete-sdg [Accessed 20 August 2022].

Haas, P.M. (2004) 'Addressing the global governance deficit', *Global Environmental Politics*, 4(4): 1–15.

Hale, T.N. and Mauzerall, D.N. (2004) 'Thinking globally and acting locally: can the Johannesburg Partnerships coordinate action on sustainable development?', *Journal of Environment and Development*, 13(3): 220–39.

Hoff, H. (2011) 'Understanding the nexus', Policy Commons. Available from: https://policycommons.net/artifacts/1359033/understanding-the-nexus/1972269/ [Accessed 20 August 2022].

IPCC (Intergovernmental Panel on Climate Change) (2018) 'Summary for policymakers', in IPCC (ed) *Global Warming of 1.5°C*, Available from: www.ipcc.ch/sr15 [Accessed 20 August 2022].

Keskinen, M., Guillaume, J., Kattelus, M., Porkka, M., Räsänen, T. and Varis, O. (2016) 'The water–energy–food nexus and the transboundary context: insights from large Asian rivers', *Water*, 8(5): 1–25

Maljean-Dubois, S. and Wemaere, M. (2017) 'Climate change and biodiversity', in E. Morgera and J. Razzaque (eds) *Biodiversity and Nature Protection Law*, Cheltenham: Edward Elgar, chapter 21.

Mert, A. and Chan, S. (2012) 'The politics of partnerships for sustainable development', in P.H. Pattberg, F. Biermann, S. Chan and A. Mert (eds) *Public–Private Partnerships for Sustainable Development*, Cheltenham: Edward Elgar.

Nilsson, M., Griggs, D. and Visbeck, M. (2016) 'Policy: map the interactions between Sustainable Development Goals', *Nature*, 534(7607): 320–22.

Ozturk, I. (2016) 'Biofuel, sustainability, and forest indicators' nexus in the panel generalized method of moments estimation: evidence from 12 developed and developing countries', *Biofuels, Bioproducts and Biorefining*, 10(2): 150–63.

Pattberg, P.H., Biermann, F., Chan, S. and Mert, A. (2012) 'Introduction: Partnerships for sustainable development', in P.H. Pattberg, F. Biermann, S. Chan and A. Mert (eds) *Public–Private Partnerships for Sustainable Development: Emergence, Influence and Legitimacy*, Cheltenham: Edward Elgar, pp 1–20.

Pattberg, P. and Widerberg, O. (2016) 'Transnational multistakeholder partnerships for sustainable development: conditions for success', *Ambio*, 45(1): 42–51.

Pörtner, H.-O., Scholes, R.J., Aard, J., Archer, E., Bai, X., Barnes, D. et al (2021) *IPBES–IPCC Co-sponsored Workshop Report on Biodiversity and Climate Change*, IPBES-IPCC.

Reppening, N., Kieffer, D. and Astor, T. (2017) 'The most underrated skill in management', *MIT Sloan Management Review*, 58(3): 39–48.

Roe, S., Streck, C., Obersteiner, M., Frank, S., Griscom, B., Drouet, L. et al (2019) 'Contribution of the land sector to a 1.5 °C world', *Nature Climate Change*, 9(11): 817–28.

Schäferhoff, M., Campe, S. and Kaan, C. (2009) 'Transnational public–private partnerships in international relations: making sense of concepts, research frameworks, and results', *International Studies Review*, 11(3): 451–74.

Schleicher, J., Schaafsma, M. and Vira, B. (2018) 'Will the Sustainable Development Goals address the links between poverty and the natural environment?', *Current Opinion in Environmental Sustainability*, 34: 43–7.

Schnurbein, G. von (ed) (2020) *Transitioning to Strong Partnerships for the Sustainable Development Goals*, Basel: MDPI.

Sharmina, M., Hoolohan, C., Bows-Larkin, A., Burgess, P.J., Colwill, J., Gilbert, P. et al (2016) 'A nexus perspective on competing land demands: wider lessons from a UK policy case study', *Environmental Science and Policy*, 59: 74–84.

Suzuki, S.C. (1992) *Listen to the children*. Speech at UNCED, Available from: www.youtube.com/watch?v=JGdS8ts63Ck [Accessed 20 August 2022].

UN (United Nations) (ed) (1992) 'Agenda 21', Available from: https://sustainabledevelopment.un.org/content/documents/Agenda21.pdf [Accessed 20 August 2022].

UN (2015) 'Transforming our world: the 2030 Agenda for Sustainable Development', A/RES/70/1, Available from: https://www.un.org/en/development/desa/population/migration/generalassembly/docs/global compact/A_RES_70_1_E.pdf [Accessed 23 June 2023].

UN (2018) '2030 Agenda requires embrace of integrated thinking, deputy secretary-general says at conference, calls for breaking down silos, forging partnerships', Press release, DSG/SM/1227-ENV/DEV/1889, New York, Available from: https://press.un.org/en/2018/dsgsm1227.doc.htm [Accessed 23 June 2023].

UNDESA (United Nations Department of Economic and Social Affairs) (2014) 'HLPF Issue Briefs 5: From silos to integrated policy making', Available from: https://sdgs.un.org/publications/hlpf-issue-briefs-5-silos-integrated-policy-making-17778 [Accessed 20 August 2022].

UNDESA (2022) 'SDG Actions Platform', Available from: https://sdgs.un.org/partnerships [Accessed 20 August 2022].

Zelinka, D. and Amadei, B. (2019) 'A systems approach for modeling interactions among the Sustainable Development Goals part 2: system dynamics', *International Journal of System Dynamics Applications*, 8(1): 41–59.

Interview with Philipp Pattberg: About Inclusive and Participatory Partnerships

Arthur Saillard and Emmanuel Dahan

How are multistakeholder partnerships different from public–private partnerships (PPPs)?

Pattberg: The short answer is: they are the same. The difference is that the MSP concept is used by academics and tends to be broader and can include NGOs, universities, local actors or even two public actors, whereas the PPP concept is narrower and more practitioner inspired.

The popularity of MSPs for the Sustainable Development Goals has increased. What are the drivers behind this increase in popularity?

Pattberg: What sets Agenda 2030 apart from previous development agendas is the emphasis on MSPs. Partnerships are no longer just a part of the SDG discourse – they are a central mechanism for delivery of the SDGs. Partnerships are becoming ever more important because they resonate with the general idea of how we should approach things in an inclusive, participatory, market-based kind of way. In short, it resonates with the overarching neoliberal configuration of global environmental and development policy making. There seems to be a good fit with the general way we understand policy making.

Research has shown that partnerships are primarily promoted by international organizations and a handful of countries in the Global North. How does this bias affect current partnerships?

Pattberg: It's correct that the partnerships we have studied are predominantly driven by Northern interests, Northern actors and Northern money. Initially, this was not so surprising given the economic capacities of the Global North, as well as the ideological fit between the concept of MSPs and the way states in the North are organized. However, this Northern bias does create problems and unintended consequences. For instance, the idea of having a separate public, privat, and non-profit sector is distinctly Western. Therefore, the very concept of MSPs doesn't travel universally. However, I think we can see some development in terms of inclusion of Southern interest and voices when you compare it to earlier iterations of partnership approaches such as the 2002 World Summit on Sustainable Development Issues. I believe that we should strive for more inclusion of affected actors in partnerships, but how we achieve that is a different question.

In your chapter you mention data limitations. What can be done to improve the monitoring of partnerships?

Pattberg: The limitation of data on partnerships is a large problem for research. If we had more data and better annual reporting, we could better assess the effectiveness of partnerships and strengthen the accountability framework. However, right now nobody knows who is working where, on which problem and with how much money. A lot of UN agencies do write comprehensive reports, but the information is not integrated and shared systematically. I believe it's a huge waste of an opportunity not to integrate the massive amounts of available data and to build a system where data can truly drive processes in a transformative way. Such a system shouldn't rely on manual inputs of static data but should feed on automated data inputs in real time to better coordinate agendas, evaluate SDG projects and ultimately generate better solutions.

14

Synthesis: The Environment in Global Sustainability Governance

Lena Partzsch

Over the past few decades, the world has witnessed many environmental summits and appeals for sustainable development. Heads of state met for the first time to address global environmental problems in Stockholm in 1972 for the United Nations Conference on the Human Environment. Another historical landmark was the Rio Earth Summit in 1992, and then the Rio+20 conference in 2012, which led to the adoption of the 2030 Agenda in 2015. The Sustainable Development Goals are the result of this global governance process, building on the Millennium Development Goals. While climate change is the environmental topic that has received most attention recently, this volume has shown that global environmental governance deals with a much broader range of problems, including deforestation (Kleinschmit et al, Chapter 3), ocean pollution (Vadrot, Chapter 4) and freshwater scarcity (Fischer et al, Chapter 5).

After the coronavirus pandemic and the Russian invasion of Ukraine, there is now a greater risk than before that already agreed environmental protection measures will be postponed or even withdrawn. Greenhouse gas emissions continue to rise (Marquardt and Schreurs, Chapter 2). Donor countries have failed to 'mobiliz[e] jointly $100 billion annually by 2020 ... to address the needs of developing countries in the context of meaningful mitigation actions', as agreed on in Agenda 2030 (target 13.a) (IPCC, 2022). None of the Agenda 2030's environmental subtargets, which were due by 2020, were accomplished. The international community did not manage to 'protect and restore water-related ecosystems, including mountains, forests, wetlands, rivers, aquifers and lakes' by 2020 (target 6.6) (UN Water, 2021). Governments failed to, by 2020, sustainably manage and protect marine and coastal ecosystems (target 14.2), effectively regulate harvesting and end overfishing (target 14.4), conserve at least 10 per cent of coastal and marine areas (target 14.5) and to prohibit subsidies that contribute to overcapacity and overfishing (target 14.6) (Maribus, 2021). Regarding life on land (SDG 15), targets to ensure ecosystem conservation, to stop deforestation and to halt biodiversity loss were also missed (WWF, 2020).

Given the multiple crises on the one hand and implementation deficits on the other, the international community needs to reform the global governance system. In this context, the authors of this volume have aimed to learn from past mistakes and to offer suggestions for more and more thorough environmental protection in global sustainability governance. The volume concludes with a synthesis along the three guiding questions that bind the chapter together:

1. How have perceptions of the environment changed in sustainability governance and research since the 1992 Earth Summit?
2. Which actors and institutions have mattered most for governance efforts over the last three decades?
3. Which alternative and innovative forms of governance exist and deserve more research attention for a transition to environmentally salient sustainability?

14.1 Perception of the environment as a global commodity

There is a great variety of perceptions of the environment in sustainability governance and research. In recent decades, the environment has increasingly been perceived as a global commodity, both within and outside the realms of sustainability governance. Authors outlined this with regard to the green goals (SDGs 6 and 13–15) in the first part of the book. Climate actions (SDG 13) are most reliant on market actors and mechanisms (as opposed to traditional command and control regulation) (Marquardt and Schreurs, Chapter 2), and this market reliance within climate change politics also affects other sectors. For example, forest protection and reforestation programs are financially compensated via the UN Framework Convention on Climate Change, leading 'life on land' (SDG 15) to thus increase in value as a global commodity (Kleinschmit et al, Chapter 3). Further, freshwater sources (SDG 6) and the oceans (SDG 14) are increasingly being commodified and developed economically (Fischer et al, Chapter 5; Mehta et al, Chapter 6; Vadrot, Chapter 4).

The perception of the environment being a global commodity is controversial in two ways. On the one hand, it is disputed whether the environment is actually a *commodity*, or whether ecosystems have value on their own and need to be preserved as a precondition for human life, including the functioning of markets (Mehta et al, Chapter 6; Vadrot, Chapter 4). Second, there is no consensus on whether the environment has a *global* value and hence its protection is an *international* responsibility. Instead, especially states with an abundance of natural resources insist on sovereignty over the environment in their territory, and often consider environmental protection to hinder national development (Kleinschmit et al, Chapter 3). The chapters have shown such tensions between national interests in socio-economic development on the one hand, and the territorial detachment of environmental concerns on the other (eg Fischer et al, Chapter 5; Mehta et al, Chapter 6; Vadrot, Chapter 4 Vadrot et al, Chapter 4). These tensions have resulted in contradictions between and within the SDGs; subtargets complement each other without representing an integrated approach (eg Dabla and Goldthau, Chapter 8). Although Agenda 2030 has been characterized as integrative (Kanie et al, 2017), its structure of goals (17), targets (169) and indicators (247) leads to its implementation in silos (Koloffon Rosas and Pattberg, Chapter 13).

The SDGs put contestations and trade-offs aside, in particular with regard to economic growth. Economic growth, now in the form of 'green' growth, is still seen as the main driver of socio-economic development (Chertkovskaya, Chapter 9; Lorek et al, Chapter 11). Back in the 1970s, Meadows et al (1972) warned of the 'limits to growth', while Brundtland's three-pillar concept assumes that dilemmas of economic growth and environmental protection can be overcome (WCED, 1987). The concept of green growth today promises that economic development can even serve environmental protection (for a critical reflection see Chertkovskaya, Chapter 9; Lorek et al, Chapter 11). This shift in perception of development is related to a shift in perception of human–nature relationships, from feminized care for nature to masculine technology management understanding (MacGregor and Mäki, Chapter 10). Scholars working on both ecofeminism and sustainable consumption and production have proposed better recognition of care work, which provides the basis for reproducing capacity for future productive activities. These researchers challenge the belief that technological innovation and diffusion through the market can solve the environmental crisis, and that sustainability governance is simply about 'rationally' managing these processes (eg Chertkovskaya, Chapter 9; Lorek et al, Chapter 11; MacGregor and Mäki, Chapter 10).

The politics of climate change is where this shift in perception is most evident, with governments more than ever relying primarily on technologies and markets to achieve emissions reduction targets. Despite agitations from the countries of the Global South, the UNFCCC, at the instigation of the US in particular, has established market-based mechanisms such as the clean development mechanism and other instruments such as payments for ecosystem services and carbon trading schemes as the main vehicle for climate action and financial redistribution between North and South (Okereke and Coventry, 2016). Proponents argue that market mechanisms offer a flexible and efficient means to reduce emissions within countries and across the world (eg Weber and Darbellay, 2011). However, in addition to the fact that the market often does not work well for the protection of the environment, critics warn that market schemes can reinforce existing social inequalities and power imbalances, thereby having a detrimental impact on local justice issues even if local communities do receive some compensation (eg Mehta et al, Chapter 6; Okereke and Coventry, 2016). Market mechanisms are reaffirming asymmetries. Given the unequal distribution of capital worldwide, a share in the profit does not imply a fairer allocation of resources in the first place. In addition, implementation does often not fulfil expectations of local populations in the Global South (Suiseeya, 2014; Okereke and Coventry, 2016).

Kleinschmit et al (Chapter 3) demonstrate how, in global forest governance, the current situation results from previous conventions, agreements and initiatives, for which the foundations were laid at the time of colonialism. In particular, the International Tropical Timber Organization still differentiates between producing and consuming countries. The first category refers to former colonies in the Global South producing (tropical) timber, which is then exported to the second category of countries in the Global North. Producing countries generally oppose environmental regulation because of the timber industry's economic interests. In consequence, the UN Forum on Forests has remained relatively weak compared to intergovernmental bodies in other sectors such as the UNFCCC (Kleinschmit et al, Chapter 3; Marquardt and Schreurs, Chapter 2). However, this weakness results from persistent inequalities in international trade between high- and low-income nation-states. The former are net importers of embodied materials, energy, land and labour, who gain a monetary trade surplus, while the latter provide resources but experience monetary trade deficits (Chertkovskaya, Chapter 9).

As in most other sectors, the international community has repeatedly failed to adopt an international forest convention due to stakeholders' adherence to the status quo (Kleinschmit et al, Chapter 3). By contrast, most actors in global ocean governance aim for a legally binding agreement (Vadrot, Chapter 4). The difference is that the high seas are still a generally unregulated territory that does not belong to a particular nation-state. Nevertheless, the underlying interests are the same for oceans as for forests and other issues of sustainability governance. Several industries have started exploring mostly untapped areas of the high and deep seas. As mainly these industries take advantage of the unregulated situation, governments have an interest in at least gaining compensation. Correspondingly, SDG 14 (Life below water) frames the oceans primarily as a commodity to be indexed, traded and used. Vadrot (Chapter 4) explains how this framing as a global commodity subsequently does not result in the protection of marine biodiversity and the respect of planetary boundaries. Instead, economic interests translate into subtargets focusing on single economic sectors, most notably the fishery sector (Vadrot, Chapter 4). Similarly, Mehta et al (Chapter 6) demonstrate how, in addition to environmental problems, treating clean water (SDG 6) as a commodity clashes with human rights approaches.

Besides the green goals, authors showed in the second part of the book that economic priorities are most apparent for the SDGs with environmental trade-offs and synergies (SDGs 2, 5, 7, 8 and 12). Dabla and Goldthau (Chapter 8) show that accelerating SDG 7 (Affordable and clean energy) is of major importance to countries of the Global South. Whether infrastructure expansion relies on renewables or on fossil energy sources remains subordinated to socio-economic evaluations and interests. Hence, planetary boundaries inevitably continue to be crossed for socio-economic development (see also Lorek et al, Chapter 11). Fischer et al (Chapter 5) demonstrate similar trade-offs for SDG 6 (Clean water and sanitation), although these take place less visibly. The authors found that environmental water issues, including decreasing water availability due to climate change, was addressed successfully only where economic opportunities were provided to the local population. Such economic opportunities included ecotourism and agri-business with local species adapted to the dry conditions of the case study area (Fischer et al, Chapter 5).

At first sight, environmental protection continues to be a concern of the Global North (eg water ecosystem protection), while the Global South is interested primarily in socio-economic development (eg water infrastructure expansion) (Kleinschmit et al, Chapter 3; Vadrot, Chapter 4). This means that the long industrialized countries still serve as an example for poorer countries to follow their model of 'development'. There is no 'mental rupture' (Sachs, 2017: 2576) in this regard. However, on further examination, when it comes to environmentally dependent livelihoods, North/South divides become much more complex. Local communities depend on an intact environment especially in countries of the Global South (Fischer et al, Chapter 5; Mehta et al, Chapter 6; Vadrot, Chapter 4). Hence, it cannot be said that the Global South per se is against environmental regulation. While debates on justice are focused on distributional outcomes in global sustainability governance, Agenda 2030 tends to neglect how Indigenous, peasant and working-class communities have been at the forefront of environmental struggles, 'caring for and sustaining their environments, and fighting injustices associated with the pursuit of growth and capital accumulation' (Chertkovskaya, Chapter 9).

Chertkovskaya as well as Lorek et al (Chapter 11) emphasize that affluence, rather than poverty, has been the crucial driver of environmental pressures. The richest 10 per cent of the global population is responsible for approximately half of total consumption-related emissions, while the poorest 50 per cent account for only about 10 per cent (Gore and Alestig, 2020). In this line, Mehta et al (Chapter 6) show that environmental problems are linked to redistributional issues.

In the third part of the book, which deals with the SDGs relevant for an environmentally sound implementation (SDGs 11 and 17), the links between environmental and social issues become most evident. Instead of addressing environmental problems in conjunction with social issues, Kosovac and Pejic (Chapter 12) highlight how, in cities and communities (SDG 11), globally orchestrated sustainability efforts generally miss the needs of the majority of the local population. There was a focus on slum development in the MDGs, which has not been entirely abandoned in the SDGs. This focus is misleading, the authors argue, as the low-income situation of slum dwellers does not lend itself to incorporating expensive sustainability construction solutions but rather relies on upgrading (or demolition) options in the bid to improve unsafe building practices, overcrowding and access to sanitation. In a similar vein, multistakeholder partnerships (SDG 17) are, first of all, meant to mobilize funding for the SDGs' implementation in the Global South (Dabla and Goldthau, Chapter 8; Koloffon Rosas and Pattberg, Chapter 13), instead of initiating a transformation to sustainable patterns of consumption and production among those populations that overuse the planet's resources, foremost, in the Global North (Lorek et al, Chapter 11). Adopting a tunnel vision (Kosovac and Pejic, Chapter 12) by focusing on poor populations can therefore have unintended consequences on the environment.

In summary, with Agenda 2030, the environment is balanced against, rather than integrated in, socio-economic development. Environmental sustainability continues to be a long-term vision. In the short term, socio-economic development is prioritized often at the expense of the environment. Paradoxically, at the core of tensions are not necessarily the planetary boundaries and conflicts between causers and sufferers of ecological overshoot, but tensions between those who possess the economic and technological means to explore and exploit the natural environment as a global commodity and those who do not. While Southern governments tend to resist environmental agreements in various sectors for the sake of sovereignty over their resources, poor people especially but not only in the Global South are most vulnerable to the consequences of persistent environmental destruction.

14.2 Actors and institutions in a fragmented and polycentric landscape

It has been shown that the SDGs reflect rather than resolve tensions between the different dimensions of sustainability. This is also seen in the fragmented and polycentric institutional landscape of global sustainability governance where, essentially, there is no centralized authority. With the High Level Political Forum, UN member states have created a body that is mandated to orchestrate the SDGs' implementation. However, this forum has no strict enforcement function (Bernstein, 2017). The SDGs demonstrate 'global governance through goal-setting' (Kanie et al, 2017), and each government is responsible for implementation in its own territory. At the same time, Agenda 2030 relies on a universally inclusive approach where all countries have committed to taking environmental action (Elder and Olsen, 2019). The principle of common but differentiated responsibilities (CBDR) in environmental governance, which the 1992 Rio Declaration enshrines (para 7), is not upheld anymore (Sachs, 2017).

The CBDR principle recognizes that, while all parties are affected by environmental degradation such as climate change, and need to address it, they differ with regard to their responsibilities and capabilities. Historically, the long industrialized countries of the Global North are responsible for the destruction of our planet (Yamin and Depledge, 2004). When the UN Environment Programme, the 'anchor institution' (Ivanova, 2020: 308) of environmental governance in the UN, was established after the Stockholm Conference in 1972, the US was the strongest proponent at the time (along with Sweden), and committed to contributing 40 per cent of the Environment Fund that provided the core resources for UNEP (Ivanova, 2020: 313). However, the chapters have shown throughout how the US has blocked most environmental agreements and systematically weakened existing institutions such as the UNFCCC (Marquardt and Schreurs, Chapter 2; Vadrot, Chapter 4).

By the time UNEP was founded in 1972, the US was the greatest polluter. With less than 6 per cent of the world's population, it produced more than one third of the global energy (Ivanova, 2020: 313–14). Today, China produces most of the world's CO_2 emission (31 per cent), followed by the United States (14 per cent) and India (7 per cent) (Statista, 2022a). Total emissions from the Global South continue to rise sharply in line with economic industrialization. The seven largest emerging economies (China, Russia, India, Brazil, Turkey, Mexico and Indonesia) have superseded the Group of Seven (G7) countries (Canada, France, Germany, Italy, Japan, the UK and the US). The creation of the G20 was a political response to the changing dynamics of the global economy. In consequence, countries of the Global South can no longer be exempt from taking environmental action (Sachs, 2017; Bodansky and Rajamani, 2018). At the same time, it should not be forgotten that US per capita emissions (14 tonnes) are still twice as high as Chinese per capita emissions (7 tonnes) and eight times higher than those of an average Indian (1.7 tonnes) (in 2019) (Statista, 2022b).

For a long time, UNEP was expected to shift UN agencies towards environmental ways and to colour its programs green (Ivanova, 2020).

Especially after the World Trade Organization was created in 1995, environmental scholars demanded that UNEP be boosted to a World Environmental Organization to allow it to balance trade and environmental issues at the international level. These scholars envisioned nation-states alike in institutions and policies at the supra-national level (eg Biermann and Simoni, 2000). The 'post-sovereign' system of the European Union, in which economic integration was followed by environmental regulation efforts, served many as the prime example of large-scale territorial governance (Sachs and Santarius, 2007; Piattoni, 2009). However, this volume has shown that environmental governance has remained fragmented and comparatively weak.

Climate change politics has brought forth the strongest institutional landscape, with the UNFCCC, the Kyoto Protocol and the Paris Agreement. A range of subsidiary bodies and global expert organizations have been established, such as in particular the Intergovernmental Panel on Climate Change (Marquardt and Schreurs, Chapter 2). Climate actors and institutions challenge the WTO and its liberal paradigm of free trade. Other environmental issues cannot rely on equivalent institutional landscapes for their defence. No other sector has an intergovernmental anchor point like the annual sessions of the Conference of the Parties to the UNFCCC. For example, Vadrot outlines in her chapter how the ocean sustainability agenda needs to be realized against the background of a fragmented institutional landscape that perpetuates ineffective action (Vadrot, Chapter 4).

In consequence to fragmentation, the implementation of Agenda 2030 depends on many actors in addition to nation-states, including subnational units such as regions and cities, businesses and CSOs (Gupta and Nilsson, 2017). The forest sector, which failed to adopt an international convention at the Rio Earth Summit, serves as a popular example of the multitude of actors and institutions. In addition to several international institutions, such as UNFF and ITTO, there are a range of national forest programs, and Western governments increasingly require legal verification for timber supply chains, including for logging activities outside their own territories (Overdevest and Zeitlin, 2014; Bartley, 2018). Moreover, there are various non-state programs such as sustainability certification by the Forest Stewardship Council, and hybrid approaches such as the Bonn Challenge, which aims to bring 350 million hectares of degraded and deforested landscapes under restoration by 2030. With the New York Forest Declaration, companies have committed to restoring a specific area of degraded forests and advertise their efforts by labelling their products as deforestation free (Kleinschmit et al, Chapter 3).

Throughout the chapters of this volume, it has become clear that, with increasing public awareness, a growing number of actors are concerned about their 'green' reputation and are hence participating in the diverse initiatives. This is not restricted to international efforts. In addition to nation-state governments, subnational units are increasingly considered crucial for the implementation of global environmental governance issues (Fischer et al, Chapter 5; see also Bansard et al, 2017). The 1992 Rio Declaration already dedicated a chapter to local authorities (chapter 28) but, with SDG 11, sustainable cities and communities have become a standalone goal. However, as has been seen in several chapters (eg Fischer et al, Chapter 5, on SDG 6; Kosovac and Pejic, Chapter 12, on SDG 11), the local level is seen as an addressee of global action, instead of integrating subnational units in decision-making processes from the very beginning. Fischer et al (Chapter 5) illustrate the relevance of subnational units for sustainable water management in Bolivia, Ecuador and Switzerland. Local authorities have considerable decision-making power in these countries, but SDG vocabulary is generally absent from their strategies. In research, the roles of subnational and local arrangements in setting priorities, determining means of implementation and adopting adequate indicators to measure progress towards greater sustainability require further empirical analysis. It needs to be better understood how subnational units participate in meaningful interaction and effectively implement global goals (Fischer et al, Chapter 5; Kosovac and Pejic, Chapter 12).

In other words, instead of further centralization and the development of large-scale territorial environmental governance, the way forward tends to be polycentrism. At the same time, the fragmented international landscape constitutes a challenge for coherent action and coordinated cooperation regarding the environment in global sustainability governance. Competing actors and institutions have been a hindrance to the overarching protection of the environment. If, as a result of multiple crises, there is now global governance reform, governments must take care that this does not happen from the top down but with the involvement of all sectors and levels.

14.3 Voluntary actions as governance alternatives

Given the acceleration of environmental change, 'governance through goal setting' (Kanie et al, 2017) has so far proven inadequate to accomplishing sustainable development. In response to governments' failure to implement goals, stakeholders have increasingly established innovative and alternative forms of governance in recent years, and voluntary actions have prevailed in all sectors to protect the environment. There is a broad variety of commitment by subnational units, business and civil society actors. City networks are an illustrative example of joint action at the subnational level (Kosovac and Pejic, Chapter 12). The greatest hope, however, lies in partnerships with business actors (eg Dabla and Goldthau; Koloffon Rosas and Pattberg, Chapter 13) and 'bottom up' civil society initiatives (Marquardt and Schreurs; MacGregor and Mäki, Chapter 8).

A popular leitmotiv has emerged, where 'cities, not states, are best equipped to deal with complex problems such as climate change' (Bansard et al, 2017: 230). Cities are beginning to act as important 'global powerhouses' (Kosovac and Pejic, Chapter 12). A growing body of research deals with city networks especially in the context of climate change (Bulkeley, 2021). The C40 Cities network consists of 96 cities that produce 25 per cent of global GDP. The Global Covenant of Mayors engages over 9,000 cities, representing nearly 800 million people or 10 per cent of the global population (Bansard et al, 2017; Nguyen et al, 2020). There are also studies on city networks focused on environmental issues other than climate change. For example, in the agri–food sector, Organic Cities encompasses more than 200 municipalities in Europe, and the Fair Trade Towns network has over 2,000 members worldwide that work towards the SDGs (Partzsch et al, 2022).

New attention given to subnational units corresponds with an increasing acknowledgement of local knowledge. There is a disconnect between global goals and local action, and several authors have highlighted the need for new environmental narratives, where the knowledge of local and Indigenous people becomes more important (Kleinschmit et al, Chapter 3; Mehta et al, Chapter 6). On the one hand, the way natural resources are governed is intractably linked to how the environment is known and represented (Vadrot et al, Chapter 4). On the other hand, people at the local level are often not even aware of global processes of decision making, nor do they have a stake in them (Fischer et al, Chapter 5). The implementation of Agenda 2030 would most certainly benefit from being more open to alternative epistemologies and governance approaches. There is a need for greater heterogeneity of (scientific) knowledge to monitor and manage unsustainable practices, including biological, geological and chemical baseline data and surveillance and control mechanisms (Mehta et al, Chapter 6; Kosovac and Pejic, Chapter 12; Vadrot, Chapter 4).

Although women tend to be more committed than men to environmental governance issues, including female parliamentarians than their male colleagues (Ramstetter and Habersack, 2019), 'the UN continues to fail women' both as decision makers and as addressees of environmental regulation (MacGregor and Mäki, Chapter 10). Given that, at the same time, women are significantly less involved in formal economies and perform their work largely unpaid (Chertkovskaya, Chapter 9; Mehta et al, Chapter 6), MacGregor and Mäki (Chapter 10) demand a more holistic view of social and ecological reproduction, and recognition of and greater value placed on both women's work and the environment. In this vein, Chertkovskaya (Chapter 9) calls for a redefinition of value, which is ultimately linked to the redefinition of knowledge. At the same time, the call for more holistic approaches leads to the consideration of multistakeholder approaches.

Several chapters of this volume outline the urgent need for multistakeholder approaches, especially partnerships (Fischer et al, Chapter 5; Kosovac and Pejic, Chapter 12; Lorek et al, Chapter 11). Agenda 2030 dedicates one goal exclusively to multistakeholder partnerships, SDG 17 (Partnerships for the goals). By doing so, Koloffon Rosas and Pattberg (Chapter 13) argue that the UN recognizes the interlinkages among the SDGs. The authors highlight how this governance approach is different from the 1992 Rio Earth Summit where environmental treaties governing climate, biodiversity and desertification were negotiated simultaneously, but then UNFCCC, the UN Convention on Biological Diversity and the UN Convention to Combat Desertification were signed independently. By contrast, Koloffon Rosas and Pattberg find evidence for partnerships being nexus facilitators, whereby partnerships combining green goals (SDGs 13–15) are most prevalent, that is, the environmental dimension is pioneering the partnership concept.

Koloffon Rosas and Pattberg also outline how partnerships are considered as the main vehicle of implementation. MSPs emerged on the scene of sustainability governance as a response to the lack of effectiveness of traditional governance approaches. It is assumed that MSPs can bring SDGs closer together to devise and implement solutions that benefit multiple issue areas simultaneously, and are effective at doing so. In this vein, for example, Kosovac and Pejic (Chapter 12) express their confidence that city networks, in combination with partnerships, can contribute to the effective implementation of SDG 11 (Sustainable cities and communities). For example, C40 Cities has been underpinned by funding from Bloomberg Philanthropies but now has a broad range of funding partners including national governments, foundations and global brands such as IKEA and Arup. Many networks also work closely with academic institutions and think tanks to develop research and policy outputs to inform cities, national governments and international organizations (Kosovac and Pejic, Chapter 12; Koloffon Rosas and Pattberg, Chapter 13).

However, despite all expectations and best practice examples, there is no evidence that the partnership community is particularly active. More than a third of the green partnerships registered on the official UN Partnership Platform do not even have a website (Koloffon Rosas and Pattberg, Chapter 13). Regardless of acknowledging trade-offs and synergies between goals and subtargets, for example, only 3 per cent of global climate finance has been devoted to initiatives that approach climate and biodiversity as a nexus (Koloffon Rosas and Pattberg, Chapter 13). To that effect, Kleinschmit et al (Chapter 3) remain generally sceptical regarding the environmental commitments of the private sector. They outline how private logging companies, which emerged from and became powerful against the backdrop of colonialism, continue to uphold principles of wood production and international markets instead of environmental conservation principles. In this vein, Vadrot (Chapter 4) warns that private companies now use partnerships to collect and own ocean data, including baseline data, to which public scientific institutions do not always enjoy access.

In parallel to the growing private sector involvement in global sustainability governance, authors highlight the significance of civil society commitment to environmental protection in all chapters of this volume. Kleinschmit et al (Chapter 3) outline how environmental movements started early on in the 1980s to pave the way for sustainability governance in the forest sector. Lorek et al (Chapter 11) emphasize that 'vast, ambitious and perhaps very bold political undertakings require partnerships with social movements seeking environmental justice and radical change to the dominant ways in which relationships of consumption and production are structured'. In this vein, the concept of degrowth, which critiques the centrality of economic growth in today's economies and societies, was developed jointly by researchers and activists (Chertkovskaya, Chapter 9). Similar ties between ecofeminist thinkers and activists can be seen, as well as a clear development of concepts over recent decades. While some ecofeminist activists in the past have used strategic arguments that supported the notion that women are inherently 'closer to nature', ecofeminists today seek to avoid conflating sex and gender and aim to dismantle essentialist thinking (MacGregor and Mäki, Chapter 10).

Intergovernmental environmental summits such as Stockholm and Rio have always been major gatherings of civil society actors. '[A] transnational world emerged, a world connected by value chains, similar consumption patterns and globalist thought' (Sachs, 2017: 2577). Before the COVID-19 pandemic, there was a massive uptake in environmental action and a variety of new movements emerged. MacGregor and Mäki (Chapter 10) point to a record number of especially young women and gender non-binary activists participating actively in the global movement for climate justice. The authors also draw hope from the growing number of 'feminist green new deals'. While both concepts of degrowth and FGND have been developed in the Global North, there has also been an increase in civil society activity in the Global South in recent years. Civil society organizations around the world are all united by an understanding that there must be a new path of development that takes planetary boundaries into account (Lorek et al, Chapter 11; Marquardt and Schreurs, Chapter 2).

14.4 Conclusion and outlook

Agenda 2030 does not resolve the tensions between perceptions of sustainable development. Given that the environment has increasingly been seen as a commodity, there are tensions at its core between those who want to economically exploit natural resources further and those who want to protect the environment for its own sake and for the sake of human survival. The latter are increasingly aligning globally to advocate ecosystem protection in global sustainability governance. While the development of centralized environmental governance has failed, a polycentric world order has emerged that includes networks of environmental pioneers (eg MacGregor and Mäki, Chapter 10; Marquardt and Schreurs, Chapter 2).

On the one hand, scholars agree that the fragmented international landscape constitutes a challenge for coherent action and coordinated cooperation to address environmental issues. On the other hand, as reform is needed in response to the multiple crises of our time, there is also a chance for bottom-up initiatives to succeed with their ideas. Voluntary initiatives at the subnational level simultaneously offer the opportunity to leave behind siloed cooperation. Multistakeholder approaches that consider the knowledge of local people, for example terms of cooperation with city networks, promise to be one of several fruitful alternatives to environmental multilateralism (Kosovac and Pejic, Chapter 12; Lorek et al, Chapter 11).

Global civil society provides many examples of the significance of individual activists in sustainability governance. Wangari Maathai initiated the Green Belt movement, an indigenous, grassroots movement of women who organized in rural Kenya to plant trees, combat deforestation and stop soil erosion. She received the Nobel Peace Prize in 2004 for her contribution to sustainable development, democracy and peace (Gorsevski, 2012). When Greta Thunberg sparked off the Fridays for Future movement in Sweden in 2018, she was in this tradition of voluntary but powerful civil society action (Marquardt, 2020). Around the world, an increasing number of individuals and groups demand that governments adopt environmental measures that meaningfully respond to the mounting evidence of environmental change (MacGregor and Mäki, Chapter 10; Marquardt and Schreurs, Chapter 2).

The present volume emerged from this context of a new awareness and recent mass protests against environmental destruction. There have now been several decades of global environmental governance. Although people are increasingly aware of the issues and are taking action, it is not enough. The chapters of this volume demonstrate that the environment continues to be neglected throughout all areas of global sustainability governance. But the last chapter has not been written yet. The authors of this volume hope to have given you as readers a better idea about the governance challenges. It is up to our generation to take on these challenges and to implement the globally agreed goals in local contexts. Together, we need to seriously consider what comes after the SDGs.

References

Bansard, J.S., Pattberg, P. and Widerberg, O. (2017) 'Cities to the rescue? Assessing the performance of transnational municipal networks in global climate governance', *International Environmental Agreements: Politics, Law and Economics*, 17(2): 229–46.

Bartley, T. (2018) *Rules Without Rights: Land, Labor, and Private Authority in the Global Economy*, Oxford: Oxford University Press.

Bernstein, S. (2017) 'The United Nations and the governance of Sustainable Development Goals', in N. Kanie and F. Biermann (eds) *Governing through Goals,* Cambridge, MA: MIT Press, pp 213–40.

Biermann, F. and Simonis, U.E. (2000) 'Institutionelle Reform der Weltumweltpolitik?: Zur politischen Debatte um die Gründung einer "Weltumweltorganisation"', *Zeitschrift für Internationale Beziehungen*, 7(1): 163–84.

Bodansky, D. and Rajamani, L. (2018) 'The evolution and governance architecture of the United Nations climate change regime', in U. Luterbacher and D.F. Sprinz (eds) *Global Climate Policy*, MIT Press, pp 13–65.

Bulkeley, H. (2021) 'Climate changed urban futures: environmental politics in the Anthropocene city', *Environmental Politics*, 36(8): 266–84.

Elder, M. and Olsen, S.H. (2019) 'The design of environmental priorities in the SDGs', *Global Policy*, 10(1): 70–82.

Gore, T. and Alestig, M. (2020) *Confronting Carbon Inequality in the European Union*, Oxford: Oxfam, Available from: www.oxfam.org/en/research/confronting-carbon-inequality-european-union [Accessed 15 June 2022].

Gorsevski, E.W. (2012) 'Wangari Maathai's emplaced rhetoric: greening global peacebuilding', *Environmental Communication*, 6(3): 290–307.

Gupta, J. and Nilsson, M. (2017) 'Toward a multi-level action framework for Sustainable Development Goals', in Kanie, N. and Biermann, F. (eds) *Governing through Goals*, Cambridge, MA: MIT Press, pp 275–94.

IPCC (Intergovernmental Panel on Climate Change) (2022) *Climate Change 2022: Impact, Adaptation and Vulnerability*, Available from: https://report.ipcc.ch/ar6wg2/pdf/IPCC_AR6_WGII_FinalDraft_FullReport.pdf [Accessed 15 March 2022].

Ivanova, M. (2020) 'Coloring the UN Environmental', *Global Governance: A Review of Multilateralism and International Organizations*, 26(2): 307–24.

Kanie, N., Bernstein, S., Biermann, F. and Haas, P.M. (2017) 'Introduction: Global governance through goal setting', in N. Kanie and F. Biermann, F. (eds) *Governing through goals*, Cambridge, MA: MIT Press, pp 1–27.

Maribus (2021) *World Ocean Review: Lebensgarant Ozean, nachhaltig nutzen, wirksam schützen* [online], Available from: https://worldoceanreview.com/wp-content/downloads/wor7/WOR7_de.pdf [Accessed 15 March 2022].

Marquardt, J. (2020) 'Fridays for Future's disruptive potential: an inconvenient youth between moderate and radical ideas', *Frontiers in Communication*, 5(4).

Meadows, D.H., Meadows, D.L., Randers, J. and Behrens, W.W. III (1972) *The Limits of Growth*, New York: Universe Books.

Nguyen, T.M.P., Davidson, K. and Coenen, L. (2020) 'Understanding how city networks are leveraging climate action: experimentation through C40', *Urban Transformations*, 2(1): 1–23.

Okereke, C. and Coventry, P. (2016) 'Climate justice and the international regime: before, during, and after Paris', *Climate Change*, 7(6): 834–51.

Overdevest, C. and Zeitlin, J. (2014) 'Assembling an experimentalist regime: transnational governance interactions in the forest sector', *Regulation & Governance*, 23(8): 22–48.

Partzsch, L., Lümmen, J. and Löhr, A.-C. (2022) 'City networks' power in global agri-food systems', *Agriculture and Human Values*, 39: 1263–75.

Piattoni, S. (2009) 'Multi-level governance: a historical and conceptual analysis', *European Integration*, 31(2): 163–80.

Ramstetter, L. and Habersack, F. (2019) 'Do women make a difference? Analysing environmental attitudes and actions of members of the European Parliament', *Environmental Politics*, 78(4): 1–22.

Sachs, W. (2017) 'The Sustainable Development Goals and *Laudato si'*: Varieties of post-development?', *Third World Quarterly*, 38(12): 2573–87.

Sachs, W. and Santarius, T. (2007) *Fair Future: Resource Conflicts, Security, and Global Justice*, London: Zed Books.

Statista (2022a) *CO2-Emissionen: Größte Länder nach Anteil am weltweiten CO2-Ausstoß im Jahr 2020*, Available from: https://de.statista.com/statis tik/daten/studie/179260/umfrage/die-zehn-groessten-c02-emittenten-weltweit [Accessed 1 July 2022].

Statista (2022b) *Energiebedingte CO_2-Emissionen pro Kopf weltweit nach ausgewählten Ländern im Jahr 2019 (in Tonnen)*, Available from: https:// de.statista.com/statistik/daten/studie/167877/umfrage/co-emissionen-nach-laendern-je-einwohner [Accessed 30 June 2022].

Suiseeya, K.R.M. (2014) 'Negotiating the Nagoya Protocol: Indigenous demands for justice', *Global Environmental Politics*, 14(3): 102–24.

UN Water (2021) *UN World Water Development Report 2021: Valuing Water* [online], Available from: www.unwater.org/publications/un-world-water-development-report-2021 [Accessed 15 March 2022].

WCED (World Commission on Environment and Development) (1987) *Our Common Future*, Oxford: Oxford University Press.

Weber, R.H. and Darbellay, A. (2011) 'The role of the financial services industry in the clean development mechanism: involving private institutions in the carbon market', *International Journal of Private Law*, 4(1): 32–53.

WWF (World Wide Fund for Nature) (2020) *Living Planet Report 2020*, Available from: https://livingplanet.panda.org [Accessed 15 March 2022].

Yamin, F. and Depledge, J. (2004) *The International Climate Change Regime: A Guide to Rules, Institutions and Procedures*, Cambridge: Cambridge University Press.

Index

References to figures appear in *italic* type; those
in **bold** type refer to tables. References to endnotes show both the
page number and the note number (235n2).